T0348748

Extracellular Matrix in Development and Disease

ADVANCES IN DEVELOPMENTAL BIOLOGY

Volume 15

Series Editor

EXTRACELLULAR MATRIX IN DEVELOPMENT AND DISEASE

Editor

Jeffrey H. Miner

Washington University School of Medicine
Renal Division
St. Louis, Missouri

2005

ELSEVIER

AMSTERDAM • BOSTON • HEIDELBERG • LONDON • NEW YORK • OXFORD
PARIS • SAN DIEGO • SAN FRANCISCO • SINGAPORE • SYDNEY • TOKYO

Elsevier
525 B Street, Suite 1900, San Diego, California 92101-4495, USA
84 Theobald's Road, London WC1X 8RR, UK

For all information on all Elsevier publications
visit our Web site at www.books.elsevier.com

ISBN-13: 978-0-444-51846-0
ISBN-10: 0-444-51846-0

Transferred to digital print 2008
Printed and bound by CPI Antony Rowe, Eastbourne

Contents

Contents

List of Contributors

Eri Arikawa-Hirasawa Research Institute for Diseases of Old Age, Department of Neurology, Juntendo University School of Medicine, Bunkyo-ku, Tokyo, Japan

Richard Belvindrah The Scripps Research Institute, Department of Cell Biology, La Jolla, California

Dominic Cosgrove National Usher Syndrome Center, Boys Town National Research Hospital, Omaha, Nebraska

Michael Anne Gratton Department of Otorhinolaryngology (Head and Neck Surgery), University of Pennsylvania, Philadelphia, Pennsylvania

Scott J. Harvey Renal Division, Washington University School of Medicine, St. Louis, Missouri

Cassandra M. Kelleher Department of Cell Biology and Physiology, Washington University School of Medicine, St. Louis, Missouri

Jie Li Department of Dermatology, University of Miami School of Medicine, Miami, Florida

Ulrich Müller The Scripps Research Institute, Department of Cell Biology, La Jolla, California

Thomas J. Mariani Division of Pulmonary and Critical Care Medicine, Brigham and Women's Hospital, Harvard Medical School, Boston, Massachusetts

M. Peter Marinkovich Program in Epithelial Biology, Department of Dermatology, Stanford University School of Medicine and Veterans' Association Palo Alto Health Care System, Palo Alto, California

Sean E. McLean Department of Cell Biology and Physiology, Washington University School of Medicine, St. Louis, Missouri

Brigham H. Mecham   Division of Pulmonary and Critical Care Medicine, Brigham and Women's Hospital, Harvard Medical School, Boston, Massachusetts

Robert P. Mecham   Department of Cell Biology and Physiology, Washington University School of Medicine, St. Louis, Missouri

Ambra Pozzi   Division of Nephrology, Departments of Medicine and Cancer Biology, Vanderbilt University Medical Center and Veterans' Affairs Hospital, Nashville, Tennessee

Paul S. Thorner   Division of Pathology, Hospital for Sick Children and Department of Laboratory Medicine and Pathobiology, University of Toronto, Toronto, Ontario, Canada

Julia Tzu   Program in Epithelial Biology, Department of Dermatology, Stanford University School of Medicine and Veterans' Association Palo Alto Health Care System, Palo Alto, California

Roy Zent   Division of Nephrology, Departments of Medicine and Cancer Biology, Vanderbilt University Medical Center and Veterans' Affairs Hospital, Nashville, Tennessee

Preface

About twelve years ago a colleague of mine lamented in a seminar that his studies of the mechanisms of action of a very famous growth factor had led him into the territory of heparan sulfate proteoglycans, which in his mind were second only to collagen in terms of boringness. Although he was certainly not alone in his opinion, which very likely extended to cover extracellular matrix molecules in general, those of us who study these proteins like to think that we have proven well beyond any reasonable doubt that a structural protein located outside of a cell can have a profound influence over cell behavior and tissue development and function. There exists a plethora of examples that demonstrate this, from simple multicellular organisms such as sponges, to our own complex species. Indeed, perhaps the most convincing argument that extracellular matrix proteins are worthy of our attention is that there are serious, relatively common human diseases that are caused by mutations in genes that encode these proteins. This has spurred a great number of researchers to study the extracellular matrix, sometimes by choice and sometimes by necessity, and much progress has been made in the last decade toward understanding what matrix proteins do and how cells interact with and respond to them.

This book is a compilation of reviews by leaders in the fields of extracellular matrix and cell/matrix interactions. Because it is beyond the scope of a book such as this to present a full picture of the extracellular matrix, I have designed it to focus on a diverse subset of matrix proteins that have been shown to be important for development, function, and disease. The book therefore both presents a broad view of the field and provides crucial details about some of the best-studied matrix molecules.

One of the most abundant proteins in the body is collagen, of which there are twenty-seven different types assembled from forty-two genetically distinct chains. The chapter by Harvey and Thorner focuses on type IV collagen, which is found ubiquitously and almost exclusively in basement membranes. Collagen IV has been found in all multicellular animal species, and for many years it was thought to serve as a requisite scaffold for basement membrane formation. However, recent data from mutant mouse embryos lacking collagen IV but containing basement membranes has forced a revision in this theory. Collagen IV also has great relevance to human health, as mutations in three of the six collagen IV chains can cause kidney failure and deafness. In addition, one of the chains harbors a cryptic epitope that in rare cases is attacked by the immune system, leading to kidney disease and lung hemorrhage.

Sulfated proteoglycans are widely deposited in the extracellular matrix. The chapter by Arikawa-Hirasawa provides an in depth discussion of perlecan, the major heparan sulfate proteoglycan in basement membranes that is also found in other extracellular matrices. Perlecan binds to many matrix molecules, growth factors, and cell surface receptors and is involved in regulating angiogenesis. Mutations in perlecan are responsible for several rare human syndromes that affect muscle and skeletal development and function. Studies in mice lacking perlecan indicate important roles in heart development and potentially in development of atherosclerosis.

Abnormal matrix accumulation in atherosclerotic vessels is certainly a serious health problem. The chapter by McLean, Mecham, and colleagues deals with extracellular matrix gene expression during normal development in the mouse aorta. These authors performed gene expression profiling at different stages of aortic development and provide a comprehensive analysis of expression of pertinent matrix proteins, including collagen, elastin, fibrillin, fibronectin, laminin, and various proteoglycans. They also profile the non-matrix-related genes expressed by the vascular smooth muscle cells present in the aortic wall. These data show that vascular smooth muscle cells exhibit a wide range of phenotypes at different stages of development that may relate to their abnormal behavior in the setting of vascular disease.

The extracellular matrix associated with the dermal-epidermal junction in the skin is crucial for maintaining the barrier function of the skin and preventing blistering. The chapter by Tzu, Li, and Marinkovich discusses the specialized nature of the matrix molecules and their receptors that comprise the dermal-epidermal junction. Interactions among these proteins maintain the integrity of the linkage between dermis and epidermis. There are a number of human skin diseases caused by mutations in these matrix components and in their receptors, and most of these result in blistering that in some cases can be so severe as to be lethal. Studies of the dermal-epidermal junction have led to an overall better understanding of the mechanisms of how cells interact with their neighboring extracellular matrix.

Cell/matrix interactions have been shown by many groups to be crucial for proliferation, differentiation, and migration of neurons and glia. The chapter by Belvindrah and Müller reviews what is known about integrin signaling and development of the central nervous system. It has become clear from studies of mice lacking either integrin receptors or their ligands that interactions between these two classes of molecules literally shape the brain, as defects in foliation are apparent in the various mutants. However, this appears to be more a product of effects on cell proliferation rather than directly on cell migration.

A complex component of the nervous system that is especially difficult to access is the auditory system of the inner ear. But its location within bone has not prevented progress toward understanding the extracellular matrix molecules that are deposited in the inner ear. These proteins are known to be important in the function of sound transmission because mutations in several of them have been found to be associated with deafness in humans. The chapter by Cosgrove and Gratton presents a detailed account of the matrix molecules expressed in the inner ear, where they are deposited, and which ones are associated with hearing loss. Mutations that affect proteins found either in connective tissue and or in basement membranes can cause impaired hearing.

Many investigators who study extracellular matrix are interested in cancer and metastasis. This is because the spread of cancer from one tissue to another requires the cancerous cells to move through one or more extracellular matrix barriers. The chapter by Zent and Pozzi concentrates on the role of extracellular matrix in the progression of cancer and how cells use matrix-degrading enzymes, primarily matrix metalloproteinases, to overcome the natural matrix barriers present in tissues. It has recently become clear that fragments of extracellular matrix proteins, which result from their cleavage by matrix-degrading enzymes, can have novel activities—both positive and negative—in regulating cancer cell migration, invasion, or recruitment of new blood vessels. Matrix components may therefore potentially be used as therapeutic tools for the treatment and prevention of cancer.

So if you are a reader who needed convincing that extracellular matrix proteins are not boring, I trust that this brief introduction has accomplished that. If you were already somewhat familiar with the activities and importance of extracellular matrix proteins in development and disease, I trust that this brief introduction has made it clear that there is so much more to learn and discover about these very interesting proteins.

JEFFREY H. MINER, Ph.D
Washington University School of Medicine

Type IV collagen: A network for development, differentiation, and disease

Scott J. Harvey[1] and Paul S. Thorner[2]

[1] *Renal Division, Washington University School of Medicine, St. Louis, Missouri*
[2] *Division of Pathology, Hospital for Sick Children and Department of Laboratory Medicine and Pathobiology, University of Toronto, Toronto, Ontario, Canada*

Contents

Advances in Developmental Biology
Volume 15 ISSN 1574-3349
DOI: 10.1016/S1574-3349(05)15001-7

1. Introduction

This chapter reviews the developmental biology of type IV collagen, a specialized network forming extracellular matrix (ECM) protein that is a major structural component of basement membranes. Similar to other collagens, it exists as a family of triple helical molecules (protomers) assembled from three α chains. With six genetically distinct α chains (α1-α6) identified in higher organisms, type IV collagen displays the greatest molecular diversity among members of the collagen superfamily. Perhaps not surprisingly, its structural and functional properties have proven equally as complex.

The first section of this chapter provides a brief overview of basement membranes, highlighting their general organization, molecular composition, and function. The four sections that follow each deal with separate aspects of the structure and assembly of type IV collagen. This begins with a review of the genes encoding type IV collagen, which display a unique pair-wise organization in the genome of nearly all species studied that distinguishes them from other collagen family members and also has important implications for their transcriptional regulation. The insight that this arrangement has provided on the evolutionary history of this truly ancient ECM component is discussed. The section that follows outlines the primary structure of type IV collagen α chains, with emphasis on domains that are of functional importance as well as common post-translational modifications. A summary of the biosynthesis and isoform composition of type IV collagen protomers is presented, followed by an overview of the organization and distribution of supramolecular networks into which these assemble.

The specific role of type IV collagen in development is then addressed, beginning with what has been learned from studies of invertebrate models. The temporal and spatial expression of type IV collagen during normal development is reviewed for several invertebrate organisms including hydra, sea urchin, nematodes, and *Drosophila*. Most invertebrates synthesize basement membranes containing only the α1 and α2 chains that are co-assembled as protomers into a single (often termed "classic") $\alpha 1/\alpha 2$-containing network. As might be predicted, experimental disruption of this network in these organisms results in profound developmental defects. The final section turns to comparable studies in higher organisms, that in addition to the "classic" network, possess a greater repertoire of type IV collagen isoforms ($\alpha 3$-$\alpha 6$) capable of assembly into "novel" networks ($\alpha 3/\alpha 4/\alpha 5$ and $\alpha 1/\alpha 2/\alpha 5/\alpha 6$) that are unique to vertebrates. Recent studies of transgenic mice are summarized that (like in invertebrates) point to the critical role for the "classic" $\alpha 1/\alpha 2$ network in embryonic development. This section also reviews developmental shifts involving the "novel" networks of type IV collagen that occur in a variety of tissues during both fetal and postnatal life. Proper assembly of $\alpha 3/\alpha 4/\alpha 5$ network is not required for early development, but is essential for the long-term maintenance of basement membrane structure and function. Disruption of this network in humans and animal models leads to Alport syndrome, a disorder characterized by progressive nephropathy, deafness, and ocular abnormalities. To date, no biologic function has been ascribed to the $\alpha 1/\alpha 2/\alpha 5/\alpha 6$ network.

2. Basement membranes

2.1. Structure

Basement membranes, or basal laminae, are sheet-like specialized ECMs that lie at the interface between cells and connective tissue stroma. They are found underlying all epithelial cells (except hepatocytes) and all endothelial cells (except those of sinusoidal capillaries in the spleen, liver, lymph nodes, and bone marrow where they are discontinuous or absent), and they ensheath individual myocytes, adipocytes, and Schwann cells (Martinez-Hernandez and Amenta, 1983; Inoue, 1989; Merker, 1994). By standard electron microscopy (EM), most "simple" basement membranes appear as uniform amorphous structures 50 to 100 nm in thickness that are in intimate contact with the adjacent cell layer. They typically have a bilaminar appearance, with an electron-dense central core termed the lamina densa and a thinner subcellular layer of lower electron density termed the lamina rara (or lamina lucida). The poorly-limited transitional region underlying the lamina densa that contacts loose connective tissue (the lamina fibroreticularis) is also formally classified as a component of the basement membrane zone. It is comprised of a variety of stromal cell-derived matrix components including tenascin, proteoglycans, and interstitial collagens. In some sites, basement membranes from opposing cell layers fuse to become a trilaminar structure that lacks a lamina fibroreticularis and contains two laminae rarae that bound a central thicker lamina densa. The glomerular and alveolar basement membranes, which both separate capillary endothelial cells from an epithelial cell layer, are examples of such "double" basement membranes and their unique organization is thought to reflect their functional specialization as filtration barriers. Other specialized basement membranes such as the lens capsule, Reichert's membrane of the parietal wall of the yolk sac, and Descemet's membrane of corneal endothelium are multilayered basement membranes of extraordinary thickness (up to $\sim$20 μm).

2.2. Molecular composition

The composition of basement membranes has been reviewed by others (Timpl, 1989; Miner, 1999; Erickson and Couchman, 2000). They are amazingly complex matrices, with $\sim$40 distinct molecular components identified to date. These can be generally classified as either extrinsic or intrinsic. Extrinsic components have a limited distribution, and are often prominent in basement membranes that serve as filtration barriers, suggesting they are synthesized elsewhere and deposited selectively in these sites. Examples include amyloid P component (Dyck et al., 1980) and fibronectin (Martinez-Hernandez et al., 1981). In contrast, intrinsic components are those that are synthesized and secreted by the adjacent cellular layer(s) and have a broad distribution among basement membranes. Their expression can vary among different tissues and developmental stages, with specific isoforms expressed early in development that are eliminated and replaced by distinct isoforms that persist at later stages.

All basement membranes have four major intrinsic components: type IV collagen, laminin, nidogen, and heparan sulfate proteoglycans (HSPGs). These molecules interact with one another to form a highly cross-linked network (Fig. 1). Type IV collagen and laminin are thought to associate in a homotypic fashion to form independent networks that are interconnected by nidogen producing a polymeric scaffold into which HSPGs and other basement membrane-associated molecules are incorporated (Yurchenco and O'Rear, 1994; Timpl and Brown, 1996). Some of the major components are multimeric proteins assembled from subunits that exist in a number of different isoforms. This is the case for type IV collagen (see section entitled Type IV collagen protomers), and also for laminin which forms the major non-collagenous component of basement membranes. Individual laminin molecules are heterotrimers of three polypeptide chains, designated α, β, and γ (Colognato and Yurchenco, 2000). Studies to date have identified 5 genetically distinct α subunits, 4 β subunits, and 3 γ subunits that co-assemble to form 15 different molecules (laminins 1-15) with additional complexity arising from alternative splicing and proteolytic processing (Libby et al., 2000; Tunggal et al., 2000). Two genetically distinct forms of nidogen exist (nidogen-1 and -2) and, although the former is more widely

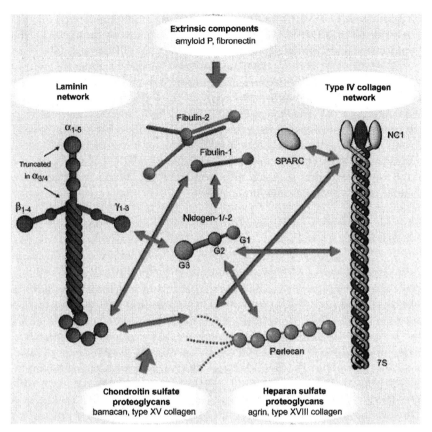

Fig. 1. Molecular composition and assembly of basement membranes. All basement membranes have four major intrinsic components: type IV collagen, laminin, nidogen, and heparan sulfate proteoglycans (HSPGs). Type IV collagen, a heterotrimer of 3 α chains, forms a network through self-interaction at both its carboxyl (NC1) and amino (7S) termini. Laminin molecules are "cross-shaped" heterotrimers consisting of an α, β, and γ chain that form a separate network through interactions between globular domains at the termini of its three short arms. These networks are bridged by nidogen, which binds to the $\gamma1$ chain of laminin through its G3 domain and also to triple helical region of type IV collagen through its G2 domain. The HSPGs such as perlecan are incorporated into this scaffold by binding to the molecules above via their core protein or heparan sulfate side chains. In some basement membranes, chondroitin sulfate proteoglycans are incorporated through similar interactions. Other intrinsic components include SPARC (which binds to type IV collagen) and members of the fibulin family (which bind to nidogen and laminin). Extrinsic molecules, such as fibronectin and amyloid P component, are not synthesized locally but are deposited in basement membranes through their affinity for the various intrinsic molecules. The components are illustrated roughly twice their scale relative to type IV collagen. Figure adapted from Current Opinion in Cell Biology, Vol. 6, No. 5, Yurchenco & O'Rear, "Basal lamina assembly," pp. 674–681, 1994, with permission from Elsevier, Ltd. (See Color Insert.)

expressed among basement membranes, they are nearly functionally equivalent (Murshed et al., 2000; Miosge et al., 2001; Schymeinsky et al., 2002). The chondroitin sulfate proteoglycans bamacan (Wu and Couchman, 1997) and type XV collagen (Li et al., 2000) are intrinsic basement membrane components. However, HSPGs predominate in most sites, and three distinct forms have been identified; perlecan (Noonan et al., 1991), agrin (Groffen et al., 1998), and type XVIII collagen (Halfter et al., 1998). Several minor intrinsic components are also recognized. These include members of the fibulin family (Timpl et al., 2003), SPARC/osteonectin (Lankat-Buttgergeit et al., 1988), and usherin (Bhattacharya et al., 2002). Finally, type VIII collagen is present in diverse ECMs but is especially prominent in Descemet's membrane (Kapoor et al., 1988; Sawada et al., 1990).

2.3. Function

Basement membranes serve an important structural role, acting both as a supportive and adhesive substrate for the cells from which they are derived, as well as a means of compartmentalizing these cells into distinct tissues. Not surprisingly, they play critical roles in development (Yurchenco et al., 2004). Their basic molecular components are first expressed and assembled in pre-implantation blastulae, with some constituents (laminin in particular) detectable as early as the two-cell stage. During the process of gastrulation, definitive basement membranes are present that segregate the embryonic germ layers (endoderm, mesoderm, and ectoderm). Their constituents, now increasing in molecular diversity, are recognized by an array of cell-surface receptors that transmit signals from the extracellular environment influencing cell growth, apoptosis, differentiation, and migration. Some tissues, such as the kidneys and lungs, are formed through branching morphogenesis, a process that requires the coordinated dissolution and *de novo* synthesis of a basement membrane at the interface of epithelial and mesenchymal cells. Basement membranes bind and sequester a variety of growth factors, and when degraded during physiologic or pathologic processes, these molecules can be released in their active form (Folkman et al., 1988; Lortat-Jacob et al., 1991; Paralkar et al., 1991; Gohring et al., 1998). Proteolytic fragments of some components, such as type IV collagen, can exert effects functionally distinct from the intact molecules (Xu et al., 2001; Ortega and Werb, 2002).

As a testament to their functional importance, mutations in the genes encoding basement membrane components have been identified in a number of human diseases. For example, those encoding the $\alpha3$-$\alpha6$ chains of type IV collagen are disrupted in Alport syndrome (see section entitled Type IV collagen in development: lessons from vertebrates), $\alpha1(XVIII)$ collagen in Knobloch syndrome (Suzuki et al., 2002), and $\alpha2(VIII)$ collagen in corneal endothelial dystrophies (Biswas et al., 2001). The subunits of laminin-5 ($\alpha3$, β3, and γ2) are disrupted in junctional epidermolysis bullosa (Nakano et al., 2002), the laminin α2 chain in congenital muscular dystrophy (Helbling-Leclerc et al., 1995), perlecan in Schwartz-Jampel syndrome (Nicole et al., 2000), and usherin in Usher syndrome (Weston et al., 2000). Acquired basement membrane diseases also exist; for example, the α3 chain of type IV collagen is the target of autoantibodies in Goodpasture syndrome (Borza et al., 2003). Finally, basement membranes play an important role in cancer, since they represent a physical barrier to the metastasis of tumor cells and must be remodeled during the process of angiogenesis, which supports tumor growth (Engbring and Kleinman, 2003; Kalluri, 2003).

3. Type IV collagen genes

Type IV collagen is one member of the collagen superfamily that in humans currently contains 27 different protomers assembled from 42 genetically distinct α chains (Gelse et al., 2003; Myllyharju and Kivirikko, 2004). These are organized into functionally distinct classes. Type IV collagen, together with types VIII and X, belong to the network-forming collagens. The remaining collagens are classified as fibrillar, microfibrillar, anchoring fibrils, transmembrane, multiplexin, and fibril-associated with interrupted triple helices (FACIT). The defining feature of all α chains are repeating units of the tripeptide Gly-Xaa-Yaa, where Xaa and Yaa may be any amino acid but are both most commonly proline. The requirement for glycine in the first position of each triplet is dictated by the triple helical structure of collagen. Briefly, the three α chains (each itself a left-handed helix) intertwine with one another about a common central axis but offset by a stagger of one residue that forms a right-handed superhelix (Bella et al., 1994). Because of this conformation, every third residue from each α-chain lies buried near the center of the molecule; glycine, the smallest amino acid, is the only residue that can occupy this position. Whereas the fibril-forming collagens are characterized by long uninterrupted stretches of repeating Gly-Xaa-Yaa triplets, type IV collagen contains numerous short interruptions along its length that are thought to confer flexibility to the molecule, allowing it to undergo network assembly.

3.1. Organization

In humans, six distinct α chains of type IV collagen exist that are designated α1-α6 based on the order in which they were identified, and each is encoded by a separate gene designated as *COL4A1-COL4A6*, respectively. These genes are larger and more complex than their fibrillar counterparts, each spanning from ~150 to ~350 kb and comprising ~50 exons. They also display an unusual genomic organization in that they are closely arranged pair-wise in head-to-head fashion, and are transcribed from opposite DNA strands under the control of a shared bidirectional promoter. The human *COL4A1-COL4A2* genes are located on chromosome 13q34 (Griffin et al., 1987; Boyd et al., 1988), *COL4A3-COL4A4* on chromosome 2q36–37 (Mariyama et al., 1992; Momota et al., 1998), and *COL4A5-COL4A6* on chromosome Xq22 (Sugimoto et al., 1994; Oohashi et al., 1995). Combining data available from public genomic databases with the published work of others reveals that orthologues of each of the human genes are represented in the genome of mice, rats, and dogs, and that their pair-wise arrangement is conserved (Kaytes et al., 1988; Thorner et al., 1996; Lu et al., 1999; Zheng et al., 1999; Lowe et al., 2003). The *COL4A5-COL4A6* complex is located on the X chromosome in all of these species. This is expected given that constraints arising from dosage compensation are thought to ensure that genes X-linked in one species of placental mammals are X-linked in all (Ohno's law). Analysis of the genomic sequence available for the chicken predicts the existence of orthologues of the vertebrate *COL4A1-COL4A6* genes that are arranged pair-wise on autosomes.

No more than two genetically distinct isoforms of type IV collagen have been identified in any invertebrate. In the sponge (*P. jarreri*), a single α chain has been identified that is encoded by the gene *PjCOL4α* (Boute et al., 1996). This isoform, designated 4α, displays features common to both the "odd" and "even" α chains identified in other species making its classification ambiguous. However, comparison of its partial genomic sequence with the human sequences suggested it was more similar to the α2, α4, or α6 chains. A single α chain has been cloned in hydra that is homologous to the α1 isoform of vertebrates (Fowler et al., 2000). In the sea urchin (*S. purpuratus*), two isoforms designated 3α and 4α have been identified that are homologues of the vertebrate α1 and α2 chains, respectively (Wessel et al., 1991; Exposito et al., 1993; Exposito et al., 1994). Their corresponding genes (*COLP3α* and *COLP4α*) may not be arranged in the urchin genome in a manner comparable to that in vertebrates. In *Drosophila*, the genes *DCg1/Cg25C* (Blumberg et al., 1988) and *viking* (Yasothornsrikul et al., 1997) that encode α chains homologous to the vertebrate α1 and α2 isoforms, respectively, are arranged pair-wise and head-to-head on chromosome 2. Both isoforms have also been cloned in other dipteran species including *S. peregrina* and *A. gambiae* (Gare et al., 2003). Homologues of the vertebrate *COL4A1* and *COL4A2* genes have been identified in *C. elegans* and are designated *emb-9* and *let-2*, respectively (Guo and Kramer, 1989; Sibley et al., 1993). The *emb-9* gene is located on chromosome III and *let-2* on chromosome X; this represents the only proven exception the usual paired arrangement of type IV collagen genes among metazoans. At ~10 kb in size and comprising between 10 and 20 exons, both genes are considerably smaller than their human counterparts, and more similar to those of *Drosophila*. Orthologues of the α1 and α2 chains of *C. elegans* are predicted in the closely related species *C. briggsae* through analysis of its genomic sequence, and an α2 isoform has been characterized in the more distantly related nematodes *B. malayi* (Caulagi and Rajan, 1995) and *A. suum* (Pettitt and Kingston, 1991).

3.2. Evolutionary history

In evolutionary terms, the molecular components of basement membranes are "living relics" among members of the ECM family (Hutter et al., 2000; Hynes and Zhao, 2000). The importance of type IV collagen and of the role it plays in the formation of basement membranes that allow for the more complex body plans of multicellular organisms is reflected by the fact that it is the only known ubiquitous collagen. It is represented throughout the metazoan kingdom, from sponges to man, a history that spans roughly 500 million years of evolution. A defining feature of all type IV collagen α chains is the globular non-collagenous domain (NC1) present at their carboxyl terminus that is crucial for their triple helical assembly into protomers and incorporation into networks. The NC1 region can be divided into two homologous subdomains with a highly conserved spatial arrangement of cysteine residues and comparable secondary structure that allows them to fold into disulfide-bonded structures of near identical topology. It seems likely that early in its evolutionary history, a portion of an ancestral gene encoding a single NC1 subdomain underwent internal duplication with the two coding regions becoming arranged in tandem.

Zhou and colleagues (1994) compared the deduced amino sequences of the human α1-α6 chains and proposed a classification system based on their degree of similarity. The α1, α3, and α5 chains were grouped as an "α1-like" class, while the α2, α4, and α6 chains formed the "α2-like" class, a division that extends to the genomic level since each gene encoding an 'α1-like' chain is paired with a neighboring gene encoding an "α2-like" chain. Several studies have speculated on the evolutionary history of vertebrate *COL4A1-COL4A6* genes based upon these criteria (Hudson et al., 1993; Zhou et al., 1994). The current model predicts a single ancestral gene that underwent duplication and inversion giving rise to a pair of genes arranged head-to-head and separated by a shared promoter region. These genes then diverged, forming the precursors of the current "α1-like" and "α2-like" genes. A second round of duplication subsequently occurred, this time involving both genes as a single unit, which gave rise to two complexes. The first represents the precursor of the *COL4A3-COL4A4* genes, while the second later underwent a final round of duplication giving rise to the *COL4A1-COL4A2* and *COL4A5-COL4A6* complexes.

The identification of homologues of the vertebrate *COL4A1* and *COL4A2* genes in a number of invertebrate genomes has allowed this model to be refined (Boute et al., 1996; Gare et al., 2003). A phylogenetic tree derived herein from analysis of the NC1 domain amino acid sequences of nearly 50 α chains from 20 different species is presented in Fig. 2 and described in the following text. In general, the deduced relationships are consistent with those that have been proposed by others, but in no case are the models identical. The point during evolution at which type IV collagen first "emerged" is

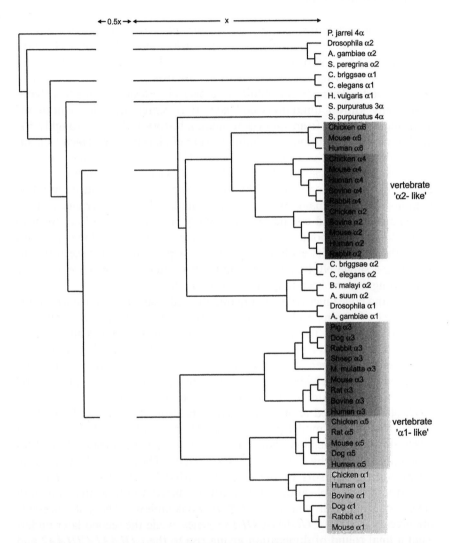

Fig. 2. Phylogeny of metazoan type IV collagen α chains. NC1 domain amino acid sequences were aligned with ClustalW, and a phylogenetic tree derived using the protein parsimony algorithm of the program protpars in the PHYLIP v3.6 software package. The tree was rooted through the sequence of the sponge (P. jarrei) which is considered ancestral. Some available sequences that were incomplete were omitted from the analysis. (See Color Insert.)

unknown. Although genetic studies were not performed, "type IV collagen-like" material has been detected immunologically in the very primitive dicyemid mesozoan *K. antartica*, where it appears to be localized intracellularly (Czaker, 2000). The body of this organism is formed of only 10 to 40 cells, the fewest among multicellular animals, and it lacks structures typical of basement membranes. The most primitive metazoan in which type IV collagen has been conclusively identified at the genetic level is the marine sponge *P. jarreri* (Boute et al., 1996). This may not be representative of other members of the phylum porifera, since only a group of sponges to which *P. jarreri* belongs (homoscleromorpha) elaborate a basement membrane. The 4α chain identified in this species is considered the most ancestral member of the type IV collagen family identified to date, since its NC1 domain displays features of both the "α1-like'" and "α2-like" classes into which all other known α chains can be assigned. Arising from the sponge sequence in the phylogenetic tree is a clade containing the α2 isoform from three dipteran species that are ancestral to all other α chains. Most closely related to these is the α1 chain of several invertebrates; first, nematodes, then hydra and sea urchin. From the sequences above emerges a common ancestor of the dipteran α1 chain and nematode and urchin α2 chain as well as all vertebrate isoforms. As expected, the vertebrate α chains ultimately emerge from this ancestor as two clades, α1-like' and "α2-like", which reflects the duplication/inversion event that gave rise to the primordial vertebrate *COL4A1-COL4A2* complex. As this complex is twice duplicated, it gives rise to three isoforms within the "α1-like" clade (α1, α3, α5) and also within the "α2-like" clade (α2, α4, α6) that are grouped together irrespective of species. Curiously, the dipteran α1 and nematode α2 chains form a separate clade, being more closely related to each other and to the vertebrate "α2-like" chains than to all other invertebrate chains. The same conclusions were reached by others through the analysis of nucleotide sequences (Gare et al., 2003).

    The metazoan kingdom is divided into two subkingdoms: the parazoa and the eumetazoa. The former, which comprises only sponges and mesozoans (such as *K. antartica*), lack true tissues and only one isoform of type IV collagen has been identified in any of its members. The eumetazoan subkingdom is divided into radiata and bilateria based on the symmetry of their body plans during early development. The radiata consist of only the phyla Ctenophora (comb-jellies) and Cnidaria (such as hydra). These organisms display radial symmetry (i.e., a top and bottom) and have only two germ layers: ectoderm and endoderm. Similar to the parazoans, only one isoform of type IV collagen has been identified in any of its members. All other animals are bilaterians. They display bilateral symmetry (i.e., a top, bottom, front, and back) and are formed of three germ layers: ectoderm, mesoderm, and endoderm. At least two isoforms of type IV collagen have been identified in all bilaterians studied. Thus, the duplication and subsequent divergence of an ancestral type IV collagen gene seems to have occurred coincident with (and was perhaps necessary for) the emergence of the more complex body plan of bilaterians. Homologues of *only* the *COL4A1-COL4A4* genes have not been identified in any organism, nor would this be expected given what is known about how their corresponding α chains are assembled into basement membranes (see section entitled Type IV collagen networks). In the accepted model on the evolution of vertebrate type IV collagen, the second duplication of the *COL4A1-COL4A2* complex that gave rise to the *COL4A5-COL4A6* genes must have occurred very soon after the first that gave rise to the *COL4A3-COL4A4* genes. Finally, essentially nothing is known about which isoforms are found in several major metazoan taxa, notably fish and amphibians. The genomes of several representative species are being sequenced and are in their early stages of annotation. Sequence data compiled from these and other organisms will allow us to complete gaps in our understanding of the evolutionary history of type IV collagen.

3.3. Comparison to other collagen genes

The vertebrate *COL4A1-COL4A6* genes show several interesting differences from their counterparts encoding fibrillar collagens. For all collagen genes, the basic organization of exons encoding collagenous portions of the molecule are similar, in that they almost invariably begin with a codon for glycine and end with a codon for an amino acid in the final position of a Gly-Xaa-Yaa triplet. In all human type IV collagen genes approximately one-half of these exons begin with split codons that start with the second nucleotide for the codon of glycine. This is in contrast to the fibrillar collagen genes, where all exons begin with a complete codon for glycine (Prockop and Kivirikko, 1995). For type IV collagen, this pattern of intron/exon boundaries only differs when the exon begins with a non-collagenous interruption. For unknown reasons, the distribution of split versus complete codons show a strong positional bias among type IV collagen genes, with exons in the 5' half starting with a complete codon, whereas the 3' exons are split. In the fibrillar collagen genes, nearly all of the exons encoding the collagenous domain are either 54 bp in length, multiples thereof (108, 162 bp), or combinations of 45 and 54 bp which has been interpreted as evidence of a conserved ancestral collagen coding unit of 54 bp (Yamada et al., 1980). Studies on the genomic structure of the mouse *COL4A1* and *COL4A2* genes first revealed that type IV collagen does not follow this "54 bp rule" (Kurkinen et al., 1985; Sakurai et al., 1986); this has since been demonstrated to be characteristic of all type IV collagen genes regardless of isoform or species. Most of the non-collagenous interruptions typical of type IV collagen are encoded at intron/exon boundaries or by entire exons of their own, which has prompted speculation that they may have evolved from introns (Buttice et al., 1990).

3.4. Transcriptional regulation

The closely apposed 5' ends of the paired type IV collagen genes are separated by an overlapping bidirectional promoter that share common elements, but which directs transcription in divergent directions in a regulated manner. This specificity is critical since the α chains encoded by each set of paired genes are usually, but not always, expressed in the same site. Bidirectional promoters are not uncommon in the human genome, and are thought to coordinate the expression of functionally related genes (Adachi and Lieber, 2002).

The *COL4A1-COL4A2* promoter has been the most extensively studied. In humans (Soininen et al., 1988) and rodents (Burbelo et al., 1988; Grande et al., 1996) this promoter is ~130 bp in size and shares many of the same features among species. The human promoter lacks canonical TATA boxes that normally fixes the orientation of RNA polymerase, but it contains A/T-rich elements ~30 bp upstream of the transcriptional start site of both genes. The intergenic promoter region contains a potential Sp1 transcription factor binding site (GC box) as well as a CCAAT box, and a total of four other GC boxes are also found within the first exon or intron of both genes. The sequence motif CCCTYCCCC (designated a CTC box) is found in the promoter and within the first intron of both genes (Fischer et al., 1993; Schmidt et al., 1993) and is recognized by a specific transcription factor termed CTCBF (Genersch et al., 1995). The CTC boxes have been identified in the promoter region of the genes encoding fibronectin, laminin $\beta 1$ and $\gamma 1$ (Bruggeman et al., 1992; Levavasseur et al., 1996) as well as all other type IV collagen chains, suggesting this motif may play an important role in regulating the coordinated expression of components of the ECM. The *COL4A1-COL4A2* promoters of humans (Pollner et al., 1997) and mice (Burbelo et al., 1988) show very limited transcriptional activity in either direction, and are highly dependent on cis-acting regulatory elements located within the first intron of both genes. In the human sequence, further control is thought to come from a silencing element in intron 3 of the *COL4A2* gene that influences transcription in both directions.

The sequence of the human (Momota et al., 1998) and mouse (Lu et al., 1999) *COL4A3-COL4A4* promoters have been determined, and in both species it displays many of the features previously described, including the lack of TATA boxes and the presence of GC, CTC, and CCAAT boxes. The human *COL4A4* gene contains two transcriptional start sites that are separated by 373 bp, with one of these located only 5 bp away from the first exon of *COL4A3*. Transcription from these sites yields two α4 mRNAs that differ in their 5' untranslated regions (UTRs). One transcript (containing exon 1) is expressed predominantly in epithelial cells, whereas the other (containing exon 1') is expressed ubiquitously at low levels. In the mouse, the intergenic region is 186 bp in length and over this region it shares 72%

identity to the human sequence. Alternative transcriptional start sites in the mouse sequence have not been reported. The *COL4A1-COL4A2* and *COL4A3-COL4A4* promoters share many putative regulatory elements, and while the α chains that both complexes encode are invariably co-expressed, all four isoforms do not have an identical distribution among basement membranes (see section entitled Type IV collagen networks). The mechanism(s) by which this specificity is dictated at the transcriptional level are poorly understood. However, the LIM-homeodomain transcription factor LMX1B, which binds to a putative enhancer region in intron 1 of the human and mouse *COL4A4* genes, might be involved in this process (Morello et al., 2001).

The *COL4A5-COL4A6* promoter has been characterized in humans, mice, and dogs (Sugimoto et al., 1994; Zheng et al., 1999; Sund et al., 2004). In the human sequence, the *COL4A6* gene contains two putative transcriptional start sites that generate mRNA transcripts differing in their 5′ UTRs, as well as partially within their encoded amino-terminal signal peptides. The transcriptional start site for the first isoform (containing exon 1′) is located 442 bp away from exon 1 of the *COL4A5* gene, whereas that of the second isoform (containing exon 1) is located an additional ∼1 kb further downstream. The α6 isoform containing exon 1′ is abundant in the placenta, whereas the transcript containing exon 1 predominates in the kidneys and lungs. The proximal intergenic region separating *COL4A5* exon 1 and *COL4A6* exon 1′ lacks TATA boxes, but contains a single CCAAT and CTC box. The same motifs are found in the more distal promoter region upstream of *COL4A6* exon 1.

4. Type IV collagen α chains

4.1. Domain structure

The complete amino acid sequence of the α1-α6 chains from humans and mice have been deduced from their cDNA sequences. A summary of these proteins is provided in Table 1. This section highlights the properties of domains that are of structural or functional importance in a prototypic α-chain (from its amino to carboxyl terminus), and summarizes its post-translational modifications.

Unprocessed α chains contain a signal peptide ∼20 to 40 residues in length at their amino terminus that is characteristically rich in hydrophobic amino acids (mainly leucine). This sequence targets the newly synthesized peptide to the endoplasmic reticulum (ER), where it is then cleaved by a signal peptidase to yield the mature α-chain. The amino terminus of mature α chains consists of a region ∼140 amino acids in length termed the 7S domain (Risteli et al., 1980), which can be further divided into two sub-regions that

Table 1
Properties of human and mouse type IV collagen α chains

Isoform	Species	RefSeq #	SP	ATNC	7S	NC2	INT	COL	NC1	Length	MW	%
α1	Human	NP_001836	27	15	135	10	21	1268	229	1669	160.6	90.7
	Mouse	NP_034061	27	15	135	10	21	1268	229	1669	160.7	
α2	Human	NP_001837	36	21	135	12	23	1302	227	1712	167.5	83.5
	Mouse	NP_034062	28	29	143	12	24	1297	227	1707	167.4	
α3	Human	NP_000082	28	14	131	11	23	1268	232	1670	161.7	78.8
	Mouse	NP_031760	28	14	131	11	24	1266	233	1669	161.8	
α4	Human	NP_000083	38	23	137	8	26	1284	231	1690	164.1	78.2
	Mouse	NP_031761	30	23	137	5	26	1279	231	1682	164.1	
α5	Human	NP_203699	26	15	135	6	22	1295	229	1691	161.6	91.2
	Mouse	NP_031762	26	15	135	6	22	1295	229	1691	160.8	
α6	Human	NP_001838	21	25	139	8	25	1295	228	1691	163.8	79.5
	Mouse	NP_444415	15	31	145	8	24	1295	228	1691	164.1	

The number of amino acids in the following domains of each isoform is indicated: SP; signal peptide; ATNC = amino-terminal non-collagenous domain; 7S = 7S domain (includes the ATNC sequence); NC2 = non-collagenous domain-2; COL = main collagenous domain; NC1 = non-collagenous domain-1. INT denotes the total number of non-collagenous interruptions, and includes the NC2 domain. The MW of the unprocessed α chains are indicated based on their deduced amino acid sequence. Interspecies comparisons are made for each pair of α chains and the overall percent identity at the amino acid level is indicated. GenBank reference sequence (RefSeq) identifiers are provided, as are the primary references where available. The data presented are specific to human α3 isoform 1, human α5 isoform 2 (renal-splice form), human α6 isoform A (containing exon 1'), and the only known mouse α6 sequence, which is homologous to human α6 isoform B (containing exon 1). Human isoforms: α1 (Soininen et al., 1987, 1989); α2 (Hostikka and Tryggvason, 1988); α3 (Mariyama et al., 1994); α4 (Leinonen et al., 1994); α5 (Zhou et al., 1992; Martin and Tryggvason, 2001); α6 (Sugimoto et al., 1994; Zhou et al., 1994). Murine isoforms: α1 (Muthukumaran et al., 1989); α2 (Saus et al., 1989);, α5 (Saito et al., 2000); α6 (Saito et al., 2000).

differ in their primary structure. The 7S domain begins with a non-collagenous stretch of ~20 to 40 residues rich in lysine that contains four cysteines which are conserved in number, but not in relative position, among all isoforms of type IV collagen regardless of species. These residues are thought to play a critical role in the formation of intramolecular and intermolecular disulfide bonds, as well as lysine-derived cross-links that are involved in the higher order assembly of protomers through their amino termini (Siebold et al., 1987). The second sub-region of the 7S domain (~100 amino acids in length) is collagenous in nature. It contains a single cysteine residue in the Xaa position of a triplet, which is exceedingly rare among collagens. This residue is conserved among all isoforms except the α4 chain of vertebrates, and is believed to participate in an intermolecular disulfide bond between protomers (Siebold et al., 1987). An early study of type IV collagen identified a small collagenase-resistant portion of the molecule that was designated the non-collagenous-2 (NC2) domain (Timpl et al., 1981). Based on biochemical and ultrastructural evidence this fragment was proposed to lie at the junction of the 7S domain and the main collagenous domain, and therefore, represents the first non-collagenous interruption in the molecule. The length of the NC2 domain in the human and murine chains varies from 5 to 12 residues. For all but the α4 chain it is perfectly conserved between species in terms of length, but not sequence (~40% to 90% identity).

All isoforms of type IV collagen are characterized by a major collagenous domain ~1,300 amino acids in length. Although this region is structurally identical to the collagenous portion of the 7S domain, the latter serves a specific role in network assembly and is functionally distinct. The six isoforms in both humans and mice each contain between 21 to 26 non-collagenous interruptions that vary in length from 1 to 24 residues, with the majority of the larger interruptions distributed preferentially in their amino-terminal half. These non-collagenous regions are believed to confer flexibility to the molecule, which may facilitate formation of a cross-linked network in basement membranes. They have also been implicated as sites for interaction with the NC1 domain of other type IV collagen molecules (Tsilibary and Charonis, 1986). Cysteines within non-collagenous interruptions may participate in either intramolecular or interchain disulfide bonds (Fessler et al., 1984; Brazel et al., 1987; Soininen et al., 1987; Brazel et al., 1988; Hostikka and Tryggvason, 1988).

At the carboxyl terminus of each α chain is a globular non-collagenous region of ~230 amino acids designated the NC1 domain (Timpl et al., 1981). The length of the NC1 domain varies between isoforms, but for a given isoform is highly conserved between species. As previously described, this region can be divided into two homologous subdomains that share ~35% identity at the amino acid level. The NC1 domain contains a total of 12 cysteine residues (6 in each subdomain) whose number and spatial arrangement is highly conserved in all isoforms. The arrangement of disulfide bonds within NC1 domain monomers was predicted initially based upon sequence analysis and has since been confirmed empirically. In a study of human placental type IV collagen, Siebold and colleagues (1988) proposed a model in which the 12 cysteine residues form 6 intramolecular disulfide bonds (3 in each subdomain) leading to multiple disulfide-bridged loops and a compact globular structure. This is supported by both biochemical (Fessler and Fessler, 1982) and crystallographic (Sundaramoorthy et al., 2002; Than et al., 2002) studies that indicate the three constituent NC1 domains of a single protomer contain intramolecular, but not intermolecular, disulfide bonds.

4.2. Post-translational modification

Type IV collagen undergoes extensive co- and post-translational modification. The hydroxylation of lysine and proline residues is a defining feature of collagens. As will be described in the following text, hydroxylysine residues serve as attachment sites for galactose or glucosylgalactose moieties, and are involved in the formation of intermolecular cross-links (Kivirikko et al., 1992). The incorporation of hydroxyproline promotes and stabilizes the triple helical conformation of collagen thereby enhancing its secretion (Berg and Prockop, 1973). In a biosynthetic pathway that requires ascorbate, O_2, Fe^{2+}, and 2-oxoglutarate, the enzymes lysyl hydroxylase and prolyl 4-hydroxylase hydroxylate lysine and proline residues in the Yaa position of Gly-Xaa-Yaa triplets, respectively (Kivirikko et al., 1992). Prolyl 3-hydroxylase acts in a similar manner on proline residues in the Xaa position of Gly-Xaa-Hyp triplets, although this is a relatively infrequent modification. Type IV collagen has a much higher content of hydroxylysine and both 4- and 3-hydroxyproline (in terms of residues/unit length) than the fibrillar collagens (Denduchis et al., 1970; Kefalides, 1971, 1972, 1973; Grant et al., 1975; Bornstein and Sage, 1980). This may reflect the fact that in the fibrillar collagens, proline and lysine residues are more evenly distributed between the variant positions of Gly-Xaa-Yaa triplets and, thus, there are quantitatively fewer residues occupying the Yaa position that may potentially be hydroxylated (Traub and Piez, 1971; Babel and Glanville, 1984). The greater content of hydroxyproline in type IV collagen might serve to compensate for the destabilizing effect of its many non-collagenous interruptions. Hydroxylysine residues serve as attachment sites for O-linked galactose or glucosylgalactose carbohydrate units (Kivirikko et al., 1992). Incorporation of galactose is catalyzed by the enzyme hydroxylysyl galactosyltransferase, which can subsequently be linked to glucose by the enzyme galactosylhydroxylysyl glucosyltransferase. Type IV collagen is glycosylated to a greater extent than the fibrillar collagens, with hexose accounting for ~12% of its dry weight, nearly all of which represents the disaccharide glucosylgalactose (Kefalides, 1971, 1973). The impact of these modifications on biologic activity of type IV collagen are unclear at present, but there is evidence that glycosylation of specific residues can modulate cell adhesion (Lauer-Fields et al., 2003).

With the exception of the human α3 chain, all isoforms of type IV collagen in both humans and mice contain a potential site for the attachment of an asparagine-linked oligosaccharide at a conserved position within the collagenous region of the 7S domain. Some isoforms also contain one to two additional sites within the main collagenous domain. Studies of the α1 and α2 chains in various species have demonstrated that the conserved asparagine residue is invariably glycosylated (Leushner, 1987; Fujiwara et al., 1991; Langeveld et al., 1991; Nayak and Spiro, 1991). The exact composition of the oligosaccharide can vary, but in general it is characterized as either a tri- or bi-antennary complex that is fucosylated at its N-acetylglucosamine core and contains N-acetyllactosamine branches variably substituted with sialic acid or galactose moieties. Steric hindrance imparted by these carbohydrates is thought to play a role in determining the maximal extent of axial overlap between protomers interacting through their amino termini (Glanville et al., 1985; Siebold et al., 1987; Langeveld et al., 1991).

Type IV collagen extracted from various sources contains reducible, lysine-derived cross-links (Tanzer and Kefalides, 1973; Heathcote et al., 1980; Wu and Cohen, 1983; Bailey et al., 1984). These are presumably formed in the extracellular compartment in the same manner as with other collagens (Eyre et al., 1984). This process is initiated by the enzyme lysyl oxidase, which deaminates the ε-amino groups of lysine or hydroxylysine residues generating the reactive aldehydes allysine and hydroxyallysine, respectively. These may spontaneously form an aldimine bond with the ε-amino group of an unmodified lysyl, hydroxylysyl, or glycosylated hydroxylysyl residue on neighboring α chains (termed a bifunctional cross-link since it is formed between two chains). Alternatively, these may form aldol condensation products with other aldehydes that can react with histidine or lysine residues forming stable, non-reducible trifunctional, and tetrafunctional cross-links. While some have suggested these more complex cross-links might exist in type IV collagen (Wu and Cohen, 1983; Bailey et al., 1984), the majority of evidence points to aldimine cross-links as the predominant form. Lysine-derived cross-links presumably contribute to the structural integrity type IV collagen networks, and along with extensive disulfide bonding account for the relative insolubility of this material in native basement membranes.

5. Type IV collagen protomers

5.1. Biosynthesis

As is typical of all collagens, type IV collagen protomers are formed from three α chains that assemble intracellularly into a triple helical conformation. The steps that occur during the biosynthesis of type IV collagen (Grant et al., 1972, 1975; Pihlajaniemi et al., 1981), appear to follow the same general principles as for the fibrillar collagens (McLaughlin and Bulleid, 1998; Gelse et al., 2003), although some distinctions are noted. As nascent fibrillar procollagen α chains are translocated into the rough ER, they are co-translationally hydroxylated and glycosylated. These reactions proceed post-translationally until sufficient hydroxyproline is incorporated to allow triple helical assembly, after which these modifications cease. It is likely that a similar sequence of events occurs during the biosynthesis of type IV collagen, since the enzymes involved are only active on non-helical α chains (Kivirikko and Myllylä, 1985). The rate at which type IV collagen is synthesized and secreted is slower compared to fibrillar collagens, a difference that has been attributed to the greater degree of post-translational modification and to a slower rate of triple helix formation (Grant et al., 1972, 1975; Pihlajaniemi et al., 1981). Type IV collagen protomers are formed through interactions between the NC1 domains of their constituent α chains that bring these molecules together allowing nucleation and subsequent propagation of triple helical structure in a zipper-like manner towards amino terminus (Tsilibary and Charonis, 1986; Dolz et al., 1988). The NC1 domain not only initiates this assembly, but it also governs the process of α-chain selection thereby ensuring that it occurs in a discriminatory manner (Boutaud et al., 2000). This interaction is non-covalent, since all cysteine residues in the NC1 domain of each α-chain are engaged intramolecular cross-links (Fessler and Fessler, 1982; Siebold et al., 1988). This is in contrast to the fibrillar collagens, where the initial association and subsequent triple helical assembly of three α chains requires the formation of interchain disulfide bonds between their carboxyl terminal propeptides.

Collagen triple helices are inherently stable structures owing to the close packing and hydrogen bonding of the three α chains—effects that are related to the high content of hydroxyproline (Bella et al., 1994; Holmgren et al., 1998). In type IV collagen, the helix is further stabilized by interchain disulfide bonds located within the main collagenous domain (Fessler et al., 1984; Brazel et al., 1987, 1988; Soininen et al., 1987; Hostikka and Tryggvason, 1988). As the cysteine residues involved are located near the amino terminus, the formation of interchain disulfide bonds must be a relatively late event during molecular assembly. Finally, as with other collagens, newly synthesized type IV collagen has been shown to interact with a number of ER resident proteins including Hsp47/colligin, a collagen-specific molecular chaperone that stabilizes the triple helical structure of collagen thereby promoting its secretion (Ferreira et al., 1996).

5.2. Isoform composition

The existence of six isoforms of type IV collagen allows for a theoretical total of 56 distinct triple helical protomers that differ with respect to the composition and stoichiometry of their α-chain subunits. However, it is clear that the six isoforms do not freely associate into trimers, but rather, they are limited to specific combinations, which then form distinct supramolecular networks that vary from tissue to tissue (Fig. 3). The α1 and α2 chains are incorporated together into a heterotrimer with the composition $\alpha1_2\alpha2$ (Crouch et al., 1980; Fessler and Fessler, 1982; Trüeb et al., 1982). The α3 and α4 chains are assembled only into a protomer with the composition α3α4α5 (Borza et al., 2002). The α5 chain also assembles into a second

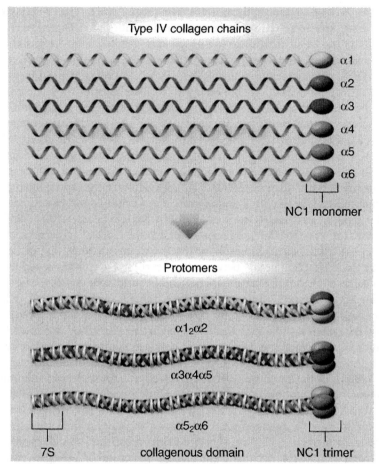

Fig. 3. Assembly of type IV collagen protomers. The six isoforms of type IV collagen assemble into three triple helical molecules (protomers) that differ in their composition. Each protomer is characterized by an amino-terminal 7S domain (~140 residues), a main collagenous domain (~1,300 residues) that contains numerous non-collagenous interruptions (white rings), and a carboxyl-terminal non-collagenous (NC1) domain. Molecular recognition events between the NC1 domains of each α chain govern their triple helical assembly into protomers. The 7S and NC1 domains are involved in cross-linking and aggregation of protomers into networks. Figure reproduced from Hudson, B., Tryggvason, K., Sundaramoorthy, M., Neilson, E. (2003). Alport's syndrome, Goodpasture's syndrome, and type IV collagen. N Engl J Med 348, 2543–2556. Copyright © 2004 Massachusetts Medical Society. All rights reserved. (See Color Insert.)

protomer together with the α6 chain that has the composition $\alpha5_2\alpha6$ (Borza et al., 2001). In general, the strongest evidence supports the concept that all type IV collagen protomers are composed of two "α1-like" chains and one "α2-like" chain.

6. Type IV collagen networks

6.1. Supramolecular assembly

In contrast to the fibrillar collagens, which require cleavage of non-collagenous propeptides at their amino and carboxyl termini prior to incorporation into fibrils, post-translational processing of type IV collagen is not required for its higher order assembly into networks within basement membranes (Minor et al., 1976; Crouch and Bornstein, 1979; Alitalo et al., 1980). The globular NC1 domain may preclude lateral association between type IV collagen protomers and, as a result, they do not form banded fibrils that are characteristic of interstitial collagens. Rather, it is the retention of the specialized 7S and NC1 domains at either end of the molecule that is critical for supramolecular assembly. The arrangement of type IV collagen in basement membranes is unique, with individual protomers self-assembling through aggregation and cross-linking of their like-ends to form a complex meshwork.

A tetramer is formed by the overlapping lateral assembly of four protomers through their 7S domains (Fig. 4), with each molecule diverging from one another outside of this region giving an "X" or "spider-like" configuration (Kühn et al., 1981; Siebold et al., 1987). Studies *in vitro* have shown this complex forms spontaneously and assembles initially as intermediates of non-covalently bound dimers and trimers, finally forming a tetramer that becomes stabilized by disulfide bonds and lysine-derived cross-links (Bächinger et al., 1982; Duncan et al., 1983). In some preparations, the 7S domain of a fifth molecule is seen interacting with the tetramer, and may be involved in bridging these aggregates (Timpl et al., 1981; Madri et al., 1983). At a position $\sim$10 nm outside their region of overlap, pairs of laterally aligned molecules diverge from one another at a mean angle of $\sim$40° at a highly flexible region that is thought to correspond to the NC2 domain (Hofmann et al., 1984).

Triple helical protomers dimerize through their carboxyl termini, and at this interface the three NC1 domains from each molecule interact to form a hexamer (Fig. 4). Collagenase digestion of basement membranes liberates the NC1 domain hexamer, that when analyzed under non-dissociative conditions has a molecular mass of $\sim$170 kDa, and by EM appears as a roughly spherical globule $\sim$11 to 17 nm in diameter (Timpl et al., 1981; Bächinger et al., 1982; Oberbäumer et al., 1982; Weber et al., 1984). The crystal structure of the $[\alpha 1_2\alpha 2]_2$ hexamer from lens capsule (Sundaramoorthy et al., 2002) and placental basement membrane (Than et al., 2002) has been determined and provides a much more precise view of this complex. The NC1 domains of a single protomer join to form a cone-shaped trimeric cap $\sim$70 Å in diameter at its base, that tapers towards the triple helical domain and encloses a hollow funnel-shaped core 12–14 Å in diameter. The NC1

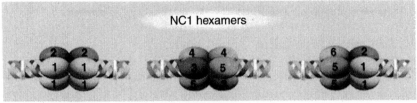

NC1 hexamers

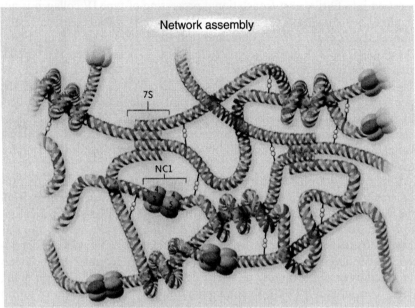

Network assembly

Fig. 4. Assembly of type IV collagen networks. (Top) Protomers dimerize through their carboxyl termini, which yields the NC1 domain as a hexamer. Three distinct populations of NC1 hexamers have been identified; two formed through homotypic assembly of $\alpha1_2\alpha2$ or $\alpha3\alpha4\alpha5$ protomers, and a third through heterotypic assembly of $\alpha1_2\alpha2$ and $\alpha5_2\alpha6$ protomers. The relative arrangement of NC1 domain monomers in each hexamer is accurately depicted. The NC1 domain monomers that oppose one another across the trimer-trimer interface may become covalently cross-linked. (Bottom) Protomers undergo supramolecular assembly into a complex network through aggregation and cross-linking of their like ends, and through supercoiling of their triple helical domains. A network comprised of $\alpha3\alpha4\alpha5$ protomers is illustrated. Individual protomers dimerize through their carboxyl-terminal NC1 domains (NC1 box) and form tetramers through their overlapping amino-terminal 7S domains (7S box). The 7S tetramer is depicted in a "stacked" alignment for simplicity, although it is thought to assume a tetragonal arrangement. The network is stabilized by interprotomer disulfide bonds between triple helical domains (white circles) that are also formed within the 7S tetramer and perhaps the NC1 hexamer (not shown). Further stabilization is introduced by lysine-derived cross-links in these regions. The bottom panel is reproduced from Hudson, B., Tryggvason, K., Sundaramoorthy, M., Neilson, E. (2003). Alport's syndrome, Goodpasture's syndrome, and type IV collagen. N Engl J Med 348, 2543–2556. Copyright © 2004 Massachusetts Medical Society. All rights reserved. (See Color Insert.)

domains are tightly packed through interchain hydrophobic interactions and hydrogen bonding, as well as through unique intramolecular and intermolecular domain-swapping interactions that act to 'clamp' the trimeric cap together. The hexamer is ellipsoid-shaped, and the trimer-trimer interface is relatively planar with each NC1 domain making close contact with a second NC1 domain on the opposing surface. Under acidic conditions, or following denaturation with agents such as SDS, urea, or guanidine, the NC1 domain hexamer dissociates into monomer and dimer species, with the latter representing 2 covalently-linked NC1 domain monomers originating from opposite protomers (Weber et al., 1984). By SDS-PAGE, the NC1 domain monomers and dimers can be resolved as a series of bands with molecular masses of ~25 and ~50 kDa, respectively, although there is variation in these values due to multiple size and charge isoforms, and interspecies differences (Langeveld et al., 1988; Gunwar et al., 1991a,b; Borza et al., 2001).

The NC1 domain monomers may be covalently linked as either heterodimers or homodimers. The strongest evidence supports the concept that the NC1 domains of 'α1-like' and 'α2-like' chains only dimerize with molecules of the same class (Borza et al., 2001, 2002). Early biochemical studies suggested the cross-links represented disulfide bonds (Fessler and Fessler, 1982; Weber et al., 1984), although many reported that a minor fraction of NC1 domain dimers were formed by non-reducible cross-links of an unknown nature (Weber et al., 1984; Langeveld et al., 1988; Siebold et al., 1988; Gunwar et al., 1991b). The precise arrangement of interprotomer cross-links remains poorly understood, and several recent studies have challenged long-held views regarding their identity. Siebold and colleagues (1988) analyzed the cyanogen bromide cleavage products of α1 NC1 domain dimers, and found the organization of intermolecular disulfide bonds was identical to intramolecular bonds within an α1 NC1 monomer. They concluded that this arrangement in the dimer could only be explained by disulfide exchange reactions between cysteines located within corresponding subdomains of the monomers. This picture has been complicated with the recent determination of the crystal structure of the $[\alpha1_2\alpha2]_2$ NC1 hexamer. Analysis of the hexamer derived from lens capsule revealed that interprotomer disulfide bonds were highly unlikely, since even the closest cysteines from opposing chains were physically too distant from one another (16 Å) to permit disulfide bonding without being severely destabilizing (Sundaramoorthy et al., 2002). The same conclusions were reached through analysis of the $[\alpha1_2\alpha2]_2$ NC1 hexamer from the placenta which is known to be highly cross-linked (Than et al., 2002). However, in this study, the crystal structure indicated a previously uncharacterized covalent thioether cross-link between specific lysine and methionine residues of opposing NC1 domains (potentially six cross-links per hexamer). Moreover, it was demonstrated biochemically that such bonds were present in dimers but not monomers. The properties of these cross-links are unknown. However, if they are exceptionally stable their presence and relative number may account for the finding that a fraction of NC1 dimers are insensitive to reduction, except under extremely harsh conditions (Reddy et al., 1993). The crystal structure of hexamers containing the remaining isoforms has yet to be determined, but should provide proof of the existence and nature of interprotomer disulfide bonds and/or lysine-derived cross-links.

In addition to interactions involving the 7S and NC1 domains, type IV collagen protomers undergo lateral association between their triple helical domains to form a highly interconnected meshwork. Reconstitution studies *in vitro* (Yurchenco and Furthmayr, 1984), as well EM studies of native basement membranes (Yurchenco and Ruben, 1987, 1988) have demonstrated that the triple helical domains of two to three protomers intertwine to form supercoiled structures.

6.2. Distribution

The dimerization of protomers through their NC1 domains is governed by specific molecular recognition events that dictate which protomers are co-assembled into the same network. At present it is unknown whether similar discriminatory interactions are involved in the assembly of the 7S tetramer. Since NC1 dimers are presumed to reflect cross-linked monomers originating from opposite protomers, determination of their subunit composition can reveal the identity of protomers that had undergone network assembly *in vivo*. At least three distinct networks have been identified using this approach. An overview of their distribution among basement membranes in humans and mice is presented in the following text, and unless indicated otherwise it is consistent between the species. The first network, $\alpha1/\alpha2$, is formed by the association of two $\alpha1_2\alpha2$ protomers (Trüeb et al., 1982; Weber et al., 1984; Siebold et al., 1988). This network is expressed ubiquitously among basement membranes, although its abundance can vary between sites. A notable example is the glomerular basement membrane (GBM), which is present in a very low abundance (Sado et al., 1995; Kalluri et al., 1997).

The second network, $\alpha3/\alpha4/\alpha5$, is formed by the association of two $\alpha3\alpha4\alpha5$ protomers (Borza et al., 2002). This network has a much more restricted distribution and is prominent in several sites that serve as a filtration barrier. It is the predominant network found in the GBM (Miner and Sanes, 1994; Sado et al., 1995), which is the only site where its existence in the exact form described in the previous text has been proven empirically. In all other tissues, the $\alpha3/\alpha4/\alpha5$ network is assumed to be present based upon co-localization of its subunit α chains. Other sites that contain the $\alpha3/\alpha4/\alpha5$ network and which serve a related functional role include the ependymal basement membrane of the choroid plexus (Kleppel et al., 1989b; Saito et al.,

2000; Urabe et al., 2002) and the alveolar basement membrane (Saito et al., 2000; Nakano et al., 2001). This network is also present in several ocular basement membranes, including the lens capsule, the corneal epithelial and endothelial (Descemet's) basement membranes, the internal limiting membrane of the retina, and in Bruch's membrane, specifically underlying retinal pigment epithelial cells (Kleppel et al., 1989a,b; Kleppel and Michael, 1990; Kelley et al., 2002; Chen et al., 2003; Ohkubo et al., 2003). The $\alpha3/\alpha4/\alpha5$ network is also present in the inner ear, and its distribution among cochlear basement membranes has been studied in several species (Cosgrove et al., 1996c; Kalluri et al., 1998; Harvey et al., 2001). This topic is reviewed by others in this series (Chapter 6). The $\alpha3/\alpha4/\alpha5$ network is prominent at the synaptic basement membrane of the skeletal neuromuscular junction (Sanes et al., 1990; Miner and Sanes, 1994). It is also found in the basement membrane of distal tubules in the kidney (Miner and Sanes, 1994; Sado et al., 1995), and underlying the epithelium at the tips of intestinal villi at various locations in the gastrointestinal tract (Saito et al., 2000; Oka et al., 2002). Studies in rodent and bovine tissues suggest the $\alpha3/\alpha4/\alpha5$ network is expressed in the basement membrane of ovarian follicles (Rodgers et al., 1998; Saito et al., 2000). In rodents it also is present in the male gonad, specifically in the seminiferous tubule basement membrane (STBM) (Frojdman et al., 1998; Saito et al., 2000). All six isoforms are found in bovine STBM, but whether an $\alpha3/\alpha4/\alpha5$ network is present in this site is unclear (Kahsai et al., 1997).

The final network, $\alpha1/\alpha2/\alpha5/\alpha6$, is formed by the association of $\alpha1_2\alpha2$ and $\alpha5_2\alpha6$ protomers, and is the only network assembled from different protomers. This network is prominent in tissues that undergo elastic changes in size. The $\alpha1/\alpha2/\alpha5/\alpha6$ network is expressed in the basement membrane of vascular smooth muscle cells in a wide variety of tissues (Seki et al., 1998; Borza et al., 2001). This is likely the case throughout the body, although exceptions have been noted for muscular vessels of small size or in certain anatomic sites that contain the $\alpha1/\alpha2$ network only (Seki et al., 1998; Urabe et al., 2002). The $\alpha1/\alpha2/\alpha5/\alpha6$ network is also expressed in the subepithelial basement membrane and surrounding visceral smooth muscle layer of the esophagus, stomach, small intestine, and colon (Heidet et al., 1997; Seki et al., 1998; Simoneau et al., 1998; Hiki et al., 2002; Oka et al., 2002). The same is true for components of the genitourinary tract including the bladder, ureter, and uterus (Seki et al., 1998; Borza et al., 2001; Kiyofuji et al., 2002), bronchial epithelium (Nakano et al., 2001), and epidermal basement membrane (Ninomiya et al., 1995).

7. Type IV collagen in development: Lessons from invertebrates

Several properties make invertebrate organisms invaluable models for developmental studies of type IV collagen. First, their basement membranes share the same basic architecture as those of higher organisms. Its framework, type IV collagen, has been conserved during metazoan evolution at the level of α chains, protomers and networks. Therefore, investigating the assembly and function of type IV collagen in "simpler" organisms can shed light on these processes in higher organisms. Moreover, invertebrates are ideally suited for experimental study. Their anatomy is well understood, and their general pattern of embryonic development mirrors, in many respects, that of vertebrates. The genomes of many invertebrates have been sequenced or are nearly completed, and techniques for genetic manipulation of these species are well established. Finally, their small size and short life cycle offers an advantage when conducting genetic studies as it allows for simple and rapid screening of mutant phenotypes.

7.1. Hydra

Cnidaria are the most primitive metazoans with defined tissue layers. One member of this phylum, *H. vulgaris* is an extremely simple creature, with a cylindrical sac-like body enclosing a gastric cavity that has a "foot" region at its basal pole and a "head" region at its apical pole consisting of a mouth opening surrounded by a crown of tentacles. Its body wall is an epithelial bilayer (ectoderm and endoderm) separated by a gelatinous layer of ECM ~300 nm in thickness termed the mesoglea. By EM, the mesoglea appears as a composite structure with amorphous basement membrane-like material concentrated at both its subepithelial aspects and a central fibrous zone that is more typical of interstitial matrices. This heterogeneity is reflected at the molecular level, as it has been found to contain some components of basement membranes including type IV collagen, laminin, and HSPG, but also other characteristics of interstitial matrices such as fibronectin and fibrillar collagen (Sarras et al., 1991a; Deutzmann et al., 2000).

Only a single isoform of type IV collagen, homologous to the α1 chain of vertebrates, has been identified in hydra (Fowler et al., 2000). Analysis of native type IV collagen extracted from mesoglea suggested the α1 chain assembles as a homotrimer. When visualized by rotary shadowing, triple helical protomers were observed that had formed dimers through their NC1 domains. However, there was no evidence that these molecules assembled as tetramers through their 7S domains, a result that was supported through biochemical analysis. This finding is unexpected, and suggests that type IV collagen of hydra might be incapable of undergoing network assembly in a manner comparable to that of higher organisms. Through *in situ* hybridization, expression of α1 mRNA was found to be restricted to the ectoderm throughout the animal, but was most intense at the base of the tentacles (Fowler et al., 2000). By immunostaining, type IV collagen is localized throughout the entire width of the mesoglea (Sarras et al., 1991a). The fact that type IV collagen was not expressed by the endoderm but was found underlying this layer implies that it is capable of assembling at sites distant from its synthesis. As described in the following text, similar findings have been made in *C. elegans*. In contrast to type IV collagen, the expression of the laminin β1 chain in hydra is restricted to the endoderm. However, it also becomes incorporated into the basement membrane-like zones underlying both the ectoderm and endoderm (Sarras et al., 1994).

When hydra are dissociated into individual cells, they reassemble and through cytodifferentiation and morphogenesis, contribute to complete regeneration of the adult body within ~96 hours. Aggregation is disrupted when this is carried out in the presence of fragments of type IV collagen or antibodies directed against the protein (Zhang et al., 1994). Purified 7S domain is the most potent inhibitor, followed by NC1 domain monomers and then NC1 hexamers. Aggregation is also blocked by agents that inhibit collagen cross-linking (Sarras et al., 1991b). Consistent with these studies, type IV collagen has been shown to also be essential for head regeneration (Fowler et al., 2000). Following decapitation, there is increased expression of α1 mRNA by cells located at the exposed surface, that becomes especially prominent in the region at the base of the tentacles as head regeneration proceeds (Fig. 5). If antisense oligonucleotides targeting the α1 transcript are first introduced into cells that will later be located basal to the decapitation site, translation of the α1 chain is disrupted and head regeneration is blocked. From these studies it is clear that type IV collagen plays a critical role in processes such as cell migration and differentiation that are essential for hydra morphogenesis.

7.2. Sea urchin

Homologues of the vertebrate α1 and α2 chains have been cloned in the sea urchin *S. purpuratus* and are designated as 3α and 4α, respectively (Exposito et al., 1993, 1994). In an early study, type IV collagen was detected along with laminin in the unfertilized egg of another species (*L. variegatus*) by immunostaining, where they co-localized to fine granules in the ooplasm (Wessel et al., 1984). It was suggested these might represent a small maternally derived pool of materials that could be recruited for basement membrane assembly prior to their *de novo* synthesis in the early embryo. As described in the following text, type IV collagen has also been reported in the unfertilized eggs of other species. Although it had yet to be proven to be a type IV collagen chain, early studies of the 3α isoform suggested it may be involved in mesenchymal differentiation. By Northern analysis, low levels of 3α mRNA were detected at the morula stage, but not in earlier 16-cell stage embryos or unfertilized eggs. Its expression peaked at the blastula stage and then diminished in gastrula and larval stages (Venkatesan et al., 1986).

α1 chain - normal adult

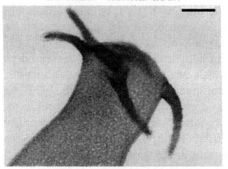

α1 chain - head regeneration

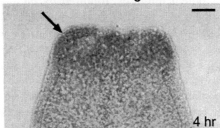

4 hr

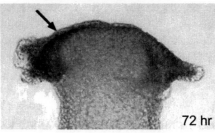

72 hr

Fig. 5. Whole mount *in situ* hybridization of type IV collagen in hydra. (Top) In normal adults, expression of the α1 chain is restricted to the ectoderm along the entire longitudinal axis of the body and is prominent at the base of the tentacles. (Bottom) The head is regenerated following decapitation. Expression of the α1 chain is upregulated in cells basal to the decapitation site by 4 hours, and becomes prominent in the region of the tentacles by 72 hours (arrows). Head regeneration is blocked if expression of the α1 chain is disrupted using antisense oligonucleotides. Scale bars: 100 μm (top) and 20 μm (bottom). Figure adapted with permission from Fowler et al., 2000 © 2004 The American Society for Biochemistry and Molecular Biology, Inc.

Others found the timing of these events were delayed, with 3α mRNA not peaking until the gastrula stage; they also reported a related transcript of smaller size (possibly 4α mRNA) that was first detected in the gastrula stage and peaked in the larval stage (Nemer and Harlow, 1988). In general, the expression patterns in the previous text are consistent with the initial formation of a basement membrane in the developing embryo, that which lines the blastocoel. It is unclear whether matrix components derived from the egg might contribute to this process, since by the morula stage when staining for type IV collagen begins to concentrate at the basal surface of cells that will form the blastocoel wall, there is already weak expression of 3α mRNA.

The expression of the 3α chain (then termed *Spcoll*) was further characterized through *in situ* hybridization (Wessel et al., 1991). It was found to be expressed by mesenchyme cells, especially those of the primary mesenchyme lineage during gastrulation and continuing through development to the larval stage. Using the same technique, expression of the 4α chain was later shown to also be restricted to primary mesenchyme cells (Exposito et al., 1994). The progenitors of primary mesenchymal cells follow a skeletogenic developmental program, and synthesize matrix components that undergo biomineralization contributing to the urchin endoskeleton. This process can be studied *in vitro*, as mesenchymal cells form spicules of mineralized matrix in culture. Spiculogenesis was inhibited when primary mesenchyme cells were cultured in the presence of an antibody to the 3α chain (Wessel et al., 1991). However, type IV collagen is not an integral component of this matrix and it was proposed that its expression by mesenchymal cells might induce their differentiation, initiating the production of other matrix components that become biomineralized.

7.3. Nematodes

Homologues of the vertebrate $\alpha 1$ and $\alpha 2$ chains have been identified in the nematode *C. elegans*, and are encoded by the genes *emb-9* and *let-2*, respectively (Guo and Kramer, 1989; Sibley et al., 1993). In the adult of this species, a thin (20-nm) basement membrane ensheathes the intestine and gonad and also lines the pseudocoelomic cavity separating the muscles of the body wall from the hypodermis (Fig. 6). The pharynx is also surrounded by a basement membrane that is approximately twice the thickness of those previously listed.

The distribution of type IV collagen during development in *C. elegans* has been studied using chain-specific antibodies, and the spatial and temporal expression patterns of both isoforms are identical (Graham et al., 1997). The stages of embryogenesis in *C. elegans* are defined by the shape of the embryo (lima or comma) and later by its length relative to its egg shell (two-fold through four-fold). Type IV collagen is first detected in body wall muscle cells immediately preceding the lima stage. At this point, which represents the onset of morphogenesis, the staining for both isoforms appears intracellular. By the 11/2-fold stage, both chains can be detected within the basement membrane that is formed between the muscle cells and the hypodermis. Weak staining is also detected among basement membranes of the

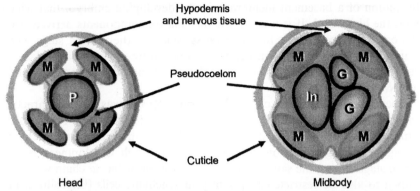

Fig. 6. Synthesis and deposition of type IV collagen in basement membranes of *C. elegans*. Schematic cross-sections through the head and midbody regions of an adult are illustrated. Basement membranes that contain type IV collagen (black) underlie the body wall muscle (M) quadrants and ensheath the intestine (In), gonad (G), and pharynx (P). A continuous basement membrane that lacks type IV collagen (grey) lines the pseudocoelom separating it from the hypodermis. Tissues that express type IV collagen (black lettering) include muscle and gonad, whereas those that do not (white lettering) include the intestine and pharynx. Figure adapted from Graham et al., The Journal of Cell Biology, 1997, Vol. 137(5), pp. 1171–1183, by copyright permission of The Rockefeller University Press, Inc.

developing pharynx and intestine, which becomes intense by the two-fold stage. From the comma through to the four-fold stage that precedes hatching, type IV collagen accumulates at the interface between the hypodermis and the body wall muscles. It can also be detected in the basement membrane of the primordial gonad and among glial-like cells in the nerve ring. This expression pattern is retained in adults. The hypodermis is separated from the pseudocoelom by a continuous basement membrane that extends into the zone between each of the muscle quadrants and is present along their pseudocoelomic face. Unexpectedly, this basement membrane fails to stain for type IV collagen in both embryos and adults. *In situ* hybridization and analysis of transgenic reporter strains revealed a discordance between the cell-specific expression of type IV collagen RNA and its localization at the protein level (Graham et al., 1997). Both isoforms are co-expressed primarily by body wall muscle and somatic cells of the gonad. Cells of the hypodermis, intestine, and pharynx do not express type IV collagen, despite the fact that these structures are ensheathed by a basement membrane. Using an epitope-tagged α1 chain whose expression was driven with a body wall muscle-specific promoter it has been shown that muscle-derived type IV collagen is incorporated into the basement membranes of the pharynx, intestine, and gonad (Graham et al., 1997). The expression of type IV collagen by muscle cells alone was sufficient for *C. elegans* to complete development. Although these embryos grew slowly and many arrested at the larval stage, some reached adulthood and were fertile.

Several groups have shown that α2 mRNA in nematodes is alternatively spliced, and that this phenomenon is developmentally regulated (Sibley et al., 1993; Pettitt and Kingston, 1994). Alternative splicing leads to two transcripts differing from one another only by the presence of a mutually exclusive exon (9 or 10); both *C. elegans* and *A. suum* are separated by an unusually small intron (<40 nt). Both exons encode five Gly-Xaa-Yaa triplets, followed by a non-collagenous interruption, then four Gly-Xaa-Yaa triplets. The identity of variant residues within the collagenous regions and the length and sequence of the interruption differs between the two exons. For exon 9, the interruption is nine amino acids in length and contains a potential glycosaminoglycan attachment site, whereas that of exon 10 is 10 amino acids in length and contains a lysine residue that may be involved in cross-linking. Greater than 90% of α2 mRNA expressed by embryos contains exon 9, whereas 80% to 90% of the transcripts in larvae and adults contain exon 10. Embryogenesis is a phase of dramatic morphologic changes, whereas the larval stage is characterized primarily by symmetric growth and enlargement of previously established tissues. These periods may require functionally distinct forms of type IV collagen that are provided through alternative splicing of the α2 chain. A temporary, perhaps more flexible collagen network assembled from exon 9-containing protomers may be later replaced by a more permanent exon 10-containing network. A comparable developmental shift occurs in specific basement membranes of higher organisms, but it involves the replacement of the "fetal" α1/α2 network by an "adult" α3/α4/α5 network.

Using a probe to the mouse α1 NC1 domain, Guo and colleagues (Guo and Kramer, 1989) identified two genes in *C. elegans* encoding basement membrane collagens that they designated *clb-1* and *clb-2*. They went on to examine several genetic loci in proximity to *clb-2* that had previously been identified in genetic screens for mutants with temperature-sensitive embryonic lethal phenotypes, and found that one of these (*emb-9*) corresponded to *clb-2* encoding the α1 chain (Guo et al., 1991). To date, 11 different alleles of *emb-9* have been described (Guo et al., 1991; Gupta et al., 1997). In general, all of these are wild-type at permissive temperatures (15°C) and larval-lethal at intermediate temperatures (20°C). If embryos are grown to the larval stage at permissive temperatures and then shifted to restrictive temperatures (25°C), they arrest development or in some cases become sterile adults. Embryos grown at restrictive temperatures arrest at the two-fold stage with gross morphologic defects including constriction and herniation of the body wall (Fig. 7). The causative mutation has been identified for 9 of the 11 alleles, and all are located in the main triple helical domain (Guo et al., 1991; Gupta et al., 1997). Six of these are missense mutations involving conserved glycine residues; one pair of these are identical. Five of the six glycine substitutions result in the embryonic lethal phenotype previously described, whereas one is lethal at the larval stage. The remaining alleles are two

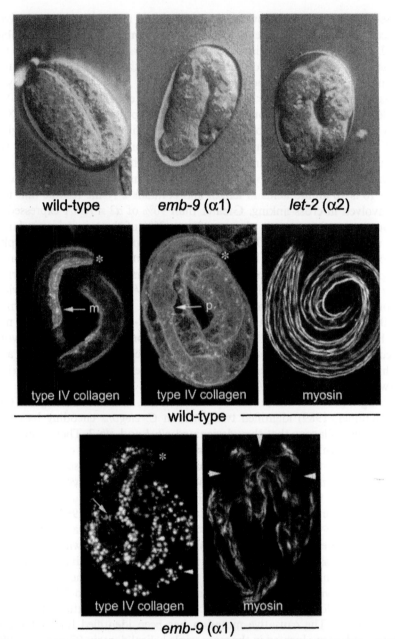

Fig. 7. Type IV collagen is required for early development in *C. elegans*. (Top) Embryos with mutations in the genes encoding the α1 or α2 chains (designated *emb-9* and *let-2*, respectively) arrest at the twofold stage and display constriction and herniation of the body wall. (Center) In wild-type embryos, type IV collagen is localized to the basement membrane separating the muscle (m) of the body wall from the hypodermis. In a different focal plane there is labeling of

different nonsense mutations and a deletion. Each of these results in a slightly milder phenotype than the glycine substitutions, causing arrest at the three-fold to four-fold stage. When present in the heterozygous state, null mutations in collagen genes generally cause a less severe phenotype than missense mutations. This is because mutant α chains can be incorporated along with normal α chains encoded by the wild-type allele into protomers which are then degraded, a classic example of a dominant negative effect that has been termed procollagen suicide (Prockop and Kivirikko, 1995). Embryos homozygous for missense mutations have a more severe phenotype than those with null mutations, suggesting that the missense proteins had other dominant effects, perhaps disrupting normal cellular biosynthetic or secretory pathways. By immunostaining, no expression of the $\alpha 1$ chain was detected in *emb-9* embryos with null mutations and the $\alpha 2$ chain was retained intracellularly. Those with missense mutations showed intracellular accumulation of both chains that was temperature-dependent. In *emb-9* mutants, immunostaining reveals prominent gaps in the distribution of myosin along the muscle quadrants (Fig. 7). Its expression is first disrupted at the onset of contractile activity in the embryo. It has been suggested that in the absence of type IV collagen, the interaction between muscle cells and the hypodermal basement membrane is altered and the mechanical stress of contraction leads to their detachment from the body wall (Gupta et al., 1997).

Sibley and colleagues (1993) recognized that a different genetic locus (*let-2*) previously identified in screens for mutants with temperature-sensitive embryonic lethal phenotypes was in proximity to the *clb-1* locus previously described. They demonstrated through transgenic rescue that *let-2* encodes the $\alpha 2$ chain, and later went on to characterize the mutation and phenotype of 17 different *let-2* alleles (Sibley et al., 1994). The phenotype of *let-2* mu-escribed for *emb-9* mutants but shows a broader range, with some alleles embryonic lethal at temperatures permissive for *emb-9*. Mutation analysis revealed that as with *emb-9*, all mutations were localized within the main

---

the pharyngeal (p) and intestinal basement membranes. The staining for myosin appears as a double row along each of the muscle quadrants. (Bottom) In *emb-9* embryos with missense mutations, the $\alpha 2$ chain is retained in muscle cells (arrowhead) and glial-like cells (arrow). The same pattern is seen for the $\alpha 1$ chain (not shown). Prominent gaps in the distribution of myosin (arrowheads) are seen in *emb-9* embryos, reflecting detachment of muscle cells from the hypodermis which occurs at the onset of contractile activity. The distribution of type IV collagen and myosin is comparable in *let-2* embryos with missense mutations (not shown). Asterisks indicate the anterior end of the embryos. Figure (top) reproduced from Sibley et al., The Journal of Cell Biology, 1993, Vol. 123(1), pp. 225–264, by copyright permission of The Rockefeller University Press, Inc. and (center and bottom) reprinted from Developmental Biology, Vol. 227, No. 2, 2000, pp. 690–705, Norman & Moerman, "The let-268 locus of ..." © 2004 with permission of Elsevier, Ltd.

collagenous domain of the α2 chain. Of the 17 alleles, 15 were missense mutations involving conserved glycine residues. Unexpectedly, three pairs of these mutations each affected the same glycine. A substitution involving an amino acid in the central position of a Gly-Xaa-Yaa triplet and another disrupting an intron splice acceptor site were also identified, and both resulted in a severe phenotype.

The important role that type IV collagen plays in *C. elegans* development is reflected in the phenotype of a mutant in which collagen processing is disrupted (Norman and Moerman, 2000). The *let-268* locus of *C. elegans* encodes the only isoform of lysyl hydroxylase identified in this species, and its spatial and temporal expression pattern mirrors that of type IV collagen. *Let-268* mutants show a strikingly similar phenotype, with developmental arrest at the twofold stage and intracellular retention of type IV collagen within muscle cells. In these mutants there is detachment of muscle cells from the hypodermis at the onset of contractile activity, which supports the proposed role of type IV collagen in maintaining the structural integrity of this interface.

The phenotype of *emb-9* and *let-2* mutants makes it clear that type IV collagen plays a critical role in embryonic development, and temperature shifting experiments involving mutant larvae also show that this protein is required throughout life. Mutation analysis has allowed the severity of the phenotype to be correlated with the specific genetic alteration, and has also raised interesting questions. It is notable that all of the mutations identified in the 26 independent *emb-9* and *let-2* alleles are located within the main collagenous domain. One would expect mutations in the 7S or NC1 domains to be equally likely. Of the nearly 1,000 conserved glycine residues that make up both α-chains, there were 4 cases in which 2 mutations affected the same glycine, which is further evidence of their non-random distribution. The genetic screens in which the *emb-9* and *let-2* alleles were originally identified were tailored to detect mutants with an embryonic lethal phenotype. The possibility exists that mutations at different positions cause milder phenotypes or result in dominant lethality or sterility and, therefore, would have gone undetected. The studies in *C. elegans* are often used to support the notion that mutations in the genes encoding the α1 and α2 chains in higher organisms must be embryonicly lethal. Both isoforms have a broad distribution among basement membranes; this is likely to be true in the setting of homozygous or compound heterozygous mutations. However, from these studies it is also clear that heterozygous mutations in either gene are not lethal, nor are they necessarily silent. Although it has not been studied in any detail, many alleles of *let-2* and *emb-9* show semi-dominant effects causing reduced viability and/or fertility in heterozygotes (Miwa et al., 1980; Wood et al., 1980; Cassada et al., 1981; Isnenghi et al., 1983; Sibley et al., 1994). It is unknown whether comparable mutations in higher organisms are disease-causing or are even compatible with life.

7.4. Drosophila

A number of potential collagen-like genes have been identified in *Drosophila* (Natzle et al., 1982; Le Parco et al., 1986a; Hynes and Zhao, 2000). However, only three genes have been characterized in any detail, and none of these encodes a prototypical fibrillar collagen. The first is *pericardin*, which was classified as being "type IV collagen-like" based on the presence of interruptions in its collagenous domain (Chartier et al., 2002). Pericardin has a non-collagenous domain at its carboxyl terminus, but shows no similarity to the highly conserved NC1 domain of type IV collagen; overall it may be more closely related to vertebrate type XVIII collagen. The remaining genes both encode type IV collagen chains. One of these, designated *DCg1/Cg25C*, is homologous to the vertebrate *COL4A1* gene encoding the α1 chain (Monson et al., 1982; Blumberg et al., 1987; Cecchini et al., 1987). The second is *viking*, which was originally identified in a screen for genes potentially involved in immunity (Rodriguez et al., 1996), and was later found to be homologous to the vertebrate *COL4A2* gene encoding the α2 chain (Yasothornsrikul et al., 1997). Orthologues of *viking* have been characterized in other dipteran species, including *S. peregrina* and *A. gambiae* (Gare et al., 2003). The α2 NC1 domain is uncharacteristically long in all of these species, containing an extension of up to 76 residues following the final conserved, cysteine. The length and sequence of this extension is not conserved, and its biologic function is unknown.

The expression of type IV collagen during development in *Drosophila* has been well documented, particularly for the α1 isoform. By immunostaining, the α1 chain has been detected in oocytes (Knibiehler et al., 1990). It was suggested that this represents maternally derived material, since no α1 mRNA was detected in the oocyte by *in situ* hybridization. The same findings have been made in sea urchin, lending support to the theory that basement membrane proteins might be stockpiled in the egg of some species. Zygotic transcription is not required for early embryogenesis, and the embryo relies instead on a stockpile of maternally supplied products (mRNA and protein) from cleavage through to a stage referred to as the mid-blastula transition. The onset of transcription differs between species and tends to occur later in lower organisms. For organisms such as the sea urchin and *Drosophila*, a maternally derived pool of type IV collagen might be required for the embryo to reach the mid-blastula transition. Granular staining for the α1 chain is detected in the yolk of pre-blastoderm embryos; later, at the syncitial blastoderm stage, there is labeling of the matrix embedding the pole cells and faint staining on the surface of the embryo. At the onset of gastrulation the staining becomes concentrated on the apical surface of cells involved in furrow formation.

By Northern blot analysis, α1 mRNA is nearly undetectable in embryos but is prominent in first and second instar larvae (Monson et al., 1982; Le

Parco et al., 1986b). However, others have detected α1 mRNA expression at early embryonic stages through *in situ* hybridization (Lunstrum et al., 1988; Mirre et al., 1988). It is first detected ∼8 to 10 hours after fertilization, which corresponds to a timepoint where there is germ band retraction and fusion of the anterior and posterior midgut. This is ∼5 hours after gastrulation has occurred, suggesting that *de novo* synthesis of type IV collagen is not required for the complex morphogenetic changes that occur during this period. In *Drosophila*, type IV collagen appears to be expressed primarily, if not exclusively, by cells of mesodermal origin. In 10-hour embryos, transcripts are detected in hemocytes that are intermingled with visceral mesoderm at sites where muscles will later form. Hemocytes are mesodermally derived cells that migrate throughout the open circulatory system of the embryo (the haemocoel). They are thought to play an important role in cellular immunity, behaving much like human macrophages. They are also considered an important source of ECM, and may deposit components such as type IV collagen into a basement membrane on the basal surface of cells lining the haemocoel. In 10-hour embryos, α1 mRNA is also expressed in the mesoderm that gives rise to the fat bodies, another component of the immune system of *Drosophila*. In 12-hour embryos through to second instar larvae, α1 mRNA is expressed primarily by fat bodies and hemocytes. In third instar larvae, expression is additionally detected within the lymph gland and cells of the imaginal discs. The latter are epithelial infoldings bounded by a basement membrane that during metamorphosis give rise to external structures of the adult (e.g., legs, wings, and antennae). At the pupal stage, there is strong expression of the α1 chain by hemocytes that have invaded larval tissues such as muscle, gut, and imaginal discs. The expression of type IV collagen diminishes at the onset of metamorphosis; during this period of major morphologic change there is remodeling of basement membranes that has been shown to involve site-specific proteolytic cleavage of the α1 chain (Fessler et al., 1993). Although not as thoroughly characterized, the spatial and temporal expression pattern of the α2 chain in embryos and larvae appears identical to that of the α1 chain (Yasothornsrikul et al., 1997).

Based solely on its localization, it has been suggested that type IV collagen may be involved in the establishment of axonal pathways during development of the central nervous system by regulating neurite outgrowth (Mirre et al., 1992). Others, by studying the localization of type IV collagen in the imaginal wing disc of late-stage larvae concluded it may be involved in establishing wing venation patterns and in the maintenance of wing integrity during metamorphosis (Murray et al., 1995). Hemocytes were implicated not only in the synthesis of type IV collagen, but also in degradation of basement membrane material as remodeling occurs during wing development. More definitive functional roles for type IV collagen have come from studies in which expression of the α1 or α2 chains has been experimentally disrupted.

Both isoforms are essential for normal embryonic development. In strains carrying mutant alleles of the *DCg1* gene (encoding the α1 chain), only 3% of embryos survive to hatching (Rodriguez et al., 1996). The small, rounded larvae that are hatched are slow-moving and short-lived. Although it was not known to encode the α2 chain at the time, strains with mutant alleles of *viking* display a similar phenotype. Some of these mutants persisted as larvae and, this, together with the fact that it was identified in a screen for genes expressed in immune system tissues (hemocytes and fat bodies) involved in both attack and defense, led to its unique name, *viking*.

Consistent with the studies described in *C. elegans*, the α1 chain has been shown to play an important role in muscle development and function (Borchiellini et al., 1996). Transgenic stains were generated in which the normal expression of the *DCg1* gene was altered using two approaches. The first involved expressing a truncated α1 chain that contained a deletion within the collagenous domain (*del* transgene). It was expected that this construct would exert a dominant negative effect and disrupt (although not abolish) expression of the native α1 and α2 chains. A construct with the intact *DCg1* sequence was used as an internal control (*col* transgene). The second approach involved transgenic expression of either sense or antisense α1 mRNA. The transgenes were expressed at the RNA level and exerted the expected effects; total collagen levels (not specific to type IV collagen) were increased in the *col* and sense mRNA lines and decreased in the *del* and antisense mRNA lines. Overexpression of the α1 chain had no effect on development, as inferred from the normal hatching rate of embryos carrying the *col* or sense transgenes. Conversely, the percentage of unhatched embryos was increased in strains carrying the *del* transgene. Unexpectedly, the hatching rate was normal in strains that expressed antisense mRNA. All *del* and antisense mRNA embryos exhibited a muscle defect, with perturbations in muscle-tendinous cell attachment. In these mutants, segmental muscles of the body wall failed to maintain their insertions at the point of attachment to tendon cells in the developing epidermis. The muscles took on a spherical morphology and appeared disorganized, lacking defined Z bands. Mutants also displayed defects in dorsal closure, germ band retraction, and nerve cord condensation. There was variability in the severity of the phenotype which the authors suggested might relate to different expression levels of the transgenes caused by position effects at their site of chromosomal integration. Some mutant embryos with less dramatic muscle defects were delayed in hatching and as larvae were hypoactive. These grew to the third instar larval stage, but had short, fat bodies and were the same size as wild-type first instar larvae. Many mutants also showed defects in visceral musculature, with longitudinal and circular muscle groups of the gut detached from one another. Some with the mildest phenotype survived to adulthood but lacked leg tendons.

8. Type IV collagen in development: Lessons from vertebrates

8.1. The α1/α2 network in mouse embryogenesis

Several early studies documented the expression of type IV collagen during mouse embryogenesis by immunostaining. While the α3-α6 chains had yet to be identified at the time these experiments were performed, the antibodies used are presumed to be directed against the $α1_2α2$ protomer. Positive staining has been reported as early as the two-cell stage (Sherman et al., 1980). However, this finding has not been supported by others using more thoroughly characterized antibodies (Adamson and Ayers, 1979; Leivo et al., 1980; Dziadek and Timpl, 1985). For example, staining for type IV collagen and laminin was detected in the unfertilized egg and four- to eight-cell stage blastocysts, but it was concluded this was non-specific (Leivo et al., 1980). In this study, expression could only be accurately assessed by the 16-cell stage, where there was granular intracellular staining for laminin but not type IV collagen.

However, type IV collagen is detected in three- to four-day pre-implantation blastocysts (32- to 64-cell stage), which coincides with the assembly of a basement membrane at the interface of the primitive endoderm and the inner cell mass (Leivo et al., 1980; Dziadek and Timpl, 1985). Although the staining was predominantly granular, suggesting an intracellular localization, some linear staining consistent with a basement membrane distribution was observed. At this developmental stage there was also patchy labeling for type IV collagen along the inner aspect of the trophectoderm, but staining for fibrillar collagens (types I and III) was negative (Leivo et al., 1980). In five-day peri-implantation blastocysts, type IV collagen is clearly detected at interface of primitive visceral endoderm and ectoderm and there is variable staining of Reichert's membrane underlying parietal endoderm cells (Adamson and Ayers, 1979; Leivo et al., 1980). In seven-day postimplantation blastocysts and at later stages there is intense expression of type IV collagen in all embryonic basement membranes (Leivo et al., 1980; Herken and Barrach, 1985). By this point, gastrulation has given rise to the three embryonic germ layers; endoderm, mesoderm, and ectoderm. The basement membrane of the ectoderm is discontinuous in the region of mesoderm formation (the primitive streak), but where present, stains for type IV collagen. Staining is also detected at the interface of the mesoderm and endoderm, and in extraembryonic basement membranes including those of the chorion, amnion, and yolk sac. The expression of type IV collagen precedes that of fibrillar collagens (types I and III). The latter are first detected in 8-day embryos, and are expressed only by mesodermally derived tissues such as mesenchymes of the head and heart (Leivo et al., 1980). As embryogenesis proceeds, the α1 and α2 chains take on their characteristic ubiquitous distribution among basement membranes. In general, the α3-α6 chains are expressed at later stages of embryonic development and in particular tissues, networks containing these isoforms are not present until the postnatal life. This phenomenon will be reviewed in the following text and is not discussed further here.

Consistent with the studies in invertebrates, the α1/α2 network has been found to be essential for embryonic development in the mouse. This was recently demonstrated through the generation of a line of transgenic mice in which the 5' ends of both the *col4a1* and *col4a2* genes and their shared promoter region were disrupted by gene targeting (Poschl et al., 2004). Quite unexpectedly, the absence of the α1/α2 network did not interfere with early embryonic development. Mutant embryos developed (essentially) normally until embryonic day 9.5 (E9.5), but died between E10.5 to E11.5 (Fig. 8). Grossly, they displayed no or only minor changes in size and body shape at E9. Extraembryonic structures such as the amnion and yolk sac showed no defects, and organ development occurred normally. Laminin and nidogen were present in both embryonic and extraembryonic basement membranes, although the staining for both was weaker and in some sites appeared discontinuous. The possibility of molecular compensation by the α3-α6 chains was ruled out, and it was concluded that type IV collagen is not required for the deposition and assembly of other molecular components into basement membranes.

Mutants showed a variable degree of growth retardation between E9.5 to E11. At E10.5, structural abnormalities of the epidermal and pial basement membranes were noted. By EM, both appeared to be of variable thickness or focally absent, and in some sites the adjacent cells had lost contact with the

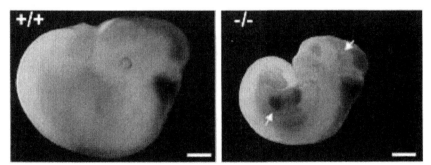

Fig. 8. The α1/α2 network is essential for mouse development. Mice lacking the α1 and α2 collagen chains (−/−) show growth retardation compared to their wild-type littermates (+/+). This is present as early as E9.5, and is accompanied by structural disruption of basement membranes in the brain, skin, and vasculature (not shown). Bleeding into the pericardium and dilatation of blood vessels (arrows) is observed in E11.5 mutants. Lethality occurs between ∼E10.5 to E11.5 due to a failure of the integrity of Reichert's membrane. Scale bars: 100 μm. Figure reproduced from Development 131: 1619–1628, 2004, Poschl et al., "Collagen IV is essential for basement membrane stability but dispensable for initiation of its assembly during early development," with the permission of The Company of Biologists, Ltd.

underlying basement membrane. At E11.5 neuronal ectopias were observed that were attributed to structural instability and local disruption of the pial basement membrane. There was also evidence that vascular basement membranes were structurally unstable. In E10.5 to 11.5 mutant embryos with beating hearts there was bleeding into the pericardium and dilatation of blood vessels. Although major elements of the vasculature had developed normally, there were subtle changes in the organization of capillary networks and blood vessels in the perineural vascular plexus. Reichert's membrane, which separates parietal endoderm from trophectoderm, appeared thinner and disorganized and showed local disruptions. As a consequence, there was detachment of the two cellular layers allowing maternal blood to enter the yolk sac cavity. Placental defects were also noted, and the authors speculated these may have impaired the exchange of nutrients to the embryo and contributed to the observed growth retardation. While the placenta was normally organized, there were defects in the labyrinth layer such that the maternal and fetal blood systems which are normally in intimate apposition were separated by cells and ECM. The fact that early events including blastocyst formation, gastrulation, and implantation occurred in mutants indicates type IV collagen is not required for early embryogenesis. Rather, it was concluded that the defects point to an important role for type IV collagen in maintaining the stability of basement membranes that are placed under increasing mechanical stress. Mice deficient in perlecan die at a similar embryonic stage (E10 to E12), although a few survive to late gestation (Costell et al., 1999). They show some similarities to mice lacking the $\alpha1/\alpha2$ collagen network, including bleeding into the pericardium and neuronal ectopias that were also attributed to the structural instability of basement membranes. Both of these mutants have a much less severe phenotype than mice lacking the laminin $\gamma1$ chain in which there is absence of embryonic basement membranes and peri-implantation lethality at $\sim$E5.5 (Smyth et al., 1999).

The phenotype of mice in which collagen processing is disrupted provides further support for the important role of type IV collagen during embryonic development. Mice lacking the collagen-specific molecular chaperone Hsp47 show structurally abnormal basement membranes and die by E11.5, findings that are consistent with a disruption in the normal biosynthesis of type IV collagen. Hsp47 mutants also show other defects (abnormally oriented epithelial tissues and ruptured blood vessels) that are likely unrelated to type IV collagen, but reflect its role in the biosynthesis of other collagen types. Very recently, mice deficient in the collagen modifying enzyme lysyl hydroxylase 3 have been described (Rautavuoma et al., 2004). Mutant mice are normal until E8.5, but then show growth retardation and die by $\sim$E9.5. Type IV collagen processing is impaired in the mutants, with the $\alpha1$ and $\alpha2$ chains displaying increased electrophoretic mobility consistent with the absence of hydoxylysine-bound carbohydrates. Basement membranes in the mutant are severely disrupted. Essentially no stretches of continuous basement membrane are found and when present, they appear as amorphous deposits underlying the cells. Type IV collagen secretion is dramatically impaired, with staining being predominantly intracellular or localized to aggregates in the extracellular space. This elegant study was the first to show a specific requirement for post-translational processing of type IV collagen that is essential for its biologic function *in vivo*. The findings reinforce the critical role that type IV collagen plays in embryonic development.

8.2. The α3/α4/α5 and α1/α2/α5/α6 networks in fetal and postnatal development

In contrast to the α1/α2 network, the α3/α4/α5 and α1/α2/α5/α6 collagen networks are not required for normal embryonic development. The expression of these "novel" networks is temporally and spatially restricted during development. In some sites they join or largely replace the α1/α2 network within basement membranes, whereas in others they may exist on their own. The α3/α4/α5 network is invariably expressed at later developmental stages than the α1/α2 network, and at least in some basement membranes this is also true for the α1/α2/α5/α6 network. While the biologic role of the α1/α2/α5/α6 network remains to be determined, the α3/α4/α5 network plays a critical role in the long-term maintenance of basement membrane structure and function. In the text that follows, the temporal and spatial distribution of these networks during development is summarized for several tissues, including kidney, testis, eye, and inner ear.

An early study of fetal human kidney revealed a distinction in the temporal and spatial expression of the α1/α2 network, and networks formed by the other (at that point undefined) isoforms during development (Kleppel and Michael, 1990). The first study that made use of well characterized antibodies to document these changes in developing kidneys was conducted in rodents (Miner and Sanes, 1994). Only the α1 and α2 chains were detected in the basement membrane of early nephric figures (i.e., vesicle, comma, and early "S-shaped" nephrons). The α3, α4, and α5 chains appeared later, together with the α1 and α2 chains in the primitive GBM at the early capillary loop stage of glomerular development. The α3, α4, and α5 isoforms were also detected in some tubular basement membranes at this stage and they continue to be expressed here into adult life. As glomeruli mature, the expression of the α3, α4, and α5 chains persists in the GBM, while expression of the α1 and α2 chains diminishes markedly and becomes prominent in the mesangium. The same transitions were subsequently described in humans (Peissel et al., 1995; Kalluri et al., 1997; Lohi et al., 1997; Kuroda et al., 1998) and dogs (Harvey et al., 1998). This can be viewed as a developmental shift, whereby the "fetal" α1/α2 collagen network is replaced by the "adult" α3/α4/α5 network in the GBM during glomerular morphogenesis. Some of the studies above included the $\alpha6$ chain, which allows the $\alpha1/\alpha2/\alpha5/\alpha6$ network to be placed into this developmental sequence. Within the glomerulus, the $\alpha6$ chain is first detected along with the $\alpha5$ chain at the pre-capillary loop stage. Both isoforms co-localize to the $\alpha1/\alpha2$-positive basement membrane of the epithelial component that is destined to become Bowman's capsule. This also reflects a developmental shift, one in which the $\alpha1/\alpha2/\alpha5/\alpha6$ network joins (but is not thought to replace) the $\alpha1/\alpha2$ network. The $\alpha5$ and $\alpha6$ chains are expressed in this site prior to the appearance of the $\alpha3$ and $\alpha4$ chains in the primitive GBM of early capillary loop stage glomeruli. Therefore, the expression of the $\alpha1/\alpha2/\alpha5/\alpha6$ network precedes that of the $\alpha3/\alpha4/\alpha5$ network during glomerular development, and both networks remain spatially separated (Fig. 9).

The $\alpha1/\alpha2$ and $\alpha1/\alpha2/\alpha5/\alpha6$ networks are present together in the basement membrane of vascular and visceral smooth muscle cells, and in epithelial basement membranes of many tissues (see section entitled Type IV collagen networks). It is unclear whether these sites also undergo a developmental shift from only an $\alpha1/\alpha2$ network to one also containing the $\alpha1/\alpha2/\alpha5/\alpha6$ network. Human fetal tissues at 18-22 weeks gestation were surveyed for expression of the $\alpha5$ and $\alpha6$ chains, and both isoforms were detected in a variety of epithelial basement membranes of the respiratory, gastrointestinal, and genitourinary tracts as well as in the skin (Peissel et al., 1995). Therefore, if a developmental shift involving the $\alpha1/\alpha2/\alpha5/\alpha6$ network took place in these sites it must have occurred prior to this timepoint. However, a study of human fetal esophagus at 20 weeks gestation suggested that such a transition may occur in visceral smooth muscle (Heidet et al., 1997).

The expression of the $\alpha1$-$\alpha6$ chains during testicular development has been studied in mice (Enders et al., 1995; Saito et al., 2000) and rats (Frojdman et al., 1998). In mice, the STBM stains only for the $\alpha1$-$\alpha5$ chains, which are presumably organized into $\alpha1/\alpha2$ and $\alpha3/\alpha4/\alpha5$ networks. The $\alpha1/\alpha2$ network is expressed in this site as early as E12.5, whereas the $\alpha3/\alpha4/\alpha5$ network is first detected ~1 week postnatally. The same developmental switch occurs in rat testis, with the $\alpha3/\alpha4/\alpha5$ network first expressed in the STBM between 2 and 6 days of age (Frojdman et al., 1998). Both networks continue to be expressed in the STBM throughout adult life, which is in contrast to the kidney where the $\alpha3/\alpha4/\alpha5$ network replaces the $\alpha1/\alpha2$ network in the GBM. There are also temporal differences in the expression of these networks between testis and kidney. All seminiferous tubules undergo this developmental transition at the same time, whereas in the kidney, the expression of specific networks varies between glomeruli that co-exist at different developmental stages. The studies in rodents noted that the $\alpha3/\alpha4/\alpha5$ network was expressed at the onset of spermatogenesis, prompting speculation that it may play a functional role in this process.

All six isoforms of type IV collagen are expressed in bovine STBM, although the exact way in which these are organized into networks is unclear

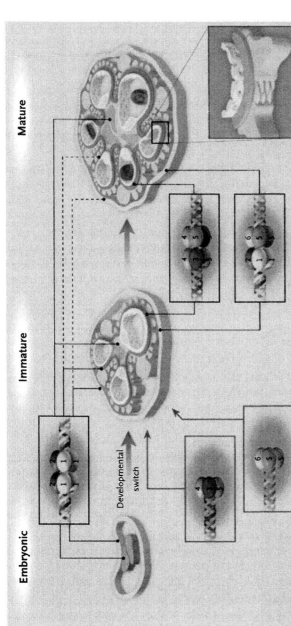

Fig. 9. Temporal and spatial distribution of type IV collagen networks during glomerular development. The basement membrane of primitive nephric figures (vesicle and 'comma' shaped nephrons) from which glomeruli arise contain only the $\alpha1/\alpha2$ collagen network (embryonic). In 'S-shaped' nephrons, the $\alpha1/\alpha2/\alpha5/\alpha6$ network is deposited in the basement membrane underlying the epithelial component destined to become Bowman's capsule (not shown). Later, in early capillary loop stage glomeruli (immature) the $\alpha3/\alpha4/\alpha5$ network becomes expressed along with the $\alpha1/\alpha2$ network in the primitive GBM (purple) that separates visceral epithelium (podocytes) (blue) from endothelium (yellow). At later stages (mature), the $\alpha1/\alpha2/\alpha5/\alpha6$ and $\alpha3/\alpha4/\alpha5$ networks appear to replace (dotted line) the $\alpha1/\alpha2$ network in Bowman's capsule and the GBM, respectively. Figure reproduced from Hudson, B., Tryggvason, K., Sundaramoorthy, M., Neilson, E. (2003). Alport's syndrome, Goodpasture's syndrome, and type IV collagen. N. Engl. J. Med. 348, 2543–2556. Copyright © 2004 Massachusetts Medical Society. All rights reserved. (See Color Insert.)

and developmental studies in this species have not been performed (Kahsai et al., 1997). Expression of the $\alpha6$ chain has not been evaluated in the rat and, therefore, it remains to be determined whether developmental shifts involving the $\alpha1/\alpha2/\alpha5/\alpha6$ network occur in testis and if so, how these might relate to testicular function. We have addressed this issue in preliminary studies of dog testis, which like bovine tissue contains all 6 isoforms. In the dog, the six α chains are expressed pair-wise in sequential fashion in the STBM during testicular development. Only the $\alpha1$ and $\alpha2$ chains are expressed in the STBM of neonatal dogs. By ~6 weeks of age this site is additionally positive for the $\alpha5$ and $\alpha6$ chains; this is followed by the $\alpha3$ and $\alpha4$ chains at ~8 weeks of age. If the protomers are assembled in dogs as they are in other species, then when interpreted at the network level this indicates that the sequence of expression is the $\alpha1/\alpha2$ network appearing first, followed by the $\alpha1/\alpha2/\alpha5/\alpha6$ network, and lastly the $\alpha3/\alpha4/\alpha5$ network. In contrast to rodents, the timing of these transitions do not correlate to the onset of spermatogenesis. In a recent study, human STBM was reported to contain only the $\alpha1$ and $\alpha2$ chains (Dobashi et al., 2003). If correct, this makes human STBM unique from that of all other species studied to date, which also contain the $\alpha3$, $\alpha4$, and $\alpha5$ chains with or without the $\alpha6$ chain.

Another site where the $\alpha1/\alpha2/\alpha5/\alpha6$ and $\alpha3/\alpha4/\alpha5$ networks co-localize is the lens capsule. Developmental studies have been carried out on this tissue in both mice and humans (Kelley et al., 2002). In mice, the $\alpha1$, $\alpha2$, $\alpha5$, and $\alpha6$ chains were detected in the lens capsule as early as E11.5. The $\alpha3$ and $\alpha4$ chains became additionally expressed in this site by ~2 weeks postnatally. The lens capsules of 54-day human embryos stained for the $\alpha1$, $\alpha2$, $\alpha5$, and $\alpha6$ chains but not the $\alpha3$ and $\alpha4$ chains, and all isoforms were present in adult tissue. This sequence indicates that expression of the $\alpha1/\alpha2/\alpha5/\alpha6$ network precedes that of the $\alpha3/\alpha4/\alpha5$ network. The authors speculated that this transition might reflect the requirement for a more elastic lens capsule comprised of the $\alpha1/\alpha2/\alpha5/\alpha6$ network during embryogenesis to allow for growth of the lens, but a more resilient structure additionally comprised of the $\alpha3/\alpha4/\alpha5$ network to withstand the mechanical stress imparted on the lens during adult life. The $\alpha1$ and $\alpha2$ chains were not expressed in the absence of the $\alpha5$ and $\alpha6$ chains in either species, which implies that either the lens capsule lacks the $\alpha1/\alpha2$ network or that a developmental shift involving the $\alpha5$ and $\alpha6$ chains occurred before the earliest timepoint examined.

Finally, studies in dogs (Harvey et al., 2001) and mice (Cosgrove et al., 1996a) have revealed that there are shifts in the expression of type IV collagen networks during cochlear development. The studies in mouse are reviewed by others in this series (Chapter 6) and will not be discussed here. In canine inner ear, the $\alpha1/\alpha2$ network is present in all vascular and perineural basement membranes. It is also localized to the inner and outer sulci, the basilar membrane, and among the root cells within the medial aspect of the spiral ligament. This expression pattern is the same for dogs of all ages. In neonatal dogs, the α3/α4/α5 network is expressed along the basilar membrane and the outer sulcus, but by one month of age it is additionally detected along the lateral aspect of the spiral ligament adjacent to the bony wall of the inner ear. The staining in this region is most intense at the base of the cochlea and corresponds to a site notably devoid of the α1/α2 network. The spatial redistribution of the α3/α4/α5 network that occurs at one month is accompanied by the expression of smooth muscle actin by cells in the lateral aspect of the spiral ligament and coincides with the acquisition of mature auditory function.

8.3. Alport syndrome

Alport syndrome is an inherited disorder characterized by progressive nephropathy, sensorineural deafness, and ocular abnormalities that is caused by mutations in the genes encoding type IV collagen. A detailed discussion of the clinicopathologic features of this disease is beyond the scope of this review and has been published elsewhere (Flinter, 1998; Kashtan, 1999; Pirson, 1999; Jais et al., 2000, 2003).

Three modes of inheritance are recognized: autosomal dominant, autosomal recessive, and X-linked (in order of increasing prevalence). Only three examples of autosomal dominant Alport syndrome caused by a mutation in the *COL4A3* gene have been reported in the literature (van der Loop et al., 2000; Pescucci et al., 2004), and the same number has been reported for the *COL4A4* gene (Ciccarese et al., 2001; Pescucci et al., 2004). Two models of this disease may exist in dogs (Hood et al., 1995, 2002). Autosomal recessive Alport syndrome accounts for ~15% of all cases (Gubler et al., 1995) and is caused by homozygous or compound heterozygous mutations in the *COL4A3* or *COL4A4* genes (Mochizuki et al., 1994). A total of ~30 different mutations in the *COL4A3* gene have been identified in families with the disease (Lemmink et al., 1994; Ding et al., 1995; Knebelmann et al., 1995; Heidet et al., 2001; Longo et al., 2002), and approximately one-half this number of mutations have been reported in the *COL4A4* gene (Boye et al., 1998; Dagher et al., 2002; Longo et al., 2002). Three different transgenic mouse models of autosomal recessive Alport syndrome exist (Cosgrove et al., 1996b; Miner and Sanes, 1996; Lu et al., 1999) and a naturally occurring model is also found in dogs (Lees et al., 1998).

Benign familial hematuria (BFH), also known as thin basement membrane nephropathy, is an autosomal dominant disorder characterized by persistent or recurrent hematuria. In contrast to Alport syndrome, BFH is not associated with deafness, significant proteinuria, or progression to renal failure (Savige et al., 2003). In ~40% of families with BFH, hematuria segregates with the *COL4A3/COL4A4* loci (Savige et al., 2003; Vega et al., 2003), and heterozygous mutations in these genes have been identified in many cases (Ozen et al., 2001; Badenas et al., 2002; Buzza et al., 2003). Thus, at least in some cases, BFH represents the carrier state of autosomal recessive Alport syndrome. It is now appreciated that mutations in the *COL4A3* or *COL4A4* genes can contribute to a clinical spectrum of disease, ranging from BFH to autosomal dominant or recessive Alport syndrome depending on the nature of the mutation and gene dosage (Longo et al., 2002; Vega et al., 2003).

X-linked Alport syndrome is the most common form of the disease, accounting for ~85% of all cases. Over 300 different mutations in the *COL4A5* gene have been identified in families with the X-linked disease. Two different models of X-linked Alport syndrome are found in dogs (Zheng et al., 1994; Lees et al., 1999), and a transgenic mouse model was recently described (Rheault et al., 2003). Mutations confined to the *COL4A6* gene have not been identified in Alport syndrome (Heiskari et al., 1996). However, this gene is involved in patients who have X-linked Alport syndrome in association with diffuse leiomyomatosis (Garcia-Torres et al., 2000). In addition to the nephropathy, deafness, and ocular abnormalities that are pathognomonic of Alport syndrome, patients with this extremely rare condition suffer from a benign diffuse or nodular proliferation of smooth muscle that usually involves the esophagus, but may also affect the tracheobronchial tree or female reproductive tract. This disorder is caused by deletions of the *COL4A5* gene that extend into but not beyond intron 2 of the *COL4A6* gene. Although these patients usually lack the $\alpha 5$ and $\alpha 6$ chains from the basement membrane of smooth muscle cells, their mere absence cannot explain the muscle phenotype. Rather, it has also been postulated that this is caused by the loss of a yet undiscovered gene in the large second intron of the *COL4A6* gene (Antignac et al., 1992; Heidet et al., 1995), loss of a regulatory sequence controlling smooth muscle proliferation (Ueki et al., 1998), or by causing dysregulation of neighboring genes through alteration of chromatin structure (Thielen et al., 2003).

In Alport syndrome, mutations in the *COL4A3*, *COL4A4*, or *COL4A5* genes interfere with the normal co-assembly of $\alpha 3$, $\alpha 4$, and $\alpha 5$ chains into a triple helical protomer and prevent or disrupt the incorporation of these protomers into an $\alpha 3/\alpha 4/\alpha 5$ network within basement membranes. Subtle defects such as missense mutations may allow for protomer assembly, but result in an improperly folded and unstable trimer that is degraded intracellularly or becomes incorporated into a network that is structurally and functionally deficient. At the most fundamental level, it is this mechanism that underlies the pathogenesis of this disease, with the consequences of this event (renal failure, hearing loss, and ocular abnormalities) arising either directly or indirectly and reflecting the specific role the $\alpha 3/\alpha 4/\alpha 5$ network plays in specialized basement membranes within different tissues.

The expression of the $\alpha 1 - \alpha 6$ chains in human Alport syndrome has been well documented (Nakanishi et al., 1994; Yoshioka et al., 1994; Peissel et al.,

1995; Naito et al., 1996; Kalluri et al., 1997; Kashtan, 1998; Mazzucco et al., 1998). The α5 and α6 chains are usually absent from all basement membranes in males with X-linked disease, such as those of Bowman's capsule and collecting ducts in the kidney, the epidermal basement membrane in skin, and vascular smooth muscle in both tissues. This can be explained by the loss of the α1/α2/α5/α6 network secondary to a defect in the α5 chain. In addition, the GBM and distal tubule basement membranes in these patients are usually comprised of the α1 and α2 chains, but lack the α3, α4, and α5 chains, which is consistent with the loss of the α3/α4/α5 network. The GBM of carrier females usually show a segmental distribution of the α3, α4, and α5 chains, a finding that presumably reflects lyonization of the normal or mutant X-chromosome in each glomerular cell. In autosomal forms of the disease, the GBM contains the α1/α2 network, but lacks the α3/α4/α5 network secondary to defects in either of the α3 or α4 chains (Gubler et al., 1995; Nomura et al., 1998). However, in contrast to X-linked Alport syndrome, the α5 chain is retained in all other sites that have been examined where it normally assembles into an α1/α2/α5/α6 network. As previously described, a GBM comprised of the α1/α2 network but not the α3/α4/α5 network such as that in Alport syndrome is transiently present in the normal developing kidney. The focal thinning and multilaminar splitting of the GBM in Alport syndrome gives it a primitive appearance reminiscent to that of normal immature glomeruli; this prompted some to speculate that Alport GBM may be in a state of developmental arrest (Spear, 1973; Rumpelt, 1987). This concept was revisited by Kalluri and colleagues (1997), who found that normal human glomeruli undergo developmental transitions in type IV collagen expression similar to those first described in rodents (Miner and Sanes, 1994). These findings prompted the authors to speculate that the molecular defects in Alport syndrome may reflect the failure of this developmental switch secondary to a mutation in the *COL4A5* gene, which leads to persistence of a "fetal" α1/α2 network in the GBM.

Several mechanisms have been proposed to account for why the α1/α2 network of Alport GBM may be insufficient for normal glomerular structure and function, and to explain the progressive nature of this disease. The α3 and α4 chains are more cysteine-rich in their 7S and triple helical regions than the α1 and α2 chains, leading to a loop structure of the triple helix covalently cross-linked by disulfide bonds (Kahsai et al., 1997; Gunwar et al., 1998). This organization may confer added structural integrity compared to basement membranes composed solely of the α1/α2 network. This additional strength would be important for basement membranes that experience high mechanical stresses such as the GBM (hydrostatic) and the inner ear (sonic). It has also been suggested that the progressive structural and functional deterioration of GBM in the Alport syndrome may reflect the requirement of the α3/α4/α5 network for conferring long-term stability to the GBM by protecting against endoproteolytic degradation (Kalluri et al., 1997).

9. Conclusions

In this chapter we sought to provide readers of this series with a current review of type IV collagen from a developmental perspective. This began with an overview of the structure, molecular composition, and function of the class of highly specialized extracellular matrices to which type IV collagen belongs—basement membranes. Genomic research has revealed that type IV collagen has been conserved throughout metazoan evolution, a finding that underscores how fundamentally important these matrices must be to the basic organization of multicellular organisms. As this field continues to advance, it will shed new light on the evolutionary history of type IV collagen and perhaps yield new surprises. Consider, for example, the unique paired organization of type IV collagen genes. Such an arrangement offers opportunities for coordinated transcriptional control for the *COL4A1-COL4A2* genes that are co-expressed in a wide variety of tissues. However, a small bidirectional promoter would seemingly present an obstacle for the *COL4A3-COL4A4* genes that are invariably co-expressed, but only among select basement membranes and also for the *COL4A5-COL4A6* genes that in some cases are expressed in different sites.

Studies of invertebrate models were the first to reveal the important role that type IV collagen plays during development. Hydra cell aggregate development and head regeneration is blocked when the expression or assembly of type IV collagen is disrupted, indicating a role in both cytodifferentiation and morphogenesis. The $\alpha1/\alpha2$ network is essential for embryonic development and plays a specific functional role in maintaining the structural stability of muscle in both *C. elegans* and *Drosophila*. In the mouse, this network is dispensable for early embryonic development, but is essential for the structural integrity of basement membranes that are placed under increasing mechanical stress as embryogenesis proceeds. The $\alpha3/\alpha4/\alpha5$ and $\alpha1/\alpha2/\alpha5/\alpha6$ networks have a restricted distribution among basement membranes and are first expressed later in fetal development or during postnatal life. The $\alpha1/\alpha2/\alpha5/\alpha6$ network is present in tissues that undergo elastic changes in size such as vascular and visceral smooth muscle. The $\alpha3/\alpha4/\alpha5$ network is present in basement membranes that serve specialized functional roles, including those that act as filtration barriers (glomerular, ependymal, and alveolar basement membranes) or perhaps as a component of a blood-tissue barrier (seminiferous tubule basement membrane). This network is also present in sites that have unique functional roles such as the lens capsule and the cochlea. The $\alpha3/\alpha4/\alpha5$ network is critical for the long-term maintenance of basement membrane structure and function. Disruption of this network results in Alport syndrome, a disorder characterized by progressive renal disease, hearing loss, and ocular abnormalities. Curiously, no specific functional role has been ascribed to the $\alpha6$ chain, or more generally, to the $\alpha1/\alpha2/\alpha5/\alpha6$ network.

Acknowledgments

S.J.H. is supported by a National Kidney Foundation Postdoctoral Research Fellowship. This review drew upon material from the doctoral thesis of S.J.H., that was supported by grants to P.S.T. from the Canada Institutes of Health Research (MOP-13254) and the National Institute of Health (P01 DK 53763–01).

References

Adachi, N., Lieber, M. 2002. Bidirectional gene organization: A common architectural feature of the human genome. Cell 109, 807–809.

Adamson, E., Ayers, S. 1979. The localization and synthesis of some collagen types in developing mouse embryos. Cell 16, 953–965.

Alitalo, K., Vaheri, A., Krieg, T., Timpl, R. 1980. Biosynthesis of two subunits of type IV procollagen and of other basement membrane proteins by a human tumor cell line. Eur. J. Biochem. 109, 247–255.

Antignac, C., Zhou, J., Sanak, M., Cochat, P., Rousel, B., Deschênes, G., Gros, F., Knebelmann, B., Hors-Cayla, M., Tryggvason, K., Gubler, M.C. 1992. Alport syndrome and diffuse leiomyomatosis: Deletions in the 5' end of the *COL4A5* collagen gene. Kidney Int. 42, 1178–1183.

Babel, W., Glanville, R. 1984. Structure of human-basement membrane (type IV) collagen. Complete amino-acid sequence of a 914 residue-long pepsin fragment from the α1(IV) chain. Eur. J. Biochem. 143, 545–556.

Bächinger, H., Fessler, L., Fessler, J. 1982. Mouse procollagen IV. Characterization and supramolecular association. J. Biol. Chem. 257, 9796–9803.

Badenas, C., Praga, M., Tazon, B., Heidet, L., Arrondel, C., Armengol, A., Andres, A., Morales, E., Camancho, J., Lens, X., Da' Vila, S., Mila, M., Antignac, C., Darnell, A., Torra, R. 2002. Mutations in the *COL4A4* and *COL4A3* genes cause familial benign hematuria. J. Am. Soc. Nephrol. 13, 1248–1254.

Bailey, A., Sims, T., Light, N. 1984. Cross-linking in type IV collagen. Biochem. J. 218, 713–723.

Bella, J., Eaton, M., Brodsky, B., Berman, H. 1994. Crystal and molecular structure of a collagen-like peptide at 1.9 A resolution. Science 266, 75–81.

Berg, R., Prockop, D. 1973. The thermal transition of a non-hydroxylated form of collagen. Evidence for a role for hydroxyproline in stabilizing the triple helix of collagen. Biochem. Biophys. Res. Commun. 52, 115–120.

Bhattacharya, G., Miller, C., Kimberling, W., Jablonski, M., Cosgrove, D. 2002. Localization and expression of usherin: A novel basement membrane protein defective in people with Usher's syndrome type IIA. Hear Res. 163, 1–11.

Biswas, S., Munier, F., Yardley, J., Hart-Holden, N., Perveen, R., Cousin, P., Sutphin, J., Noble, B., Batterbury, M., Kielty, C., Hackett, A., Bonshek, R., Ridgeway, A., McLeod, D., Sheffield, V., Stone, E., Schorderet, D., Black, G. 2001. Missense mutations in *COL8A2*, the gene encoding the α2 chain of type VIII collagen, cause two forms of corneal endothelial dystrophy. Hum. Mol. Genet. 10, 2415–2423.

Blumberg, B., MacKrell, A., Fessler, J. 1988. *Drosophila* basement membrane procollagen α1 (IV). II. Complete cDNA sequence, genomic structure, and general implications for supramolecular assemblies. J. Biol. Chem. 263, 18328–18337.

Blumberg, B., MacKrell, A., Olson, P., Kurkinen, M., Monson, J., Natzle, J., Fessler, J. 1987. Basement membrane procollagen IV and its specialized carboxyl domain are conserved in *Drosophila*, mouse, and human. J. Biol. Chem. 262, 5947–5950.

Borchiellini, C., Coulon, J., Le Parco, Y. 1996. The function of type IV collagen during *Drosophila* muscle development. Mech. Dev. 58, 179–191.

Bornstein, P., Sage, H. 1980. Structurally distinct collagen types. Ann. Rev. Biochem. 49, 957–1003.

Borza, D., Neilson, E., Hudson, B. 2003. Pathogenesis of Goodpasture syndrome: A molecular perspective. Semin. Nephrol. 23, 522–531.

Borza, D., Bondar, O., Ninomiya, Y., Sado, Y., Naito, I., Todd, P., Hudson, B. 2001. The NC1 domain of collagen IV encodes a novel network composed of the α1, α2, α5, and α6 chains in smooth muscle basement membranes. J. Biol. Chem. 276, 28532–28540.

Borza, D., Bondar, O., Todd, P., Sundaramoorthy, M., Sado, Y., Ninomiya, Y., Hudson, B. 2002. Quaternary organization of the Goodpasture antigen, the α3(IV) collagen chain: Sequestration of two cryptic epitopes by intra-protomer interactions with the α4 and α5 NC1 domains. J. Biol. Chem. 277, 40075–40083.

Boutaud, A., Borza, D., Bondar, O., Gunwar, S., Netzer, K.O., Singh, N., Ninomiya, Y., Sado, Y., Noelken, M., Hudson, B. 2000. Type IV collagen of the glomerular basement membrane. Evidence that the chain specificity of network assembly is encoded by the noncollagenous NC1 domain. J. Biol. Chem. 275, 30716–30724.

Boute, N., Exposito, J., Boury-Esnault, N., Vacelet, J., Noro, N., Miyazaki, K., Yoshizato, K., Garrone, R. 1996. Type IV collagen in sponges, the missing link in basement membrane ubiquity. Biol. Cell. 88, 37–44.

Boyd, C., Toth-Fejel, S., Gadi, I., Litt, M., Condon, M., Kolbe, M., Hagen, I., Kurkinen, M., MacKenzie, J., Magenis, E. 1988. The genes encoding for human pro α1(IV) collagen and pro α2(IV) collagen are both located at the end of the long arm of chromosome 13. Am. J. Hum. Genet. 42, 309–314.

Boye, E., Mollet, G., Forestier, L., Cohen-Solal, L., Heidet, L., Cochat, P., Grünfeld, J.P., Palcoux, J., Gubler, M.C., Antignac, C. 1998. Determination of the genomic structure of the *COL4A4* gene and of novel mutations causing autosomal recessive Alport syndrome. Am. J. Hum. Genet. 63, 1329–1340.

Brazel, D., Oberbäumer, I., Dieringer, H., Babel, W., Glanville, R., Deutzmann, R., Kühn, K. 1987. Completion of the amino acid sequence of the α1(IV) chain of human basement membrane collagen (type IV) reveals 21 non-triplet interruptions located within the collagenous domain. Eur. J. Biochem. 168, 529–536.

Brazel, D., Pollner, R., Oberbäumer, I., Kühn, K. 1988. Human basement membrane collagen (type IV). The amino acid sequence of the α2(IV) chain and its comparison with the α1(IV) chain reveals deletions in the α1(IV) chain. Eur. J. Biochem. 172, 35–42.

Bruggeman, L., Burbelo, P., Yamada, Y., Klotman, P. 1992. A novel sequence in the type IV collagen promoter binds nuclear proteins from Engelbreth-Holm-Swarm tumor. Oncogene 7, 1497–1502.

Burbelo, P., Martin, G., Yamada, Y. 1988. α1(IV) and α2(IV) collagen genes are regulated by a bidirectional promoter and a shared enhancer. Proc. Natl. Acad. Sci. USA 85, 9679–9682.

Buttice, G., Kaytes, P., D'Armiento, J., Vogeli, G., Kurkinen, M. 1990. Evolution of collagen IV genes from a 54-base pair exon: A role for introns in gene evolution. J. Mol. Evol. 30, 479–488.

Buzza, M., Dagher, H., Wang, Y., Wilson, D., Babon, J., Cotton, R., Savige, J. 2003. Mutations in the *COL4A4* gene in thin basement membrane disease. Kidney Int. 63, 447–453.

Cassada, R., Isnenghi, E., Culotti, M., von Ehrenstein, G. 1981. Genetic analysis of temperature-sensitive embryogenesis mutants in *Caenorhabditis elegans*. Dev. Biol. 84, 193–205.

Caulagi, V., Rajan, T. 1995. The structural organization of an α2 (type IV) basement membrane collagen gene from the filarial nematode *Brugia malayi*. Mol. Biochem. Parasitol. 70, 227–229.

Cecchini, J., Knibiehler, B., Mirre, C., Le Parco, Y. 1987. Evidence for a type-IV-related collagen in *Drosophila melanogaster*. Evolutionary constancy of the carboxyl-terminal non-collagenous domain. Eur. J. Biochem. 165, 587–593.

Chartier, A., Zaffran, S., Astier, M., Semeriva, M., Gratecos, D. 2002. Pericardin, a *Drosophila* type IV collagen-like protein is involved in the morphogenesis and maintenance of the heart epithelium during dorsal ectoderm closure. Development 129, 3241–3253.

Chen, L., Miyamira, N., Ninomiya, Y., Honda, T. 2003. Distribution of the collagen IV isoforms in human Bruch's membrane. Br. J. Ophthalmol. 87, 212–215.

Ciccarese, M., Casu, D., Ki Wong, F., Faedda, R., Arvidsson, S., Tonolo, G., Luthman, H., Satta, A. 2001. Identification of a new mutation in the α4(IV) collagen gene in a family with autosomal dominant Alport syndrome and hypercholesterolaemia. Nephrol. Dial. Transpl. 16, 2008–2012.

Colognato, H., Yurchenco, P. 2000. Form and function: The laminin family of heterotrimers. Dev. Dyn. 218, 213–234.

Cosgrove, D., Kornak, J., Samuelson, G. 1996a. Expression of basement membrane type IV collagen chains during postnatal development in the murine cochlea. Hearing Res. 100, 21–32.

Cosgrove, D., Meehan, D., Grunkemeyer, J., Kornak, J., Sayers, R., Hunter, W., Samuelson, G. 1996b. Collagen *COL4A3* knockout: A mouse model for Alport syndrome. Genes Dev. 10, 2981–2992.

Cosgrove, D., Samuelson, G., Pinnt, J. 1996c. Immunohistochemical localization of basement membrane collagens and associated proteins in the murine cochlea. Hearing Res. 97, 54–65.

Costell, M., Gustafsson, E., A, A., Morgelin, M., Bloch, W., Hunziker, E., Addicks, K., Timpl, R., Fassler, R. 1999. Perlecan maintains the integrity of cartilage and some basement membranes. J. Cell. Biol. 147, 1109–1122.

Crouch, E., Bornstein, P. 1979. Characterization of a type IV procollagen synthesized by human amniotic fluid cells in culture. J. Biol. Chem. 254, 4197–4204.

Crouch, E., Sage, H., Bornstein, P. 1980. Structural basis for apparent heterogeneity of collagens in human basement membranes: Type IV collagen contains two distinct chains. Proc. Natl. Acad. Sci. USA 77, 745–749.

Czaker, R. 2000. Extracellular matrix (ECM) components in a very primitive multicellular animal, the Dicyemid Mesozoan *Kantharella antartica*. Anat. Rec. 259, 52–59.

Dagher, H., Yan Wang, Y., Fassett, R., Savige, J. 2002. Three novel *COL4A4* mutations resulting in stop codons and their clinical effects in autosomal recessive Alport syndrome. Hum. Mutat. 20, 321–322.

Denduchis, B., Kefalides, N., Bezkorovainy, A. 1970. The chemistry of sheep anterior lens capsule. Arch. Biochem. Biophys. 138, 582–589.

Deutzmann, R., Fowler, S., Zhang, X., Boone, K., Dexter, S., Boot-Handford, R., Rachal, R., Sarras, M., Jr. 2000. Molecular, biochemical and functional analysis of a novel and developmentally important fibrillar collagen (Hcol-1) in hydra. Development 127, 4669–4680.

Ding, J., Stitzel, J., Berry, P., Hawkins, E., Kashtan, C. 1995. Autosomal recessive Alport syndrome: Mutation in the *COL4A3* gene in a woman with Alport syndrome and post-transplant antiglomerular basement membrane nephritis. J. Am. Soc. Nephrol. 5, 1714–1717.

Dobashi, M., Fujisawa, M., Naito, I., Yamazaki, T., Okada, H., Kamidono, S. 2003. Distribution of type IV collagen subtypes in human testes and their association with spermatogenesis. Fertil. Steril. 80, 755–760.

Dolz, R., Engel, J., Kühn, K. 1988. Folding of collagen IV. Eur. J. Biochem. 178, 357–366.

Duncan, K., Fessler, L., Bächinger, H., Fessler, J. 1983. Procollagen IV: Association to tetramers. J. Biol. Chem. 258, 5869–5877.

Dyck, R., Lockwood, C., Kershaw, M., McHugh, N., Duance, V., Baltz, M., Pepys, M. 1980. Amyloid P-component is a constituent of normal human glomerular basement membrane. J. Exp. Med. 152, 1162–1174.

Dziadek, M., Timpl, R. 1985. Expression of nidogen and laminin in basement membranes during mouse embryogenesis and in teratocarcinoma cells. Dev. Biol. 111, 372–382.

Enders, G., Kahsai, T., Lian, G., Funabiki, K., Killen, P., Hudson, B. 1995. Developmental changes in seminiferous tubule extracellular matrix components of the mouse testis: α3(IV) collagen chain expressed at the initiation of spermatogenesis. Biol. Reprod. 53, 1489–1499.

Engbring, J., Kleinman, H. 2003. The basement membrane matrix in malignancy. J. Pathol. 200, 465–470.

Erickson, A., Couchman, J. 2000. Still more complexity in mammalian basement membranes. J. Histol. Cytochem. 48, 1291–1306.

Exposito, J., D' Alessio, M., Di Liberto, M., Ramirez, F. 1993. Complete primary structure of a sea urchin type IV collagen a chain and analysis of the 5′ end of its gene. J. Biol. Chem. 268, 5249–5254.

Exposito, J., Suzuki, H., Geourjon, C., Garrone, R., Solurah, M., Ramirez, F. 1994. Identification of a cell lineage-specific gene coding for a sea urchin α2(IV)-like collagen chain. J. Biol. Chem. 269, 13167–13171.

Eyre, D., Paz, M., Gallop, P. 1984. Cross-linking in collagen and elastin. Ann. Rev. Biochem. 53, 717–748.

Ferreira, L., Norris, K., Smith, T., Hebert, C., Sauk, J. 1996. Hsp47 and other ER-resident molecular chaperones form heterocomplexes with each other and with collagen type IV chains. Conn. Tiss. Res. 33, 265–273.

Fessler, L., Condic, M., Nelson, R., Fessler, J., Fristrom, J. 1993. Site-specific cleavage of basement membrane collagen IV during Drosophila metamorphosis. Development 117, 1061–1069.

Fessler, L., Duncan, K., Fessler, J., Salo, T., Tryggvason, K. 1984. Characterization of the procollagen IV cleavage products produced by a specific tumor collagenase. J. Biol. Chem. 259, 9783–9789.

Fessler, L., Fessler, J. 1982. Identification of the carboxyl peptides of mouse procollagen IV and its implications for the assembly and structure of basement membrane procollagen. J. Biol. Chem. 257, 9804–9810.

Fischer, G., Schmidt, C., Opitz, J., Cully, Z., Kühn, K., Poschl, E. 1993. Identification of a novel sequence element in the common promoter region of human collagen type IV genes, involved in the regulation of divergent transcription. Biochem. J. 292, 687–695.

Flinter, F. 1998. Disorders of the basement membrane: Hereditary nephritis. In: Inherited disorders of the kidney (S. Morgan, J.P. Grünfeld, Eds.), : Oxford University press, pp. 192–214.

Folkman, J., Klagsbrun, M., Sasse, J., Wadzinski, M., Ingber, D., Vlodavsky, I. 1988. A heparin-binding angiogenic protein -basic fibroblast growth factor - is stored within basement membrane. Am. J. Pathol. 130, 393–400.

Fowler, S., Jose, S., Zhang, X., Deutzmann, R., Sarras, M., Jr., Boot-Handford, R. 2000. Characterization of Hydra type IV collagen. Type IV collagen is essential for head regeneration and its expression is up-regulated upon exposure to glucose. J. Biol. Chem. 275, 39589–39599.

Frojdman, K., Pelliniemi, L., Virtanen, I. 1998. Differential distribution of type IV collagen chains in the developing rat testis and ovary. Differentiation 63, 125–130.

Fujiwara, S., Shinkai, H., Timpl, R. 1991. Structure of N-linked oligosaccharide chains in the triple helical domains of human type VI and mouse type IV collagen. Matrix 11, 307–312.

Garcia-Torres, R., Cruz, D., Orozco, L., Heidet, L., Gubler, M.C. 2000. Alport syndrome and diffuse leiomyomatosis. Clinical aspects, pathology, molecular biology and extracellular matrix studies. A synthesis. Nephrologie 21, 9–12.

Gare, D., Piertney, S., Billingsley, P. 2003. Anopheles gambiae collagen IV genes: Cloning, phylogeny and midgut expression associated with blood feeding and Plasmodium infection. Int. J. Parasitol. 33, 681–690.

Gelse, K., Poschl, E., Aigner, T. 2003. Collagens-structure, function and biosynthesis. Adv. Drug. Deliv. Rev. 55, 1531–1546.

Genersch, E., Eckerskorn, C., Lottspeich, F., Herzog, C., Kühn, K., Poschl, E. 1995. Purification of the sequence-specific transcription factor CTCBF, involved in the control of human collagen IV genes: Subunits with homology to Ku antigen. EMBO J. 14, 791–800.

Glanville, R., Qian, R., Siebold, B., Risteli, J., Kühn, K. 1985. Amino acid sequence of the N-terminal aggregation and cross-linking region (7S domain) of the α1(IV) chain of human basement membrane collagen. Eur. J. Biochem. 152, 213–219.

Gohring, W., Sasaki, T., Heldin, C., Timpl, R. 1998. Mapping of the binding of platelet-derived growth factor to distinct domains of the basement membrane proteins BM-40 and perlecan and distinction from the BM-40 collagen-binding epitope. Eur. J. Biochem. 255, 60–66.

Graham, P., Johnson, J., Wang, S., Sibley, M., Gupta, M., Kramer, J. 1997. Type IV collagen is detectable in most, but not all, basement membranes of *Caenorhabditis elegans* and assembles on tissues that do not express it. J. Cell. Biol. 137, 1171–1183.

Grande, J., Melder, D., Kluge, D., Wieben, E. 1996. Structure of the rat collagen IV promoter. Biochim. Biophys. Acta 1309, 85–88.

Grant, M., Harwood, R., Williams, I. 1975. The biosynthesis of basement-membrane collagen by isolated rat glomeruli. Eur. J. Biochem. 54, 531–540.

Grant, M., Kefalides, N., Prockop, D. 1972. The biosynthesis of basement membrane collagen in embryonic chick lens: Delay between the synthesis of polypeptide chains and the secretion of collagen by matrix-free cells. J. Biol. Chem. 247, 3539–3544.

Griffin, C., Emanuel, B., Hansen, J., Cavenee, W., Myers, J. 1987. Human collagen genes encoding basement membrane α1(IV) and α2(IV) chains map to the distal long arm of chromosome 13. Proc. Natl. Acad. Sci. USA 84, 512–516.

Groffen, A., Buskens, C., van Kuppevelt, T., Veerkamp, J., Monnens, L., van den Heuvel, L. 1998. Primary structure and high expression of human agrin in basement membranes of adult lung and kidney. Eur. J. Biochem. 254, 123–128.

Gubler, MC., Knebelmann, B., Beziau, A., Broyer, M., Pirson, Y., Haddoum, F., Kleppel, M., Antignac, C. 1995. Autosomal recessive Alport syndrome: Immunohistochemical study of type IV collagen chain distribution. Kidney Int. 47, 1142–1147.

Gunwar, S., Ballester, F., Kalluri, R., Timoneda, J., Chonko, A., Edwards, S., Noelken, M., Hudson, B. 1991a. Glomerular basement membrane. Identification of dimeric subunits of the noncollagenous domain (hexamer) of collagen IV and the Goodpasture antigen. J. Biol. Chem. 266, 15318–15324.

Gunwar, S., Ballester, F., Noelken, M., Sado, Y., Ninomiya, Y., Hudson, B. 1998. Glomerular basement membrane. Identification of a novel disulfide-cross-linked network of α3, α4 and α5 chains of type IV collagen and its implications for the pathogenesis of Alport syndrome. J. Biol. Chem. 273, 8767–8775.

Gunwar, S., Bejarano, P., Kalluri, R., Langeveld, J., Wisdom, B., Noelken, M., Hudson, B. 1991b. Alveolar basement membrane: Molecular properties of the noncollagenous domain (hexamer) of collagen IV and its reactivity with Goodpasture autoantibodies. Am. J. Respir. Cell Mol. Biol. 5, 107–112.

Guo, X., Johnson, J., Kramer, J. 1991. Embryonic lethality caused by mutations in basement membrane collagen of *C. elegans*. Nature 349, 707–709.

Guo, X., Kramer, J. 1989. The two *Caenorhabditis elegans* basement membrane (type IV) collagen genes are located on separate chromosomes. J. Biol. Chem. 264, 17574–17582.

Gupta, M., Graham, P., Kramer, J. 1997. Characterization of a1(IV) collagen mutations in *Caenorhabditis elegans* and the effects of α1 and α2(IV) mutations on type IV collagen distribution. J. Cell. Biol. 137, 1185–1196.

Halfter, W., Dong, S., Schurer, B., Cole, G. 1998. Collagen XVIII is a basement membrane heparan sulfate proteoglycan. J. Biol. Chem. 273, 25404–25412.

Harvey, S., Mount, R., Sado, Y., Naito, I., Ninomiya, Y., Jefferson, B., Jacobs, R., Thorner, P. 2001. The inner ear of dogs with X-linked nephritis provides clues to the pathogenesis of hearing loss in X-linked Alport syndrome. Am. J. Pathol. 159, 1097–1104.

Harvey, S., Zheng, K., Sado, Y., Naito, I., Ninomiya, Y., Jacobs, R., Hudson, B., Thorner, P. 1998. The role of distinct type IV collagen networks in glomerular development and function. Kidney Int. 54, 1857–1866.

Heathcote, J., Muhammed, S., Smith, E., Grant, M. 1980. Biosynthetic studies on the collagenous components of basement membranes. Ren. Physiol. 3, 36–40.

Heidet, L., Arrondel, C., Forestier, L., Cohen-Solal, L., Mollet, G., Gutierrez, B., Stavrou, C., Gubler, M.C., Antignac, C. 2001. Structure of the human type IV collagen gene COL4A3 and mutations in autosomal Alport syndrome. J. Am. Soc. Nephrol. 12, 97–106.

Heidet, L., Cai, Y., Sado, Y., Ninomiya, Y., Thorner, P., Guicharnaud, L., Boye, E., Chauvet, V., Solal, L., Beziau, A., Torres, R., Antignac, C., Gubler, M.C. 1997. Diffuse leiomyomatosis associated with X-linked Alport syndrome: Extracellular matrix study using immunohistochemistry and in situ hybridization. Lab. Invest. 76, 233–243.

Heidet, L., Dahan, K., Zhou, J., Xu, Z., Cochat, P., Gould, J., Leppig, K., Proesmans, W., Guyot, C., Guillot, M., Roussel, B., Tryggvason, K., Grünfeld, J.P., Gubler, MC., Antignac, C. 1995. Deletions of both α5(IV) and α6(IV) collagen genes in Alport syndrome and in Alport syndrome associated with smooth muscle tumours. Hum. Mol. Genet. 4, 99–108.

Heiskari, N., Zhang, X., Zhou, J., Leinonen, A., Barker, D., Gregory, M., Atkin, C., Netzer, KO., Weber, M., Reeders, S., Grönhagen-Riska, C., Neumann, H., Trembath, R., Tryggvason, K. 1996. Identification of 17 mutations in ten exons in the COL4A5 collagen gene, but no mutations found in four exons in COL4A6: A study of 250 patients with hematuria and suspected of having Alport syndrome. J. Am. Soc. Nephrol. 7, 702–709.

Helbling-Leclerc, A., Zhang, X., Topaloglu, H., Cruaud, C., Tesson, F., Weissenbach, J., Tome, F., Schwartz, K., Fardeau, M., Tryggvason, K., Guicheney, P. 1995. Mutations in the laminin α2-chain gene (LAMA2) cause merosin-deficient congenital muscular dystrophy. Nat. Genet. 11, 216–218.

Herken, R., Barrach, H. 1985. Ultrastructural localization of type IV collagen and laminin in the seven-day-old mouse embryo. Anat. Embryol. 171, 365–371.

Hiki, Y., Iyama, K., Tsuruta, J., Egami, H., Kamio, T., Suko, S., Naito, I., Sado, Y., Ninomiya, Y., Ogawa, M. 2002. Differential distribution of basement membrane type IV collagen α1(IV), α2(IV), α5(IV) and α6(IV) in colorectal epithelial tumors. Pathol. Int. 52, 224–233.

Hofmann, H., Voss, T., Kühn, K. 1984. Localization of flexible sites in thread-like molecules from electron micrographs. J. Mol. Biol. 172, 325–343.

Holmgren, S., Taylor, K., Bretscher, L., Raines, R. 1998. Code for collagen's stability deciphered. Nature 392, 666–677.

Hood, J., Huxtable, C., Naito, I., Smith, C., Sinclair, R., Savige, J. 2002. A novel model of autosomal dominant Alport syndrome in Dalmatian dogs. Nephrol. Dial. Transplant. 17, 2094–2098.

Hood, J., Savige, J., Hendtlass, A., Kleppel, M., Huxtable, C., Robinson, W. 1995. Bull terrier hereditary nephritis: A model for autosomal dominant Alport syndrome. Kidney Int. 47, 758–765.

Hostikka, S., Tryggvason, K. 1988. The complete primary structure of the a2(IV) chain of human type IV collagen and comparison with the α1(IV) chain. J. Biol. Chem. 263, 19488–19493.

Hudson, B., Reeders, S., Tryggvason, K. 1993. Type IV collagen: Structure, gene organization and role in human diseases: Molecular basis of Goodpasture and Alport syndromes and diffuse leiomyomatosis. J. Biol. Chem. 268, 26033–26036.

Hudson, B., Tryggvason, K., Sundaramoorthy, M., Neilson, E. 2003. Alport's syndrome, Goodpasture's syndrome, and type IV collagen. N. Engl. J. Med. 348, 2543–2556.

Hutter, H., Vogel, B., Plenefisch, J., Norris, C., Proenca, R., Spieth, J., Guo, C., Mastwal, S., Zhu, X., Scheel, J., Hedgecock, E. 2000. Conservation and novelty in the evolution of cell adhesion and extracellular matrix genes. Science 287, 989–994.

Hynes, R., Zhao, Q. 2000. The evolution of cell adhesion. J. Cell Biol. 150, F89–F95.

Inoue, S. 1989. Ultrastructure of basement membranes. Int. Rev. Cytol. 117, 57–98.

Isnenghi, E., Cassada, R., Smith, K., Denich, K., Radnia, K., von Ehrenstein, G. 1983. Maternal effects and temperature-sensitive period of mutations affecting embryogenesis in *Caenorhabditis elegans*. Dev. Biol. 98, 465–480.

Jais, J., Knebelmann, B., Giatras, I., De Marchi, M., Rizzoni, G., Renieri, A., Weber, M., Gross, O., Netzer, K.O., Flinter, F., Pirson, Y., Dahan, K., Weislander, J., Persson, U., Tryggvason, K., Martin, P., Hertz, J., Schroder, C., Sanak, M., Carvalho, M., Saus, J., Antignac, C., Smeets, H., Gubler, M.C. 2003. X-linked Alport syndrome: Natural history and genotype-phenotype correlations in girls and women belonging to 195 families: A "European community Alport Syndrome Concerted Action" study J. Am. Soc. Nephrol. 14, 2603–2610.

Jais, J., Knebelmann, B., Giatras, I., De Marchi, M., Rizzoni, G., Renieri, A., Weber, M., Gross, O., Netzer, K.O., Flinter, F., Pirson, Y., Verellen, C., Wieslander, J., Persson, U., Tryggvason, K., Martin, P., Hertz, J., Schroder, C., Sanak, M., Krejcova, S., Carvalho, M., Saus, J., Antignac, C., Smeets, H., Gubler, M.C. 2000. X-linked Alport syndrome: Natural history in 195 families and genotype-phenotype correlations in males. J. Am. Soc. Nephrol. 11, 649–657.

Kahsai, T., Enders, G., Gunwar, S., Brunmark, C., Wieslander, J., Kalluri, R., Zhou, J., Noelken, M., Hudson, B. 1997. Seminiferous tubule membrane. Composition and organization of type IV collagen chains, and the linkage of $\alpha3(IV)$ and $\alpha5(IV)$ chains. J. Biol. Chem. 272, 17023–17032.

Kalluri, R. 2003. Basement membranes: Structure, assembly and role in tumour angiogenesis. Nat. Rev. Cancer 3, 422–433.

Kalluri, R., Gattone, V., Hudson, B. 1998. Identification and localization of type IV collagen chains in the inner ear cochlea. Conn. Tissue. Res. 37, 143–150.

Kalluri, R., Shield, C., Todd, P., Hudson, B., Neilson, E. 1997. Isoform switching of type IV collagen is developmentally arrested in X-linked Alport syndrome leading to increased susceptibility of renal basement membranes to endoproteolysis. J. Clin. Invest. 99, 2470–2478.

Kapoor, R., Sakai, L., Funk, S., Roux, E., Bornstein, P., Sage, E. 1988. Type VIII collagen has a restricted distribution in specialized extracellular matrices. J. Cell. Biol. 107, 721–730.

Kashtan, C. 1998. Alport syndrome and thin glomerular basement membrane disease. J. Am. Soc. Nephrol. 9, 1736–1750.

Kashtan, C. 1999. Alport syndrome. An inherited disorder of renal, ocular, and cochlear basement membranes. Medicine 78, 338–360.

Kaytes, P., Wood, L., Theriault, N., Kurkinen, M., Vogeli, G. 1988. Head-to-head arrangement of murine type IV collagen genes. J. Biol. Chem. 263, 19274–19277.

Kefalides, N. 1971. Isolation of a collagen from basement membranes containing three identical a-chains. Biochem. Biophys. Res. Commun. 45, 226–234.

Kefalides, N. 1972. Isolation and characterization of cyanogen bromide peptides from basement membrane collagen. Biochem. Biophys. Res. Commun. 47, 1151–1158.

Kefalides, N. 1973. Structure and biosynthesis of basement membranes. Int. Rev. Connect. Tissue Res. 6, 63–104.

Kelley, P., Sado, Y., Duncan, M. 2002. Collagen IV in the developing lens capsule. Matrix Biol. 21, 415–423.

Kivirikko, K., Myllylä, R. 1985. Post-translational processing of procollagens. Ann. NY Acad. Sci. 460, 187–201.

Kivirikko, K., Myllylä, R., Pihlajaniemi, T. 1992. Hydroxylation of proline and lysine residues in collagens and other animal and plant proteins. In: *Post-Translational Modifications of Proteins* (J. Harding, M. Crabbe, Eds.), CRC press, pp. 1–51.

Kiyofuji, M., Iyama, K., Kitaoka, M., Sado, Y., Ninomiya, Y., Ueda, S. 2002. Quantitative analysis of type IV collagen a chains in the basement membrane of human urogenital epithelium. Histochem. J. 34, 479–486.

Kleppel, M., Kashtan, C., Santi, P., Wieslander, J., Michael, A. 1989. Distribution of familial nephritis antigen in normal tissue and renal basement membranes of patients with homozygous and heterozygous Alport familial nephritis. Relationship of familial nephritis and Goodpasture antigens to novel collagen chains and type IV collagen. Lab. Invest. 61, 278–289.

Kleppel, M., Michael, A. 1990. Expression of novel basement membrane components in the developing human kidney and eye. Am. J. Anat. 187, 165–174.

Kleppel, M., Santi, P., Cameron, J., Wieslander, J., Michael, A. 1989. Human tissue distribution of novel basement membrane collagen. Am. J. Pathol. 134, 813–825.

Knebelmann, B., Forestier, L., Drouot, L., Quinones, S., Chuet, C., Benessy, F., Saus, J., Antignac, C. 1995. Splice-mediated insertion of an Alu sequence in the *COL4A3* mRNA causing autosomal recessive Alport syndrome. Hum. Mol. Genet. 4, 675–679.

Knibiehler, B., Mirre, C., Le Parco, Y. 1990. Collagen type IV of *Drosophila* is stockpiled in the growing oocyte and differentially located during early stages of embryogenesis. Cell. Differ. Dev. 30, 147–157.

Kühn, K., Wiedemann, H., Timpl, R., Risteli, J., Dieringer, H., Voss, T., Glanville, R. 1981. Macromolecular structure of basement membrane collagens: Identification of 7S collagen as a cross-linking domain of type IV collagen. FEBS Lett. 125, 123–128.

Kurkinen, M., Bernard, M., Barlow, D., Chow, L. 1985. Characterization of 64-, 123- and 182-base-pair exons in the mouse alpha 2(IV) collagen gene. Nature 317, 177–179.

Kuroda, N., Yoshikawa, N., Nakanishi, K., Iijima, K., Hanioka, K., Hayashi, Y., Imai, Y., Sado, Y., Nakayama, M., Itoh, H. 1998. Expression of type IV collagen in the developing human kidney. Pediatr. Nephrol. 12, 554–558.

Langeveld, J., Noelken, M., Hard, K., Todd, P., Vliegenthart, J., Rouse, J., Hudson, B. 1991. Bovine glomerular basement membrane. Location and structure of the asparagine-linked oligosaccharide units and their potential role in the assembly of the 7 S collagen IV tetramer. J. Biol. Chem. 266, 2622–2631.

Langeveld, J., Wieslander, J., Timoneda, J., McKinney, P., Butkowski, R., Wisdom, B., Hudson, B. 1988. Structural heterogeneity of the noncollagenous domain of basement membrane collagen. J. Biol. Chem. 263, 10481–10488.

Lankat-Buttgergeit, B., Mann, K., Deutzmann, R., Timpl, R., Krieg, T. 1988. Cloning and complete amino acid sequences of human and murine basement membrane protein BM-40 (SPARC, osteonectin). FEBS Lett. 236, 352–356.

Lauer-Fields, J., Malkar, N., Richert, G., Drauz, K., Fields, G. 2003. Melanoma cell CD44 interaction with the α1(IV)1263–1277 region from basement membrane collagen is modulated by ligand glycosylation. J. Biol. Chem. 278, 14321–14330.

Le Parco, Y., Cecchini, J., Knibiehler, B., Mirre, C. 1986. Characterization and expression of collagen-like genes in *Drosophila melanogaster*. Biol. Cell 56, 217–226.

Le Parco, Y., Knibiehler, B., Cecchini, J., Mirre, C. 1986. Stage- and tissue-specific expression of a collagen gene during *Drosophila melanogaster* development. Exp. Cell Res. 163, 405–412.

Lees, G., Helman, R., Kashtan, C., Michael, A., Homco, L., Millichamp, N., Camacho, Z., Templeton, J., Ninomiya, Y., Sado, Y., Naito, I., Kim, Y. 1999. New form of X-linked dominant hereditary nephritis in dogs. Am. J. Vet. Res. 60, 373–383.

Lees, G., Helman, R., Kashtan, C., Michael, A., Homco, L., Millichamp, N., Ninomiya, Y., Sado, Y., Naito, I., Kim, Y. 1998. A model of autosomal recessive Alport syndrome in English cocker spaniel dogs. Kidney Int. 54, 706–719.

Leinonen, A., Mariyama, M., Mochizuki, T., Tryggvason, K., Reeders, S. 1994. Complete primary structure of the human type IV collagen α4(IV) chain. Comparison with structure and expression of the other α(IV) chains. J. Biol. Chem. 269, 26172–26177.

Leivo, I., Vaheri, A., Timpl, R., Wartiovaara, J. 1980. Appearance and distribution of collagens and laminin in the early mouse embryo. Dev. Biol. 76, 100–114.

Lemmink, H., Mochizuki, T., van den Heuvel, L., Schröder, C., Barrientos, A., Monnens, L., van Oost, B., Brunner, H., Reeders, S., Smeets, H. 1994. Mutations in the type IV collagen α3 (*COL4A3*) gene in autosomal recessive Alport syndrome. Hum. Mol. Genet. 3, 1269–1273.

Leushner, J. 1987. Partial characterization of the heteropolysaccharide associated with the 7S domain of type IV collagen from placenta. Biochem. Cell. Biol. 65, 501–506.

Levavasseur, F., Lietard, J., Ogawa, K., Theret, N., Burbelo, P., Yamada, Y., Guillouzo, A., Clement, B. 1996. Expression of laminin γ1 in cultured hepatocytes involves repeated CTC and GC elements in the *LAMC1* promoter. Biochem. J. 313, 745–752.

Li, D., Clark, C., Myers, J. 2000. Basement membrane zone type XV collagen is a disulfide-bonded chondroitin sulfate proteoglycan in human tissues and cultured cells. J. Biol. Chem. 275, 22339–22347.

Libby, R., Champliaud, M., Claudepierre, T., Xu, Y., Gibbons, E., Koch, M., Burgeson, R., Hunter, D., Brunken, W. 2000. Laminin expression in adult and developing retinae: Evidence of two novel CNS laminins. J. Neurosci. 20, 6517–6528.

Lohi, J., Korhonen, M., Leivo, I., Kangas, L., Tani, T., Kalluri, R., Miner, J., Lehto, V.-P., Virtanen, I. 1997. Expression of type IV collagen α1(IV)-α6(IV) polypeptides in normal and developing human kidney and in renal cell carcinomas and oncocytomas. Int. J. Cancer 72, 43–49.

Longo, I., Porcedda, P., Mari, F., Giachino, D., Meloni, I., Deplano, C., Brusco, A., Bosio, M., Massella, L., Lavoratti, G., Roccatello, D., Frasca, G., Mazzucco, G., Muda, A., Conti, M., Fasciolo, F., Arrondel, C., Heidet, L., Renieri, A., De Marchi, M. 2002. *COL4A3/COL4A4* mutations: From familial hematuria to autosomal dominant or recessive Alport syndrome. Kidney Int. 61, 1947–1956.

Lortat-Jacob, H., Kleinman, H., Grimaud, J. 1991. High-affinity binding of interferon-γ to a basement membrane complex (matrigel). J. Clin. Invest. 87, 878–883.

Lowe, J., Guyon, R., Cox, M., Mitchell, D., Lonkar, A., Lingaas, F., Andre, C., Gailbert, F., Ostrander, E., Murphy, K. 2003. Radiation hybrid mapping of the canine type I and type IV collagen gene subfamilies. Funct. Integr. Genomics 3, 112–116.

Lu, W., Phillips, C., Killen, P., Hlaing, T., Harrison, W., Elder, F., Miner, J., Overbeek, P., Meisler, M. 1999. Insertional mutation of the collagen genes Col4a3 and Col4a4 in a mouse model of Alport syndrome. Genomics 61, 113–124.

Lunstrum, G., Bächinger, H., Fessler, L., Duncan, K., Nelson, R., Fessler, J. 1988. *Drosophila* basement membrane procollagen IV. I. Protein characterization and distribution. J. Biol. Chem. 263, 18318–18327.

Madri, J., Foellmer, H., Furthmayr, H. 1983. Ultrastructural morphology and domain structure of a unique collagenous component of basement membranes. Biochemistry 22, 2797–2804.

Mariyama, M., Leinonen, A., Mochizuki, T., Tryggvason, K., Reeders, S. 1994. Complete primary structure of the human α3(IV) collagen chain. Coexpression of the α3(IV) and α4 (IV) collagen chains in human tissues. J. Biol. Chem. 269, 23103–23107.

Mariyama, M., Zheng, K., Yang-Feng, T., Reeders, S. 1992. Colocalization of the genes for the α3(IV) and α4(IV) chains of type IV collagen to chromosome 2 bands q35-q37. Genomics 13, 809–813.

Martin, P., Tryggvason, K. 2001. Two novel alternatively spliced 9-bp exons in the *COL4A5* gene. Pediatr. Nephrol. 16, 41–44.

Martinez-Hernandez, A., Amenta, P. 1983. The basement membrane in pathology. Lab. Invest. 48, 656–677.

Martinez-Hernandez, A., Marsh, C., Clark, C., Macarak, E., Brownell, A. 1981. Fibronectin: Its relationship to basement membranes. II. Ultrastructural studies in rat kidney. Coll. Relat. Res. 1, 405–418.

Mazzucco, G., Barsotti, P., Muda, A., Fortunato, M., Mihatsch, M., Torri-Tarelli, L., Renieri, A., Faraggiana, T., de Marchi, M., Monga, G. 1998. Ultrastructural and immuno-histochemical findings in Alport's syndrome: A study of 108 patients from 97 Italian families with particular emphasis on *COL4A5* mutation correlations. J. Am. Soc. Nephrol. 9, 1023–1031.

McLaughlin, S., Bulleid, N. 1998. Molecular recognition in procollagen chain assembly. Matrix. Biol. 16, 369–377.

Merker, H.J. 1994. Morphology of the basement membrane. Microsc. Res. Tech. 28, 95–124.

Miner, J. 1999. Renal basement membrane components. Kidney Int. 56, 2016–2024.

Miner, J., Sanes, J. 1994. Collagen IV α3, α4, and α5 chains in rodent basal laminae: Sequence, distribution, association with laminins, and developmental switches. J. Cell. Biol. 127, 879–891.

Miner, J., Sanes, J. 1996. Molecular and functional defects in kidneys of mice lacking collagen α3(IV): Implications for Alport syndrome. J. Cell. Biol. 135, 1403–1413.

Minor, R., Clark, C., Strause, E., Koszalka, T., Brent, R., Kefalides, N. 1976. Basement membrane procollagen is not converted to collagen in organ cultures of parietal yolk sac endoderm. J. Biol. Chem. 251, 1789–1794.

Miosge, N., Holzhausen, S., Zelent, C., Sprysch, P., Herken, R. 2001. Nidogen-1 and nidogen-2 are found in basement membranes during embryonic development. Histochem. J. 33, 523–530.

Mirre, C., Cecchini, J., Le Parco, Y., Knibiehler, B. 1988. *De novo* expression of a type IV collagen gene in *Drosophila* embryos is restricted to mesodermal derivatives and occurs at germ band shortening. Development 102, 369–376.

Mirre, C., Le Parco, Y., Knibiehler, B. 1992. Collagen IV is present in the developing CNS during *Drosophila* neurogenesis. J. Neurosci. Res. 31, 146–155.

Miwa, J., Schierenberg, E., Miwa, S., von Ehrenstein, G. 1980. Genetics and mode of expression of temperature-sensitive mutations arresting embryonic development in *Caenorhabditis elegans*. Dev. Biol. 76, 160–174.

Mochizuki, T., Lemmink, H., Mariyama, M., Antignac, C., Gubler, MC., Pirson, Y., Verellen-Dumoulin, C., Chan, B., Schröder, C., Smeets, H., Reeders, S. 1994. Identification of mutations in the α3(IV) and α4(IV) collagen genes in autosomal recessive Alport syndrome. Nature Genet. 8, 77–82.

Momota, R., Sugimoto, M., Oohashi, T., Kigasawa, K., Yoshioka, H., Ninomiya, Y. 1998. Two genes, *COL4A3* and *COL4A4* coding for the human α3(IV) and α4(IV) collagen chains are arranged head-to-head on chromosome 2q36. FEBS Lett. 424, 11–16.

Monson, J., Natzle, J., Friedman, J., McCarthy, B. 1982. Expression and novel structure of a collagen gene in *Drosophila*. Proc. Natl. Acad. Sci. USA 79, 1761–1765.

Morello, R., Zhou, G., Dreyer, S., Harvey, S., Ninomiya, Y., Thorner, P., Miner, J., Cole, W., Winterpacht, A., Zabel, B., Oberg, K., Lee, B. 2001. Regulation of glomerular basement membrane collagen expression by *LMX1B* contributes to renal disease in nail patella syndrome. Nature Genet. 27, 205–208.

Murray, M., Fessler, L., Palka, J. 1995. Changing distributions of extracellular matrix components during early wing morphogenesis in *Drosophila*. Dev. Biol. 168, 150–165.

Murshed, M., Smyth, N., Miosge, N., Karolat, J., Krieg, T., Paulsson, M., Nischt, R. 2000. The absence of nidogen 1 does not affect murine basement membrane formation. Mol. Cell. Biol. 20, 7007–7012.

Muthukumaran, G., Blumberg, B., Kurkinen, M. 1989. The complete primary structure for the α1-chain of mouse collagen IV. Differential evolution of collagen IV domains. J. Biol. Chem. 264, 6310–6317.

Myllyharju, J., Kivirikko, K. 2004. Collagens, modifying enzymes and their mutations in humans, flies and worms. Trends Genet. 20, 33–43.

Naito, I., Kawai, S., Nomura, S., Sado, Y., Osawa, G. 1996. Relationship between *COL4A5* gene mutation and distribution of type IV collagen in male X-linked Alport syndrome. Kidney Int. 50, 304–311.

Nakanishi, K., Yoshikawa, N., Iijima, K., Kitagawa, K., Nakamura, H., Ito, H., Yoshioka, K., Kagawa, M., Sado, Y. 1994. Immunohistochemical study of α1–5 chains of type IV collagen in hereditary nephritis. Kidney Int. 46, 1413–1421.

Nakano, A., Chao, S., Pulkkinen, L., Murrell, D., Bruckner-Tuderman, L., Pfendner, E., Uitto, J. 2002. Laminin 5 mutations in junctional epidermolysis bullosa: Molecular basis of Herlitz vs. non-Herlitz phenotypes. Hum. Genet. 110, 41–51.

Nakano, K., Iyama, K., Mori, T., Yoshioka, M., Hiraoka, T., Sado, Y., Ninomiya, Y. 2001. Loss of alveolar basement membrane type IV collagen α3, α4 and α5 chains in bronchioalveolar carcinoma of the lung. J. Pathol. 194, 420–427.

Natzle, J., Monson, J., McCarthy, B. 1982. Cytogenetic location and expression of collagen-like genes in *Drosophila*. Nature 296, 368–371.

Nayak, B., Spiro, R. 1991. Localization and structure of the asparagine-linked oligosaccharides of type IV collagen from glomerular basement membrane and lens capsule. J. Biol. Chem. 266, 13978–13987.

Nemer, M., Harlow, P. 1988. Sea-urchin RNAs displaying differences in developmental regulation and in complementarity to a collagen exon probe. Biochim. Biophys. Acta 950, 445–449.

Nicole, S., Davoine, C., Topaloglu, H., Cattolico, L., Barral, D., Beighton, P., Hamida, C., Hammouda, H., Cruaud, C., White, P., Samson, D., Urtizberea, J., Lehmann, H.F., Weissenbach, J., Hentati, F., Fontaine, B. 2000. Perlecan, the major proteoglycan of basement membranes, is altered in patients with Schwartz-Jampel syndrome (chondrodystrophic myotonia). Nat. Genet. 26, 480–483.

Ninomiya, Y., Kagawa, M., Iyama, K., Naito, I., Kishiro, Y., Seyer, J., Sugimoto, M., Oohashi, T., Sado, Y. 1995. Differential expression of two basement membrane collagen genes, *COL4A6* and *COL4A5*, demonstrated by immunofluorescence staining using peptide-specific monoclonal antibodies. J. Cell. Biol. 130, 1219–1229.

Nomura, S., Naito, I., Fukushima, T., Tokura, T., Kataoka, N., Tanaka, I., Tanaka, H., Osawa, G. 1998. Molecular genetic and immunohistological study of autosomal recessive Alport syndrome. Am. J. Kidney Dis. 31, E4.

Noonan, D., Fulle, A., Valente, P., Cai, S., Horigan, E., Sasaki, M., Yamada, Y., Hassell, J. 1991. The complete sequence of perlecan, a basement membrane heparan sulfate proteoglycan, reveals extensive similarity with laminin A chain, low density lipoprotein receptor, and the neural cell adhesion molecule. J. Biol. Chem. 266, 22939–22947.

Norman, K., Moerman, D. 2000. The *let-268* locus of *Caenorhabditis elegans* encodes a procollagen lysyl hydroxylase that is essential for type IV collagen secretion. Dev. Biol. 227, 690–705.

Oberbäumer, I., Wiedemann, H., Timpl, R., Kühn, K. 1982. Shape and assembly of type IV procollagen obtained from cell culture. EMBO J. 1, 805–810.

Ohkubo, S., Takeda, H., Higashide, T., Ito, M., Sakurai, M., Shirao, Y., Yanagida, T., Oda, Y., Sado, Y. 2003. Immunohistochemical and molecular genetic evidence for type IV collagen α5 chain abnormality in the anterior lenticonus associated with Alport syndrome. Arch. Ophthalmol. 121, 846–850.

Oka, Y., Naito, I., Manabe, K., Sado, Y., Matsushima, H., Ninomiya, Y., Mizuno, M., Tsuji, T. 2002. Distribution of collagen type IV α1–6 chains in human normal colorectum and colorectal cancer demonstrated by immunofluorescence staining using chain-specific epitope-defined monoclonal antibodies. J. Gastroenterol. Hepatol. 17, 980–986.

Oohashi, T., Ueki, Y., Sugimoto, M., Ninomiya, Y. 1995. Isolation and structure of the *COL4A6* gene encoding the human α6(IV) collagen chain and comparison with other type IV collagen genes. J. Biol. Chem. 270, 26863–26867.

Ortega, N., Werb, Z. 2002. New functions for non-collagenous domains of basement membrane collagens. J. Cell. Biol. 115, 4201–4214.

Ozen, S., Ertoy, D., Heidet, L., Solal, L., Ozen, H., Besbas, N., Bakkaoglu, A., Antignac, C. 2001. Benign familial hematuria associated with a novel *COL4A4* mutation. Pediatr. Nephrol. 16.

Paralkar, V., Vukicevic, S., Reddi, A. 1991. Transforming growth factor β type I binds to collagen IV of basement membrane matrix: Implications for development. Dev. Biol. 143, 303–308.

Peissel, B., Geng, L., Kalluri, R., Kashtan, C., Rennke, H., Gallo, G., Yoshioka, K., Sun, M., Hudson, B., Neilson, E., Zhou, J. 1995. Comparative distribution of the α1(IV), α5(IV), and α6(IV) collagen chains in normal human adult and fetal tissues and in kidneys from X-linked Alport syndrome patients. J. Clin. Invest. 96, 1948–1957.

Pescucci, C., Mari, F., Longo, I., Vogiatzi, P., Caselli, R., Scala, E., Abaterusso, C., Gusmano, R., Seri, M., Miglietti, N., Bresin, R., Renieri, A. 2004. Autosomal-dominant Alport syndrome: Natural history of a disease due to *COL4A3* or *COL4A4* gene. Kidney Int. 65, 1598–1603.

Pettitt, J., Kingston, I. 1991. The complete primary structure of a nematode α2(IV) collagen and the partial structural organization of its gene. J. Biol. Chem. 266, 16149–16156.

Pettitt, J., Kingston, I. 1994. Developmentally regulated alternative splicing of a nematode type IV collagen gene. Dev. Biol. 161, 22–29.

Pihlajaniemi, T., Myllylä, R., Alitalo, K., Vaheri, A., Kivirikko, K. 1981. Posttranslational modifications in the biosynthesis of type IV collagen by a human tumor cell line. Biochemistry 20, 7409–7415.

Pirson, Y. 1999. Making the diagnosis of Alport's syndrome. Kidney Int 56, 760–775.

Pollner, R., Schmidt, C., Fischer, G., Kühn, K., Poschl, E. 1997. Cooperative and competitive interactions of regulatory elements are involved in the control of divergent transcription of human *COL4A1* and *COL4A2* genes. FEBS Lett. 405, 31–36.

Poschl, E., Schlotzer-Schrehardt, U., Brachvogel, B., Saito, K., Ninomiya, Y., Mayer, U. 2004. Collagen IV is essential for basement membrane stability but dispensable for initiation of its assembly during early development. Development 131, 1619–1628.

Prockop, D., Kivirikko, K. 1995. Collagens: Molecular biology, diseases and potentials for therapy. Annu. Rev. Biochem. 64, 403–434.

Rautavuoma, K., Takaluoma, K., Sormunen, R., Myllyharju, J., Kivirikko, K., Soininen, R. 2004. Premature aggregation of type IV collagen and early embryonic lethality in lysyl hydroxylase 3 null mice. Proc. Natl. Acad. Sci. USA 101, 14120–14125.

Reddy, G., Hudson, B., Bailey, A., Noelken, M. 1993. Reductive cleavage of the disulfide bonds of the collagen IV noncollagenous domain in aqueous sodium dodecyl sulfate: Absence of intermolecular nondisulfide cross-links. Biochem. Biophys. Res. Commun. 190, 277–82.

Rheault, M., Kren, S., Thielen, B., Mesa, H., Crosson, J., Thomas, W., Sado, Y., Kashtan, C., Segal, Y. 2003. Mouse model of X-linked Alport syndrome. J. Am. Soc. Nephrol. 15, 1466–1474.

Risteli, J., Bächinger, H., Engel, J., Furthmayr, H., Timpl, R. 1980. 7-S collagen: Characterization of an unusual basement membrane structure. Eur. J. Biochem. 108, 239–250.

Rodgers, H., Irvine, C., van Wezel, I., Lavranos, T., Luck, M., Sado, Y., Ninomiya, Y., Rodgers, R. 1998. Distribution of the α1 to α6 chains of type IV collagen in bovine follicles. Biol. Reprod. 59, 1334–1341.

Rodriguez, A., Zhou, Z., Tang, M., Meller, S., Chen, J., Bellen, H., Kimbrell, D. 1996. Identification of immune system and response genes, and novel mutations causing melanotic tumor formation in *Drosophila melanogaster*. Genetics 143, 929–940.

Rumpelt, H. 1987. Alport's syndrome: Specificity and pathogenesis of glomerular basement membrane alterations. Pediatr. Nephrol. 1, 422–427.

Sado, Y., Kagawa, M., Kishiro, Y., Sugihara, K., Naito, I., Seyer, J., Sugimoto, M., Oohashi, T., Ninomiya, Y. 1995. Establishment by the rat lymph node method of epitope-defined monoclonal antibodies recognizing the six different a chains of human type IV collagen. Histochem. Cell. Biol. 104, 267–275.

Saito, K., Naito, I., Seki, T., Oohashi, T., Kimura, E., Momota, R., Kishiro, Y., Sado, Y., Yoshioka, H., Ninomiya, Y. 2000. Differential expression of mouse α5(IV) and α6(IV) collagen genes in epithelial basement membranes. J. Biochem. 128, 427–434.

Sakurai, Y., Sullivan, M., Yamada, Y. 1986. α1 type IV collagen gene evolved differently from fibrillar collagen genes. J. Biol. Chem. 261, 6654–6657.

Sanes, J., Engvall, E., Butkowski, P., Hunter, D. 1990. Molecular heterogeneity of basal laminae: Isoforms of laminin and collagen IV at the neuromuscular junction and elsewhere. J. Cell. Biol. 111, 1685–1699.

Sarras, M., Jr., Madden, M., Zhang, X., Gunwar, S., Huff, J., Hudson, B. 1991a. Extracellular matrix (mesoglea) of *Hydra vulgaris*. I. Isolation and characterization. Dev. Biol. 148, 481–494.

Sarras, M., Jr., Meador, D., Zhang, X. 1991b. Extracellular matrix (mesoglea) of *Hydra vulgaris*. II. Influence of collagen and proteoglycan components on head regeneration. Dev. Biol. 148, 495–500.

Sarras, M., Jr., Yan, L., Grens, A., Zhang, X., Agbas, A., Huff, J., St. John, P., Abrahamson, D. 1994. Cloning and biological function of laminin in *Hydra vulgaris*. Dev. Biol. 164, 312–324.

Saus, J., Quinones, S., Mac Krell, A., Blumberg, B., Muthukumaran, G., Pihlajaniemi, T., Kurkinen, M. 1989. The complete primary structure of mouse α2(IV) collagen. Alignment with mouse α1(IV) collagen. J. Biol. Chem. 264, 6318–6324.

Savige, J., Rana, K., Tonna, S., Buzza, M., Dagher, H., Wang, Y. 2003. Thin basement membrane nephropathy. Kidney Int. 64, 1169–1178.

Sawada, H., Konomi, H., Hirosawa, K. 1990. Characterization of the collagen in the hexagonal lattice of Descemet's membrane: Its relation to type VIII collagen. J. Cell. Biol. 110, 219–227.

Schmidt, C., Fischer, G., Kadner, H., Gernersch, E., Kühn, K., Poschl, E. 1993. Differential effects of DNA-binding proteins on bidirectional transcription from the common promoter region of human collagen type IV genes *COL4A1* and *COL4A2*. Biochim. Biophys. Acta 1174, 1–10.

Schymeinsky, J., Nedbal, S., Miosge, N., Poschl, E., Rao, C., Breier, D., Skarnes, W., Timpl, R., Bader, B. 2002. Gene structure and functional analysis of the mouse nidogen-2 gene: Nidogen-2 is not essential for basement membrane formation in mice. Mol. Cell. Biol. 22, 6820–6830.

Seki, T., Naito, I., Oohashi, T., Sado, Y., Ninomiya, Y. 1998. Differential expression of type IV collagen isoforms, α5(IV) and α6(IV) chains, in basement membranes surrounding smooth muscle cells. Histochem. Cell. Biol. 110, 359–366.

Sherman, M., Gay, R., Gay, S., Miller, E. 1980. Association of collagen with preimplantation and peri-implantation mouse embryos. Dev. Biol. 74, 470–478.

Sibley, M., Graham, P., von Mende, N., Kramer, J. 1994. Mutations in the α2(IV) basement membrane collagen gene of *Caenorhabditis elegans* produce phenotypes of differing severities. EMBO J. 13, 3278–3285.

Sibley, M., Johnson, J., Mello, C., Kramer, J. 1993. Genetic identification, sequence, and alternative splicing of the *Caenorhabditis elegans* α2(IV) collagen gene. J. Cell. Biol. 123, 255–264.

Siebold, B., Deutzmann, R., Kühn, K. 1988. The arrangement of intra- and intermolecular disulfide bonds in the carboxyterminal, non-collagenous aggregation and cross-linking domain of basement-membrane type IV collagen. Eur. J. Biochem. 176, 617–624.

Siebold, B., Qian, R., Glanville, R., Hofmann, H., Deutzmann, R., Kühn, K. 1987. Construction of a model for the aggregation and cross-linking region (7S domain) of type IV collagen based upon an evaluation of the primary structure of the α1 and α2 chains in this region. Eur. J. Biochem. 168, 569–575.

Simoneau, A., Herring-Gillam, F., Vachon, P., Perreault, N., Basora, N., Bouatrouss, Y., Pageot, L., Zhou, J., Beaulieu, J. 1998. Identification, distribution and tissular origin of the α5(IV) and α6(IV) collagen chains in the developing human intestine. Dev. Dyn. 212, 437–447.

Smyth, N., Vatansever, H., Murray, P., Meyer, M., Frie, C., Paulsson, M., Edgar, D. 1999. Absence of basement membranes after targeting the *LAMC1* gene results in embryonic lethality die to failure of endoderm differentiation. J. Cell. Biol. 144, 151–160.

Soininen, R., Haka-Risku, T., Prockop, D., Tryggvason, K. 1987. Complete primary structure of the α1(IV)-chain of human basement membrane (type IV) collagen. FEBS Lett. 225, 188–194.

Soininen, R., Huotari, M., Ganguly, A., Prockop, D., Tryggvason, K. 1989. Structural organization of the gene for the a1 chain of human type IV collagen. J. Biol. Chem. 264, 13565–13571.

Soininen, R., Huotari, M., Hostikka, S., Prockop, D., Tryggvason, K. 1988. The structural genes for α1 and α2 chains of human type IV collagen are divergently encoded on opposite DNA strands and have an overlapping promoter region. J. Biol. Chem. 263, 17217–17220.

Spear, G. 1973. Alport's syndrome: A consideration of pathogenesis. Clin. Nephrol. 1, 336–337.

Sugimoto, M., Oohashi, T., Ninomiya, Y. 1994. The genes *COL4A5* and *COL4A6*, coding for basement membrane collagen chains α5(IV) and α6(IV), are located in head-to-head in close proximity on human chromosome Xq22 and *COL4A6* is transcribed from two alternative promoters. Proc. Natl. Acad. Sci. USA 91, 11679–11683.

Sund, M., Maeshima, Y., Kalluri, R. 2005. Bi-functional promoter of type IV collagen *COL4A5* and *COL4A6* genes regulates the expression of α5 and α6 chains in a distinct cell specific fashion. Biochem. J. 387, 755–761.

Sundaramoorthy, M., Meiyappan, M., Todd, P., Hudson, B. 2002. Crystal structure of NC1 domains. Structural basis for type IV collagen assembly in basement membranes. J. Biol. Chem. 277, 31142–31153.

Suzuki, O., Sertie, A., Der Kaloustian, V., Kok, F., Carpenter, M., Murray, J., Czeizel, A., Klienmann, S., Rosemberg, S., Monteiro, M.B.R.O., Passos-Bueno, M. 2002. Molecular analysis of collagen XVIII reveals novel mutations, presence of a third isoform, and possible genetic heterogeneity in Knobloch syndrome. Am. J. Hum. Genet. 71, 1320–1329.

Tanzer, M., Kefalides, N. 1973. Collagen crosslinks: Occurrence in basement membrane collagens. Biochem. Biophys. Res. Commun. 51, 775–780.

Than, M., Henrich, S., Huber, R., Ries, A., Mann, K., Kühn, K., Timpl, R., Bourenkov, G., Bartunik, H., Bode, W. 2002. The 1.9A crystal structure of the noncollagenous (NC1) domain of human placental collagen IV shows stabilization via a novel type of covalent Met-Lys cross-link. Proc. Natl. Acad. Sci. USA 99, 6607–6612.

Thielen, B., Barker, D., Nelson, R., Zhou, J., Kren, S., Segal, Y. 2003. Deletion mapping in Alport syndrome and Alport syndrome-diffuse leiomyomatosis reveals potential mechanisms of visceral smooth muscle overgrowth. Hum. Mutat. 22, 419–426.

Thorner, P., Zheng, K., Kalluri, R., Jacobs, R., Hudson, B. 1996. Coordinate gene expression of the α3, α4 and α5 chains of collagen type IV: Evidence from a canine model of X-linked nephritis with a *COL4A5* gene mutation. J. Biol. Chem. 271, 13821–13828.

Timpl, R. 1989. Structure and biological activity of basement membrane proteins. Eur. J. Biochem. 180, 487–502.

Timpl, R., Brown, J. 1996. Supramolecular assembly of basement membranes. Bioessays 18, 123–132.

Timpl, R., Sasaki, T., Kostka, G., Chu, M.L. 2003. Fibulins: A versatile family of extracellular matrix proteins. Nat. Rev. Mol. Cell. Biol. 4, 479–489.

Timpl, R., Wiedemann, H., van Delden, V., Furthmayr, H., Kühn, K. 1981. A network model for the organization of type IV collagen molecules in basement membranes. Eur. J. Biochem. 120, 203–211.

Traub, W., Piez, K. 1971. The chemistry and structure of collagen. Adv. Protein. Chem. 25, 243–352.

Trüeb, B., Gröbli, B., Spiess, M., Dermatt, B., Winterhalter, K. 1982. Basement membrane (type IV) collagen is a heteropolymer. J. Biol. Chem. 257, 5239–5245.

Tsilibary, E., Charonis, A. 1986. The role of the main noncollagenous domain (NC1) in type IV collagen self assembly. J. Cell. Biol. 103, 2467–2473.

Tunggal, P., Smyth, N., Paulsson, M., Ott, M. 2000. Laminins: Structure and genetic regulation. Microsc. Res. Tech. 51, 214–227.

Ueki, Y., Naito, I., Oohashi, T., Sugimoto, M., Seki, T., Yoshioka, H., Sado, Y., Sato, H., Sawai, T., Sasaki, F., Matsuoka, M., Fukuda, S., Ninomiya, Y. 1998. Topoisomerase I and II consensus sequences in a 17 kb deletion junction of the *COL4A5* and *COL4A6* genes and immunohistochemical analysis of esophageal leiomyomatosis associated with Alport syndrome. Am. J. Hum. Genet. 62, 253–262.

Urabe, N., Naito, I., Saito, K., Yonezawa, T., Sado, Y., Yoshioka, H., Kusachi, S., Tsuji, T., Ohtsuka, A., Taguchi, T., Murakami, T., Ninomiya, Y. 2002. Basement membrane type IV collagen molecules in the choroid plexus, pia mater and capillaries in the mouse brain. Arch. Histol. Cytol. 65, 133–143.

van der Loop, F., Heidet, L., Timmer, E., van den Bosch, B., Leinonen, A., Antignac, C., Jefferson, J., Maxwell, A., Monnens, L., Schroder, C., Smeets, H. 2000. Autosomal dominant Alport syndrome caused by a *COL4A3* splice site mutation. Kidney Int. 58, 1870–1875.

Vega, B., Badenas, C., Ars, E., Lens, X., Mila, M., Darnell, A., Torra, R. 2003. Autosomal recessive Alport's syndrome and benign familial hematuria are collagen type IV diseases. Am. J. Kidney Dis. 42, 952–959.

Venkatesan, M., De Pablo, F., Vogeli, G., Simpson, R. 1986. Structure and developmentally regulated expression of a *Stronglyocentrotus purpuratus* collagen gene. Proc. Natl. Acad. Sci. USA 83, 3351–3355.

Weber, S., Engel, J., Wiedemann, H., Glanville, R., Timpl, R. 1984. Subunit structure and assembly of the globular domain of basement-membrane collagen type IV. Eur. J. Biochem. 139, 401–410.

Wessel, G., Etkin, M., Benson, S. 1991. Primary mesenchyme cells of the sea urchin embryo require an autonomously produced, nonfibrillar collagen for spiculogenesis. Dev. Biol. 148, 261–272.

Wessel, G., Marchase, R., McClay, D. 1984. Ontogeny of the basal lamina in the sea urchin embryo. Dev. Biol. 103, 235–245.

Weston, M., Eudy, J., Fujita, S., Yao, S., Usami, S., Cremers, C., Greenberg, J., Ramesar, R., Martini, A., Moller, C., Smith, R., Sumegi, J., Kimberling, W., Greenberg, J. 2000. Genomic structure and identification of novel mutations in usherin, the gene responsible for Usher syndrome type IIa. Am. J. Hum. Genet. 66, 1199–1210.

Wood, W., Hecht, R., Carr, S., Vanderslice, R., Wolf, N., Hirsh, D. 1980. Parental effects and phenotypic characterization of mutations that affect early development in *Caenorhabditis elegans*. Dev. Biol. 74, 446–469.

Wu, R., Couchman, J. 1997. cDNA cloning of the basement membrane chondroitin sulfate proteoglycan core protein, bamacan: A five domain structure including coiled-coil motifs. J. Cell. Biol. 136, 433–444.

Wu, V., Cohen, M. 1983. Intermolecular reducible cross-links in rat glomerular basement membrane. Renal. Physiol. 6, 232–239.

Xu, J., Rodriguez, D., Petitclerc, E., Kim, J., Hangai, M., Moon, Y., Davis, G., Brooks, P., Yuen, S. 2001. Proteolytic exposure of a cryptic site within collagen type IV is required for angiogenesis and tumor growth *in vivo*. J. Cell. Biol. 154, 1069–1079.

Yamada, Y., Avvedimento, V., Mudryj, M., Ohkubo, H., Vogeli, G., Irani, M., Pastan, I., de Crombrugghe, B. 1980. The collagen gene: Evidence for its evolutionary assembly by amplification of a DNA segment containing an exon of 54 bp. Cell 22, 887–892.

Yasothornsrikul, S., Davis, W., Cramer, G., Kimbrell, D., Dearolf, C. 1997. *viking*: Identification and characterization of a second type IV collagen in *Drosophila*. Gene 198, 17–25.

Yoshioka, K., Hino, S., Takemura, T., Maki, S., Wieslander, J., Takekoshi, Y., Makino, H., Kagawa, M., Sado, Y., Kashtan, C. 1994. Type IV collagen α5 chain: Normal distribution and abnormalities in X-linked Alport syndrome revealed by monoclonal antibody. Am. J. Pathol. 144, 986–996.

Yurchenco, P., Amenta, P., Patton, B. 2004. Basement membrane assembly, stability and activities observed through a developmental lens. Matrix Biol. 22, 521–538.

Yurchenco, P., Furthmayr, H. 1984. Self-assembly of basement membrane collagen. Biochemistry 23, 1839–1850.

Yurchenco, P., O' Rear, J. 1994. Basal lamina assembly. Curr Opin Cell Biol 6, 674–681.

Yurchenco, P., Ruben, G. 1987. Basement membrane structure *in situ*: Evidence for lateral associations in the collagen type IV network. J. Cell. Biol. 105, 2559–2568.

Yurchenco, P., Ruben, G. 1988. Type IV collagen lateral associations in the EHS tumor matrix. Comparison with amniotic and *in vitro* networks. Am. J. Pathol. 132, 278–291.

Zhang, X., Hudson, B., Sarras, M., Jr. 1994. Hydra cell aggregate development is blocked by selective fragments of fibronectin and type IV collagen. Dev. Biol. 164, 10–23.

Zheng, K., Harvey, S., Sado, Y., Naito, I., Ninomiya, Y., Jacobs, R., Thorner, P. 1999. Absence of the α6(IV) chain of collagen type IV in Alport syndrome is related to a failure at the protein assembly level and does not result in diffuse leiomyomatosis. Am. J. Pathol. 154, 1883–1892.

Zheng, K., Thorner, P., Marrano, P., Baumal, R., McInnes, R. 1994. Canine X chromosome-linked hereditary nephritis: A genetic model for human X-linked hereditary nephritis resulting from a single base mutation in the gene encoding the α5 chain of collagen type IV. Proc. Natl. Acad. Sci. USA 91, 3989–3993.

Zhou, J., Ding, M., Zhao, Z., Reeders, S. 1994. Complete primary structure of the sixth chain of human basement membrane collagen α6(IV). Isolation of the cDNAs for α6(IV) and comparison with five other type IV collagen chains. J. Biol. Chem. 269, 13193–13199.

Zhou, J., Hertz, J., Leinonen, A., Tryggvason, K. 1992. Complete amino acid sequence of the human α5(IV) collagen chain and identification of a single-base mutation in exon 23 converting glycine 521 in the collagenous domain to cysteine in an Alport syndrome patient. J. Biol. Chem. 267, 12475–12481.

Role of perlecan in development and diseases

Eri Arikawa-Hirasawa

Research Institute for Diseases of Old Age, Department of Neurology, Juntendo University School of Medicine, Bunkyo-ku, Tokyo, Japan

Contents

Advances in Developmental Biology
Volume 15 ISSN 1574-3349
DOI: 10.1016/S1574-3349(05)15002-9

1. Introduction

Perlecan is a large heparan sulfate proteoglycan (HSPG) that is expressed in all basement membranes but also in other connective tissues as well. The core protein consists of five distinctive structural domains that can be substituted primarily with heparan sulfate (HS) chains. Perlecan binds to many matrix molecules, growth factors, and cell surface receptors. Studies with gene knockout mice and human genetic diseases have demonstrated that perlecan is essential for development and that the lack of perlecan results in embryonic and perinatal lethality. In addition, it has been shown that perlecan has a number of complex biological effects in tissue homeostasis and diseases that depend on cell types and the environment. In this chapter, the structure and function of perlecan and related diseases will be reviewed.

2. Structure of perlecan

The basement membrane is a thin extracellular matrix (ECM) that forms in the interface between parenchymal cells and their surrounding connective tissues and is implicated in development and diseases. Proteoglycans are a class of glycosylated proteins that have covalent attachment of one or more glycosaminoglycan (GAG) chains (reviewed in Bernfield et al., 1999; Kramer et al., 2003). The GAG chains are modified by sulfation and epimerization during synthesis in the Golgi. These modifications result in specific structural patterns and protein-binding abilities of the GAG chains. A specific class of proteoglycans, known as HSPGs, is emerging as a key molecule class, governing crucial events in embryonic development, inflammation, wound repair, and cancer. To date, the crucial roles of HSPGs in regulating key developmental signaling pathways such as Wnt, Hedgehog, transforming growth factor-β (TGF-β), and fibroblast growth factor (FGF) pathways have been demonstrated (Lin et al., 1999; Esko and Lindahl, 2001; Perrimon and Bernfield, 2001; Lin, 2004; Perrimon and Hacker, 2004).

There are many different genes and gene families for the core proteins of HSPGs that include the four members of the syndecan family, the glypican family, perlecan, agrin, and collagen XVIII (Bernfield et al., 1999; Iozzo, 2001; Nakato and Kimata, 2002; Lin, 2004). Perlecan is an HSPG that is present in all basement membranes and in the cartilage matrix (Murdoch et al., 1994). It contains an estimated 400-kDa core protein with two to three HS side chains at the N-terminus and one or two HS/chondroitin sulfate (CS) chains at the C-terminus. On some occasions the HS side chains may be substituted with other glycosaminoglycan chains such as CS, dermatan sulfate (DS), hybrid HS/CS, CS/DS, or secreted as GAG-free glycoprotein (reviewed in Jiang et al., 2003).

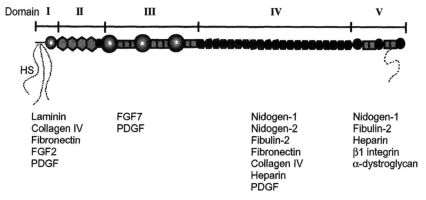

Domain I | II | III | IV | V

HS

Laminin	FGF7	Nidogen-1	Nidogen-1
Collagen IV	PDGF	Nidogen-2	Fibulin-2
Fibronectin		Fibulin-2	Heparin
FGF2		Fibronectin	β1 integrin
PDGF		Collagen IV	α-dystroglycan
		Heparin	
		PDGF	

I: HS attachment II: LDL receptor III: Laminin short arm

IV: N-CAM (IgG) V: Laminin α-chain G and EGF

Fig. 1. Schematic structure of perlecan and binding molecules.

The amino acid sequence of murine perlecan deduced from complementary deoxyribonucleic acid (cDNA) cloning (Noonan et al., 1988, 1991) indicates that the protein core is 369 kDa in size and consists of five distinct domains (Fig. 1). Domain I contains an SEA (sperm protein, enterokinase, agrin) module and three SerGlyAsp peptide sequences, in which the serine residue is a site for HS attachment (Dolan et al., 1997); domain II is similar to the cholesterol-binding region of the low-density lipoprotein (LDL) receptor; domain III is similar to the short arm of laminin α chains; domain IV is the largest domain and contains 14 repeats of immunoglobulin (Ig)G-like motifs similar to those in the neural cell adhesion molecule; and domain V has three globular subdomains, similar to the LG subdomain of laminin α chains, with interruptions by four EGF-like motifs, and can be substituted with HS and/or CS chains (Brown et al., 1997; Tapanadechopone et al., 1999; Friedrich, 2000). These domains bind ECMs (Murdoch et al., 1992; Noonan and Hassell, 1993; Rogalski et al., 1993; Iozzo, 1994; Moerman et al., 1996; Tapanadechopone et al., 1999; Friedrich et al., 2000), growth factors, and cell surface receptors (Iozzo, 1994).

The structure of human perlecan is essentially identical to murine perlecan except that domain IV contains 21 repeats of the IgG-like motifs, which increases the core protein to 466 kDa (Murdoch et al., 1992). A longer form of perlecan was shown to be expressed in mice as an alternately spliced form of perlecan (Noonan and Hassell, 1993). A homologue of perlecan is also present in the basement membranes of *C. elegans* (Rogalski et al., 1993; Moerman et al., 1996) and *Drosophila* (Friedrich et al., 2000; Park et al., 2003). Mutations in the alternatively spliced perlecan gene in *C. elegans* produce defective muscle development and cause paralysis (Rogalski et al., 2001). Perlecan may contain only HS chains (Rogalski et al., 2001) or both HS and CS chains (SundarRaj et al., 1995).

The size of the GAG chains also varies according to tissue and cell type (Hassell et al., 1980; Molist et al., 1998). Biological activity of perlecan may be varied by modifications of these sugar chains. Genetically modified mice, in which the HS attachment sites at the N terminus of perlecan are removed, survive embryonic development and are viable and fertile but develop small eyes (Rossi et al., 2003). However, mutant mice exhibit significantly delayed wound healing, retarded FGF-2–induced tumor growth, and defective angiogenesis (Zhou et al., 2004). Although these results suggest that the HS chains of perlecan positively regulate angiogenesis *in vivo*, gene knockout of the whole perlecan molecule may result in different phenotypes.

3. Interactions of perlecan with extracellular molecules and cell surface receptors

Perlecan is implicated in supporting basement membrane structure, since it interacts with basement membrane components. The HS chains interact with collagen IV and laminin (reviewed in Bernfield et al., 1999). The core protein also binds to collagen IV, nidogen/entactin–1 and –2, and fibulin (Talts et al., 1999; Kramer and Yost, 2003). Perlecan can self-aggregate into dimers and multimers via domain V. Domain V interacts with cell surface receptors such as $\beta 1$ integrin and α-dystroglycan (Brown et al., 1997; Friedrich et al., 1999). Perlecan's ability to interact with many proteins and receptors suggests that perlecan provides structural support for the basement membrane and bridges matrix and cell interactions for the stable formation of tissue architecture.

4. Expression of perlecan in development

Expression of perlecan is detected from a very early stage of embryogenesis. Perlecan expression on the external trophectodermal cell surfaces of mouse blastocysts increases in the peri-implantation period (Smith et al., 1997). Perlecan expression appears early in tissues of vasculogenesis including heart, pericardium, and major blood vessels (Handler et al., 1997). Deposition of perlecan occurs within the developing cartilage, especially the cartilage undergoing endochondral ossification, where it remains elevated throughout all developmental stages up to adulthood. During later stages of development (embryonic day [E]13 to E17.5), perlecan messenger ribonucleic acid (mRNA) levels progressively increase, and its expression correlates with the onset of tissue differentiation of various parenchymal organs, including the developing kidneys, lungs, liver, spleen, and gastrointestinal tract. The central nervous system showed no perlecan expression with the exception of the calvaria and choroid plexus.

5. Functional-null mutation of perlecan gene in mice

The perlecan-null mutation in mice showed perinatal lethal chondrodysplasia. Some of the perlecan $-/-$ mice died at around E10.5 with defective cephalic development and defective cardiomyocyte basement membranes (Arikawa-Hirasawa et al., 1999; Costell et al., 1999). The remaining perlecan $-/-$ mice go on to develop skeletal dysplasia characterized by shortened long bones and craniofacial abnormalities, and die shortly after birth (Figs. 2 and 3)

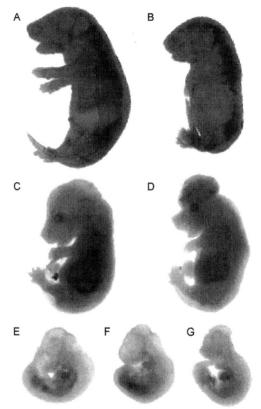

Fig. 2. Gross appearance of wild-type (A, C, E) and Hspg2$-/-$ mice (B, D, F, G) (Arikawa-Hirasawa et al., 1999). (A) Newborn wild-type mouse. (B) Newborn Hspg2$-/-$ mouse. (C) E16.5 wild-type mouse. (D) E16.5 Hspg2$-/-$ mouse with exencephaly. (E) E10.5 wild-type mouse. (F, G) E10.5 Hspg2$-/-$ mice. Approximately 60% of Hspg2$-/-$ mice show no apparent abnormalities at this stage (F), but develop dwarfism at a later stage and survive until birth. The remaining 40% of Hspg2$-/-$ mice show defective cephalic development (G) with poor development of the forebrain and hindbrain, and most die at ~E10.5. A few mice survive and develop exencephaly (D). Total percent of Hspg2$-/-$ mice, including both those that die at ~E10.5 and the surviving mice, is ~25% of the total embryos (~500) examined, consistent with mendelian inheritance. The partial penetrance of the cephalic defects in Hspg2$-/-$ mice is similar to other mouse models with exencephaly and is probably due to the mixed genetic backgrounds of the mouse strains. Hspg 2 = perlecan.

E. Arikawa-Hirasawa

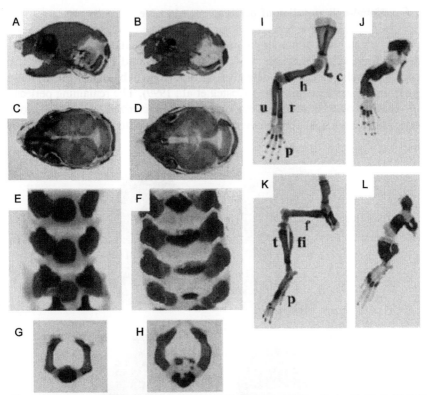

Fig. 3. Skeletal preparations of wild-type (A, C, E, G, I, K) and Hspg2−/− (B, D, F, H, J, L) newborn mice stained with Alcian blue and Alizarin red (Arikawa-Hirasawa et al., 1999). Lateral view of Hspg2−/−skull (B) shows shorter and thicker mandibula and delayed ossification of bony ossicles of middle ear. Dilated sutures and brachycephaly are observed in the skull vault of Hspg2−/− mice (D). Shorter and wider vertebrae (F, H) of Hspg2−/− mice are shown. Multiple ossification centers were observed at the periphery of the vertebral body (H). The scapula, humerus, radius, and ulna of Hspg2−/− forelimb (J) and ilium, femur, tibia and fibula of the hindlimb (L) are shorter and broader. The size of the phalanges in Hspg2−/− mice is relatively normal. c = clavicle; h = humerus; p = phalanx; r = radius; u = ulna; f = femur; fi = fibula; t = tibia; Hspg2 = perlecan. (See Color Insert.)

(Arikawa-Hirasawa et al., 1999). Histological evaluation of the perlecan −/− mice revealed a disorganized growth plate with reduced chondrocyte proliferation and differentiation and defective endochondral ossification (Arikawa-Hirasawa et al., 1999). The columnar structure of the hypertrophic chondrocytes is absent in the mutant growth plate. The vertebral bodies of the mutant mice have decreased height and increased width and contain multiple ossification centers at the periphery.

The abnormalities of cartilage are not obvious until approximately E14.5, at which time endochondral ossification begins; after that, disruption of cartilage matrix becomes progressively more severe. Perlecan-null cartilage matrix contains reduced and disorganized collagen fibrils and glycosamino-glycans, suggesting that perlecan plays an important role in cartilage matrix formation. Perlecan-null growth plate also shows reduced chondrocyte pro-liferation and disorganized expression of Indian hedgehog. Since FGF/FGFR3c signaling is a key regulator for chondrocyte differentiation and proliferation in the growth plate, these results may suggest a role for perlecan in the modulation of FGF signaling in cartilage development. The pheno-type of perlecan-null mice is similar to that caused by the constitutive activation mutation of *FGFR3* in humans (Rousseau et al., 1994, 1996; Wilcox et al., 1998) and is opposite to that of *Fgfr3*-null mice (Deng et al., 1996; Ornitz, 2001; Brodie and Deng, 2003). These results suggest that perlecan may be a negative regulator of FGFR3 signaling in the growth plate. In this regard, perlecan may function by sequestering ligands/growth factors in the matrix. We also observed an abnormal vascular invasion of the growth plate (Arikawa-Hirasawa, unpublished data, 2004) where vascular endothelial growth factor (VEGF) was more immunostained in perlecan-null cartilage compared to normal growth plate. This result may support the potential role of perlecan as a VEGF "sink" in the matrix.

6. Functional-null mutations in humans with dyssegmental dysplasia, Silverman-Handmaker type

Similarities of growth plate histology and X-ray findings for skeletal abnormalities of perlecan-null mice led to the identification of a human dis-order called Dyssegmental dysplasia, Silverman-Handmaker type (DDSH) (Fig. 4) (Arikawa-Hirasawa et al., 2001). DDSH is a rare lethal autosomal recessive skeletal dysplasia characterized by anisospondyly and micromelia (Aleck et al., 1987). The endochondral growth plate of DDSH patients is short and contains large and unfused calcospherites similar to that of the perlecan mutant mice. Unlike the mice, humans have patchy mucoid degen-eration of the resting cartilage and intracellular inclusions in chondrocytes. Immunostaining with an antibody to the perlecan core protein showed strong staining of the pericellular cartilage matrix in normal controls. In contrast, there was little staining in the DDSH cartilage matrix, but there was significant staining of inclusion bodies within the cells. Similar immu-nostaining results were observed with cultured fibroblasts from DDSH patients. Biochemical analysis revealed that normal-sized perlecan core pro-tein was not detected in the cell media from DDSH fibroblasts, but several smaller-sized proteins were detected in the cell lysates. Three patients studied were identified to have perlecan mutations (Arikawa-Hirasawa et al., 2001). A homozygous 89-bp duplication mutation was found in exon 36 of the perlecan gene in a pair of siblings with DDSH. Heterozygous point muta-tions were identified at the donor site of exon 54 and at the middle of exon 73

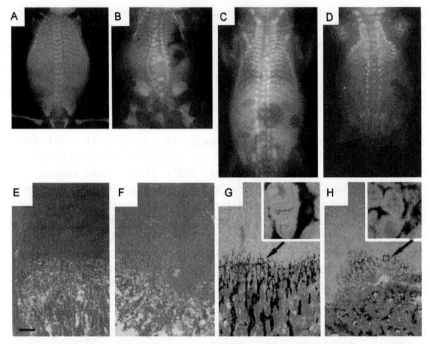

Fig. 4. Radiographic and histologic features of DDSH and the perlecan-null mouse. (Arikawa-Hirasawa et al., 2001). Radiographs are shown of a normal newborn human (A), a newborn patient with DDSH (B), a normal newborn mouse (C), and an Hspg2-null newborn mouse (D). Note the similarity in the radiographic appearance between the human and the null mouse. In both cases, there is anisospondly, a small chest, and short, bent long bones of the limbs. Profound dyssegmental ossification is observed in the spine in case 1 and the Hspg2-null mouse. Also shown is the histology of the cartilage of the femur: (E, F) Toluidine blue stain, ×6.25, scale bar, 500 μm; (G, H) von Kossa trichrome stain, ×6.25. Inset, higher magnification, ×50. (E, G), Normal human. (F, H) A patient with DDSH. In DDSH cartilage, reduced matrix staining and endochondral ossification with disorganized columnar structures of hypertrophic chondrocytes are seen. The calcospherites are unfused and show granular appearances in the DDSH hypertrophic zone, whereas they are fused in normal tissue. Hspg2 = perlecan. (See Color Insert.)

in the third, unrelated patient, causing exon skipping. These mutations are predicted to produce premature termination codons. Immunostaining reveals that truncated proteins are synthesized but not secreted. Thus, DDSH is caused by functional null mutations of the perlecan gene similar to the gene knockout mice.

7. Neuromuscular junction (NMJ) of knockout mice

Perlecan is present in muscle basement membranes and is enriched at the NMJ (Arikawa-Hirasawa et al., 2002b). The NMJ forms a highly specialized structure where a unique set of molecules (Sanes and Lichtman, 1999) such

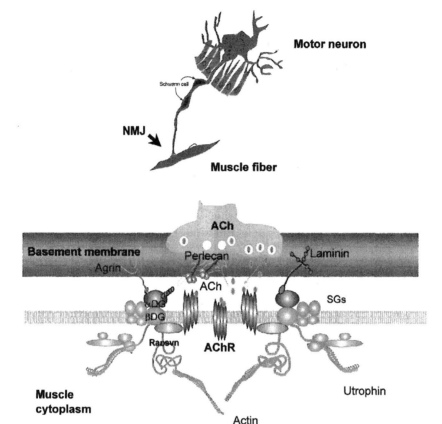

Fig. 5. In skeletal muscles, acetylcholine (ACh) is released from the nerve terminal and acts as a neurotransmitter. Ach binds AchR on the muscle cell surface, activates Na channels, and induces muscle contraction. The NMJ forms a highly specialized structure where a unique set of molecules, such as acetylcholinesterase (AchE), acetylcholine receptors (AchR), agrin, dystroglycans, rapsyn, and utrophine are clustered. Perlecan is also enriched at the NMJ. AchE rapidly inactivates Ach activity for muscle relaxation and recycling of Ach. Perlecan is essential to target AchE at the NMJ.

as acetylcholinesterase (AChE), acetylcholine receptors (AChR), agrin, dystroglycans, rapsyn, and utrophin are clustered (Fig. 5). In skeletal muscles, acetylcholine (ACh) is released from the nerve terminal and acts as a neurotransmitter. ACh binds AChR on the muscle cell surface, activates Na channels, and induces muscle contraction. AChE rapidly inactivates this process for muscle relaxation and recycling of ACh. The collagen-tail form (ColQ) of AChE is preferentially expressed in innervated regions of muscles and is shown to bind perlecan *in vitro* (Rotundo, 2003; Steen and Froehner, 2003). In the perlecan knockout mice, muscle development and differentiation appear to be normal, and the nerve terminals are normally formed at birth (Arikawa-Hirasawa et al., 2002b). Clustering molecules are present at

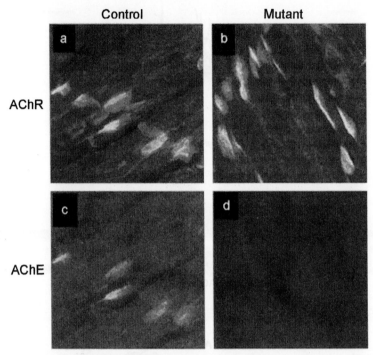

Fig. 6. AChE is absent from the NMJs of perlecan-null mice (Arikawa-Hirasawa et al., 2002b). Whole mounts of embryo limb muscles, as well as intercostal muscle and diaphragm (data not shown), were stained for AChE using fluorescent OR-Fas2 as a probe and for AChR using Alexa–594-conjugated α-bungarotoxin. Although there was strong AChR clustering in both normal heterozygous and perlecan-null muscle, AChE clustering was completely absent from perlecan-null muscle. Thus, perlecan is essential for attachment of the AChE to the synaptic basal lamina.

the NMJ of the mutant mouse muscles. However, AChE is absent at the newborn NMJ, although the AChE protein is synthesized normally (Fig. 6) (Arikawa-Hirasawa et al., 2002b). Thus, perlecan is essential for high-density localization of AChE at the NMJ.

8. Partially functional mutations in humans with Schwartz-Jampel syndrome

Schwartz-Jampel syndrome (SJS) was initially identified for perlecan mutations by positional cloning and gene linkage (Nicole et al., 2000; Arikawa-Hirasawa et al., 2002a). SJS is characterized by a unique combination of myotonia and chondrodysplasia and, unlike DDSH, patients with

SJS survive. SJS is a rare autosomal recessive skeletal dysplasia associated with myotonia. This disorder is characterized by short stature, osteochondrodysplasia, myotonia, and a characteristic face with a fixed facial expression, blepharophimosis, pursed lips, and sometimes low-set ears and myopia. Skeletal abnormalities include kyphoscoliosis, platyspondyly with coronal clefts in the vertebrae, delayed and irregular epiphyseal ossification, and joint contractures. Phenotypic differences in DDSH and SJS appear to reflect differences in the amount of functional perlecan in the matrix. Characterization of the perlecan gene in these patients showed that it contained homozygous missense and splicing mutations that could result in either truncated perlecan proteins missing part of domain IV and the whole domain V or molecules defective in disulfide bonds in domain III. Recent studies on additional SJS patients revealed different types of mutations that result in various altered forms of perlecan. For example, heterozygous mutations produce either truncated perlecan proteins that lack domain V or significantly reduced levels of wild-type perlecan. A homozygous 7-kb deletion resulted in reduced amounts of nearly full-length perlecan. Unlike DDSH, SJS mutations result in different, abnormal forms of the perlecan molecule in reduced levels that are secreted into the ECM and are partially functional. These findings indicate an important role of perlecan in neuromuscular function and cartilage formation.

Recently, Spranger et al. (2000) reported that other skeletal dysplasias previously identified as chondrodysplasias, such as kyphomelic chondrodysplasia, micromelic chondrodysplasia, and Burton's disease, could be reclassified based on a clinical examination as SJS. Mutation analyses of the perlecan gene will be required for the diagnoses of chondrodysplasia and myotonic myopathy. These mutation searches may reveal a wider spectrum of the SJS phenotype in the near future.

9. Role of perlecan in cardiovascular development and atherosclerosis

Deletion of the perlecan gene causes myocardial abnormality as well as cephalic and cartilage abnormalities, including transposition of the great vessels (Costell et al., 2002). Perlecan-deficient embryos demonstrate abnormally abundant mesenchymal cells expressing the smooth muscle cell (SMC)-specific alpha-actin isoform in the left ventricular outflow tract, which can lead to varying levels of outflow tract obstruction.

In relation to cardiovascular biology, perlecan has been identified in the atherosclerotic intimal lesions of apoE–/– and LDLR–/– mice (Kunjathoor et al., 2002) and may contribute to the retention of lipoproteins at the earliest stage of atherosclerosis. Perlecan is associated with intermediate and advanced lesions of hypercholesterolemic non-human primates and in cultures of medial SMC from human atherosclerotic tissue (Evanko et al.,

1998). The exact role of perlecan in the arterial wall remains unclear. Perlecan has a module domain that shares homology with the binding domain of the LDLR, suggesting that perlecan could potentially bind lipoproteins. Atherosclerosis was studied in mice with a heterozygous deletion for the perlecan gene made by crossing the defect into apolipoprotein E (apoE) and LDL receptor knockout backgrounds. At 12 weeks, chow-fed apoE-null mice with a heterozygous deletion had less atherosclerosis, but this difference was not significant at 24 weeks (Vikramadithyan et al., 2004). These results suggest that the loss of perlecan leads to less atherosclerosis in early lesions.

10. Angiogenesis and tumor formation

Perlecan has been implicated in angiogenesis and tumor formation and metastasis (reviewed in Jiang et al., 2003). Perlecan promotes angiogenesis in an *in vivo* angiogenesis model (Aviezer et al., 1994) and its expression is associated with vessel formation (Guelstein et al., 1993; Roskams et al., 1998). Recently, domain V of perlecan was shown to inhibit angiogenesis (Mongiat et al., 2003). This portion of perlecan is called endorepellin because of its ability to bind collagen XVIII, which contains endostatin, a potent antiangiogenic factor (Mongiat et al., 2003). Endorepellin counteracts the antiangiogenic activity of endostatin and itself inhibits angiogenesis at nanomolar concentrations. Most of endorepellin's activity resides on the LG3 subdomain of domain V of perlecan. The LG3 protein was detected in the urine of patients with end-stage renal disease and in the amniotic fluid of pregnant women who had premature rupture. In addition, specific metalloproteases of the BMP-1/Tolloid family can cleave LG3 from perlecan and endorepellin (Gonzalez et al., 2005). More recently, endorepellin was found to regulate endothelial cell activity in a calcium-dependent but HS-independent manner by disrupting actin stress fibers and focal adhesions through an interaction with $\alpha 2\beta 1$ integrin (Bix and Iozzo, 2005). These results suggest that the C-terminal region of perlecan indeed plays a significant role in biological functions *in vivo* as both a proteolytic fragment and as a portion of the intact perlecan molecule.

Perlecan has been reported to regulate tumor growth and metastasis (reviewed in Jiang et al., 2003). Expression of perlecan is often increased in many tumor types *in vivo* and *in vitro*. Antisense suppression of perlecan expression inhibited the invasion of melanoma cells, while similar antisense suppression of perlecan increased tumor growth of human fibrosarcoma cells (Mathiak et al., 1997) and Kaposi's sarcoma cells (Marchisone et al., 2000). These differences may be dependent on tumor cell types and tissues. The significant role of the HS chains of perlecan in angiogenesis and tumor formation was demonstrated using genetically engineered mutant mice that express endogenous perlecan lacking the HS chains (Zhou et al.,

2004). Angiogenesis and tumor formation were inhibited in these mice, suggesting that the HS chains enhance tumor growth and new blood vessel formation.

11. Conclusions

Perlecan has multiple functions in cell growth and differentiation and tissue organization. Recent studies with gene knockout and mutant mice and human genetic disorders have shed light on the *in vivo* functions of perlecan. The complete loss of functional perlecan results in embryonic and perinatal lethality with defective cartilage and cephalic development and cardiac failure. Perlecan is essential for clustering of AChE at the NMJ, and partially functional mutated versions of perlecan result in myotonia and milder chondrodysplasia. The deletion of a short coding sequence including the HS attachment sites of the N-terminus of perlecan in mice revealed that the HS chains play a role for supporting the integrity of the lens capsule and for promoting angiogenesis and tumor growth. In vitro studies indicate that many extracellular molecules interact with the HS chains and core protein of perlecan. Perlecan may work as a modulator for growth factor signaling as a ligand reservoir. Different effects of perlecan on different cells may be due to the HS GAG side chains that directly affect cell signaling or through interactions with heparin-binding growth factors. The significance of these interactions and structure and function relationships *in vivo* can be addressed in mouse models.

References

Aleck, K.A., Grix, A., Clericuzio, C., Kaplan, P., Adomian, G.E., Lachman, R., Rimoin, D.L. 1987. Dyssegmental dysplasias: Clinical, radiographic, and morphologic evidence of heterogeneity. Am. J. Med. Genet. 27, 295–312.

Arikawa-Hirasawa, E., Le, A.H., Nishino, I., Nonaka, I., Ho, N.C., Francomano, C.A., Govindraj, P., Hassell, J.R., Devaney, J.M., Spranger, J., Stevenson, R.E., Iannaccone, S., Dalakas, M.C., Yamada, Y. 2002a. Structural and functional mutations of the perlecan gene cause Schwartz-Jampel syndrome, with myotonic myopathy and chondrodysplasia. Am. J. Hum. Genet. 70, 1368–1375.

Arikawa-Hirasawa, E., Rossi, S.G., Rotundo, R.L., Yamada, Y. 2002b. Absence of acetylcholinesterase at the neuromuscular junctions of perlecan-null mice. Nat. Neurosci. 5, 119–123.

Arikawa-Hirasawa, E., Watanabe, H., Takami, H., Hassell, J.R., Yamada, Y. 1999. Perlecan is essential for cartilage and cephalic development. Nat. Genet. 23, 354–358.

Arikawa-Hirasawa, E., Wilcox, W.R., Le, A.H., Silverman, N., Govindraj, P., Hassell, J.R., Yamada, Y. 2001. Dyssegmental dysplasia, Silverman-Handmaker type, is caused by functional null mutations of the perlecan gene. Nat. Genet. 27, 431–434.

Aviezer, D., Hecht, D., Safran, M., Eisinger, M., David, G., Yayon, A. 1994. Perlecan, basal lamina proteoglycan, promotes basic fibroblast growth factor-receptor binding, mitogenesis, and angiogenesis. Cell 79, 1005–1013.

Bernfield, M., Gotte, M., Park, P.W., Reizes, O., Fitzgerald, M.L., Lincecum, J., Zako, M. 1999. Functions of cell surface heparan sulfate proteoglycans. Annu. Rev. Biochem. 68, 729–777.

Bix, G., Iozzo, R.V. 2005. Matrix revolutions: "tails" of basement-membrane components with angiostatic functions. Trends Cell. Biol. 15, 52–60.

Brodie, S.G., Deng, C.X. 2003. Mouse models orthologous to FGFR3-related skeletal dysplasias. Pediatr. Pathol. Mol. Med. 22, 87–103.

Brown, J.C., Sasaki, T., Gohring, W., Yamada, Y., Timpl, R. 1997. The C-terminal domain V of perlecan promotes beta1 integrin-mediated cell adhesion, binds heparin, nidogen and fibulin-2 and can be modified by glycosaminoglycans. Eur. J. Biochem. 250, 39–46.

Costell, M., Carmona, R., Gustafsson, E., Gonzalez-Iriarte, M., Fassler, R., Munoz-Chapuli, R. 2002. Hyperplastic conotruncal endocardial cushions and transposition of great arteries in perlecan-null mice. Circ. Res. 91, 158–164.

Costell, M., Gustafsson, E., Aszodi, A., Morgelin, M., Bloch, W., Hunziker, E., Addicks, K., Timpl, R., Fassler, R. 1999. Perlecan maintains the integrity of cartilage and some basement membranes. J. Cell. Biol. 147, 1109–1122.

Deng, C., Wynshaw-Boris, A., Zhou, F., Kuo, A., Leder, P. 1996. Fibroblast growth factor receptor 3 is a negative regulator of bone growth. Cell 84, 911–921.

Dolan, M., Horchar, T., Rigatti, B., Hassell, J.R. 1997. Identification of sites in domain I of perlecan that regulate heparan sulfate synthesis. J. Biol. Chem. 272, 4316–4322.

Esko, J.D., Lindahl, U. 2001. Molecular diversity of heparan sulfate. J. Clin. Invest. 108, 169–173.

Evanko, S.P., Raines, E.W., Ross, R., Gold, L.I., Wight, T.N. 1998. Proteoglycan distribution in lesions of atherosclerosis depends on lesion severity, structural characteristics, and the proximity of platelet-derived growth factor and transforming growth factor-beta. Am. J. Pathol. 152, 533–546.

Friedrich, M.V., Gohring, W., Morgelin, M., Brancaccio, A., David, G., Timpl, R. 1999. Structural basis of glycosaminoglycan modification and of heterotypic interactions of perlecan domain V. J. Mol. Biol. 294, 259–270.

Friedrich, M.V., Schneider, M., Timpl, R., Baumgartner, S. 2000. Perlecan domain V of *Drosophila melanogaster*. Sequence, recombinant analysis and tissue expression. Eur. J. Biochem. 267, 3149–3159.

Gonzalez, E.M., Reed, C.C., Bix, G., Fu, J., Zhang, Y., Gopalakrishnan, B., Greenspan, D.S., Iozzo, R.V. 2005. BMP-1/tolloid-like metalloproteases process endorepellin, the angiostatic C-terminal fragment of perlecan. J. Biol. Chem. 280, 7080–7087

Guelstein, V.I., Tchypysheva, T.A., Ermilova, V.D., Ljubimov, A.V. 1993. Myoepithelial and basement membrane antigens in benign and malignant human breast tumors. Int. J. Cancer. 53, 269–277.

Handler, M., Yurchenco, P.D., Iozzo, R.V. 1997. Developmental expression of perlecan during murine embryogenesis. Dev. Dyn. 210, 130–145.

Hassell, J.R., Robey, P.G., Barrach, H.J., Wilczek, J., Rennard, S.I., Martin, G.R. 1980. Isolation of a heparan sulfate-containing proteoglycan from basement membrane. Proc. Natl. Acad. Sci. USA 77, 4494–4498.

Iozzo, R.V. 1994. Perlecan: A gem of a proteoglycan. Matrix Biol. 14, 203–208.

Iozzo, R.V. 2001. Heparan sulfate proteoglycans: Intricate molecules with intriguing functions. J. Clin. Invest. 108, 165–167.

Jiang, X., Couchman, J.R. 2003. Perlecan and tumor angiogenesis. J. Histochem. Cytochem. 51, 1393–1410.

Kramer, K.L., Yost, H.J. 2003. Heparan sulfate core proteins in cell-cell signaling. Annu. Rev. Genet. 37, 461–484.

Kunjathoor, V.V., Chiu, D.S., O'Brien, K.D., LeBoeuf, R.C. 2002. Accumulation of biglycan and perlecan, but not versican, in lesions of murine models of atherosclerosis. Arterioscler. Thromb. Vasc. Biol. 22, 462–468.

Lin, X. 2004. Functions of heparan sulfate proteoglycans in cell signaling during development. Development 131, 6009–6021.

Lin, X., Buff, E.M., Perrimon, N., Michelson, A.M. 1999. Heparan sulfate proteoglycans are essential for FGF receptor signaling during *Drosophila* embryonic development. Development 126, 3715–3723.

Marchisone, C., Del Grosso, F., Masiello, L., Prat, M., Santi, L., Noonan, D.M. 2000. Phenotypic alterations in Kaposi's sarcoma cells by antisense reduction of perlecan. Pathol. Oncol. Res. 6, 10–17.

Mathiak, M., Yenisey, C., Grant, D.S., Sharma, B., Iozzo, R.V. 1997. A role for perlecan in the suppression of growth and invasion in fibrosarcoma cells. Cancer Res. 57, 2130–2136.

Moerman, D.G., Hutter, H., Mullen, G.P., Schnabel, R. 1996. Cell autonomous expression of perlecan and plasticity of cell shape in embryonic muscle of Caenorhabditis elegans. Dev. Biol. 173, 228–242.

Molist, A., Romaris, M., Lindahl, U., Villena, J., Touab, M., Bassols, A. 1998. Changes in glycosaminoglycan structure and composition of the main heparan sulphate proteoglycan from human colon carcinoma cells (perlecan) during cell differentiation. Eur. J. Biochem. 254, 371–377.

Mongiat, M., Fu, J., Oldershaw, R., Greenhalgh, R., Gown, A.M., Iozzo, R.V. 2003. Perlecan protein core interacts with extracellular matrix protein 1 (ECM1), a glycoprotein involved in bone formation and angiogenesis. J. Biol. Chem. 278, 17491–17499.

Murdoch, A.D., Dodge, G.R., Cohen, I., Tuan, R.S., Iozzo, R.V. 1992. Primary structure of the human heparan sulfate proteoglycan from basement membrane (HSPG2/perlecan). A chimeric molecule with multiple domains homologous to the low density lipoprotein receptor, laminin, neural cell adhesion molecules, and epidermal growth factor. J. Biol. Chem. 267, 8544–8557.

Murdoch, A.D., Liu, B., Schwarting, R., Tuan, R.S., Iozzo, R.V. 1994. Widespread expression of perlecan proteoglycan in basement membranes and extracellular matrices of human tissues as detected by a novel monoclonal antibody against domain III and by *in situ* hybridization. J. Histochem. Cytochem. 42, 239–249.

Nakato, H., Kimata, K. 2002. Heparan sulfate fine structure and specificity of proteoglycan functions. Biochim. Biophys. Acta 1573, 312–318.

Nicole, S., Davoine, C.S., Topaloglu, H., Cattolico, L., Barral, D., Beighton, P., Hamida, C.B., Hammouda, H., Cruaud, C., White, P.S., Samson, D., Urtizberea, J.A., Lehmann-Horn, F., Weissenbach, J., Hentati, F., Fontaine, B. 2000. Perlecan, the major proteoglycan of basement membranes, is altered in patients with Schwartz-Jampel syndrome (chondrodystrophic myotonia). Nat. Genet. 26, 480–483.

Noonan, D.M., Fulle, A., Valente, P., Cai, S., Horigan, E., Sasaki, M., Yamada, Y., Hassell, J.R. 1991. The complete sequence of perlecan, a basement membrane heparan sulfate proteoglycan, reveals extensive similarity with laminin A chain, low density lipoprotein-receptor, and the neural cell adhesion molecule. J. Biol. Chem. 266, 22939–22947.

Noonan, D.M., Hassell, J.R. 1993. Perlecan, the large low-density proteoglycan of basement membranes: Structure and variant forms. Kidney Int. 43, 53–60.

Noonan, D.M., Horigan, E.A., Ledbetter, S.R., Vogeli, G., Sasaki, M., Yamada, Y., Hassell, J.R. 1988. Identification of cDNA clones encoding different domains of the basement membrane heparan sulfate proteoglycan. J. Biol. Chem. 263, 16379–16387.

Ornitz, D.M. 2001. Regulation of chondrocyte growth and differentiation by fibroblast growth factor receptor 3. Novartis. Found. Symp. 232, 63–76; discussion 76–80, 272–282.

Park, Y., Rangel, C., Reynolds, M.M., Caldwell, M.C., Johns, M., Nayak, M., Welsh, C.J., McDermott, S., Datta, S. 2003. *Drosophila* perlecan modulates FGF and hedgehog signals to activate neural stem cell division. Dev. Biol. 253, 247–257.

Perrimon, N., Bernfield, M. 2001. Cellular functions of proteoglycans—an overview. Semin. Cell. Dev. Biol. 12, 65–67.

Perrimon, N., Hacker, U. 2004. Wingless, hedgehog and heparan sulfate proteoglycans. Development 131, 2509–2511; author reply 2511–2503.

Rogalski, T.M., Mullen, G.P., Bush, J.A., Gilchrist, E.J., Moerman, D.G. 2001. UNC–52/ perlecan isoform diversity and function in *Caenorhabditis elegans*. Biochem. Soc. Trans. 29, 171–176.

Rogalski, T.M., Williams, B.D., Mullen, G.P., Moerman, D.G. 1993. Products of the unc–52 gene in Caenorhabditis elegans are homologous to the core protein of the mammalian basement membrane heparan sulfate proteoglycan. Genes Dev. 7, 1471–1484.

Roskams, T., De Vos, R., David, G., Van Damme, B., Desmet, V. 1998. Heparan sulphate proteoglycan expression in human primary liver tumours. J. Pathol. 185, 290–297.

Rossi, M., Morita, H., Sormunen, R., Airenne, S., Kreivi, M., Wang, L., Fukai, N., Olsen, B.R., Tryggvason, K., Soininen, R. 2003. Heparan sulfate chains of perlecan are indispensable in the lens capsule but not in the kidney. EMBO J. 22, 236–245.

Rotundo, R.L. 2003. Expression and localization of acetylcholinesterase at the neuromuscular junction. J. Neurocytol. 32, 743–766.

Rousseau, F., Bonaventure, J., Legeai-Mallet, L., Pelet, A., Rozet, J.M., Maroteaux, P., Le Merrer, M., Munnich, A. 1994. Mutations in the gene encoding fibroblast growth factor receptor-3 in achondroplasia. Nature 371, 252–254.

Rousseau, F., Bonaventure, J., Legeai-Mallet, L., Pelet, A., Rozet, J.M., Maroteaux, P., Le Merrer, M., Munnich, A. 1996. Mutations of the fibroblast growth factor receptor-3 gene in achondroplasia. Horm. Res. 45, 108–110.

Sanes, J.R., Lichtman, J.W. 1999. Development of the vertebrate neuromuscular junction. Annu. Rev. Neurosci. 22, 389–442.

Smith, S.E., French, M.M., Julian, J., Paria, B.C., Dey, S.K., Carson, D.D. 1997. Expression of heparan sulfate proteoglycan (perlecan) in the mouse blastocyst is regulated during normal and delayed implantation. Dev. Biol. 184, 38–47.

Spranger, J., Hall, B.D., Hane, B., Srivastava, A., Stevenson, R.E. 2000. Spectrum of Schwartz-Jampel syndrome includes micromelic chondrodysplasia, kyphomelic dysplasia, and Burton disease. Am. J. Med. Genet. 94, 287–295.

Steen, M.S., Froehner, S.C. 2003. PerleCan fix your muscle AChEs. Trends Neurosci. 26, 241–242.

SundarRaj, N., Fite, D., Ledbetter, S., Chakravarti, S., Hassell, J.R. 1995. Perlecan is a component of cartilage matrix and promotes chondrocyte attachment. J. Cell. Sci. 108(Pt. 7), 2663–2672.

Talts, J.F., Andac, Z., Gohring, W., Brancaccio, A., Timpl, R. 1999. Binding of the G domains of laminin alpha1 and alpha2 chains and perlecan to heparin, sulfatides, alpha-dystroglycan and several extracellular matrix proteins. EMBO J. 18, 863–870.

Tapanadechopone, P., Hassell, J.R., Rigatti, B., Couchman, J.R. 1999. Localization of glycosaminoglycan substitution sites on domain V of mouse perlecan. Biochem. Biophys. Res. Commun. 265, 680–690.

Vikramadithyan, R.K., Kako, Y., Chen, G., Hu, Y., Arikawa-Hirasawa, E., Yamada, Y., Goldberg, I.J. 2004. Atherosclerosis in perlecan heterozygous mice. J. Lipid. Res. 45, 1806–1812.

Wilcox, W.R., Tavormina, P.L., Krakow, D., Kitoh, H., Lachman, R.S., Wasmuth, J.J., Thompson, L.M., Rimoin, D.L. 1998. Molecular, radiologic, and histopathologic correlations in thanatophoric dysplasia. Am. J. Med. Genet. 78, 274–281.

Zhou, Z., Wang, J., Cao, R., Morita, H., Soininen, R., Chan, K.M., Liu, B., Cao, Y., Tryggvason, K. 2004. Impaired angiogenesis, delayed wound healing and retarded tumor growth in perlecan heparan sulfate-deficient mice. Cancer Res. 64, 4699–4702.

Extracellular matrix gene expression in the developing mouse aorta

Sean E. McLean,[1] Brigham H. Mecham,[2]
Cassandra M. Kelleher,[1] Thomas J. Mariani[2]
and Robert P. Mecham[1]

[1]*Department of Cell Biology and Physiology, Washington University School of Medicine,
St. Louis, Missouri*
[2]*Division of Pulmonary and Critical Care Medicine, Brigham and Women's Hospital,
Harvard Medical School, Boston, Massachusetts*

Contents

Advances in Developmental Biology
Volume 15 ISSN 1574-3349
DOI: 10.1016/S1574-3349(05)15003-0

1. Introduction

Embryonic development requires the establishment of a functional circulatory system early in embryogenesis. The general outline of the forming vascular network is established in the absence of blood flow by endothelial cells through angiogenic or vasculogenic processes. With the initiation of blood flow, recruitment and differentiation of cells that make up the vascular wall begins in a process that is highly sensitive to local hemodynamic forces (Langille, 1996).

The general histological form of the large blood vessels includes three compartments: the *tunica intima*, consisting of a single layer of endothelial cells that sit directly on the internal elastic lamina (IEL); the *tunica media*, consisting of concentric layers of smooth muscle cells (SMCs) between sheets of elastin (the elastic laminae); and the *tunica adventitia*, made up of myofibroblasts that produce mainly collagen fibers. Within the medial layer, the collagen and elastin fibers are arranged to form a "two-phase" system, in which circumferentially aligned collagen fibers of high tensile strength and elastic modulus bear most of the stressing force at and above physiologic blood pressure. Elastin, which is distensible and has a low tensile strength, functions primarily as an elastic reservoir and distributes stress evenly throughout the wall and onto collagen fibers (Wolinsky and Glagov, 1967; Berry et al., 1972; Gerrity and Cliff, 1975). As the vessel wall matures, the SMCs go through multiple overlapping phenotypic transitions, characterized broadly by cellular proliferation, matrix production, and the assembly of an appropriate contractile apparatus within the cell cytoplasm. Defining the functional characteristics of developing vascular wall cells is difficult because of the transient nature of many marker proteins that are characteristic of the SMC phenotype. Expression of most known SMC differentiation markers in other cell types either during development or in response to injury is also a problem with cell-type determination. Further complicating our understanding of the vascular SMC is the cellular heterogeneity (Frid et al., 1994; Gittenberger-de Groot et al., 1999) and phenotypic plasticity (Schwartz and Mecham, 1995) observed during embryogenesis and vessel maturation.

In medium and large vessels, a major function of the SMC is to synthesize and organize the unique extracellular matrix (ECM) responsible for the mechanical properties of the wall. Unlike cells in the small muscular and resistance vessels, the SMCs of the elastic conducting vessels contribute little to the static mechanical properties of the wall. Hence, their ability to produce ECM can be considered to be their "differentiated" phenotype. Because the formation of a functional ECM must occur in an organized sequence, the "matrix phenotype" is changing throughout the entire period of vessel wall development.

In addition to providing the structural and mechanical properties required for vessel function, the ECM provides instructional signals that induce, define, and stabilize smooth muscle phenotypes. There are many examples of ECM molecules playing critical roles in the regulation of gene expression by interacting with specific matrix receptors on cells and by binding and storing growth factors that influence cellular function. This reciprocal instructive interaction between the cell and its ECM is important in directing the developmental transitions that occur in embryogenesis, postnatal development, and in response to injury. How vascular cells interpret these regulatory signals is a major area of research today.

This chapter will discuss the vascular SMC phenotype and ECM molecules made by vessel wall cells during vascular development, with the primary focus on the developing mouse aorta. Several excellent reviews have summarized our current understanding of SMC phenotypes based on expression of cytoskeletal and other marker proteins (Glukhova and Koteliansky, 1995; Owens, 1995; Hungerford et al., 1996; Owens et al., 2004). There are also numerous ultrastructural studies documenting the architecture of the developing vessel wall (Pease and Paule, 1960; Karrer, 1961; Paule, 1963; Haust et al., 1965; Albert, 1972; Gerrity and Cliff, 1975; Thyberg et al., 1979). Extensive information on the vascular smooth muscle cell and a still timely discussion of questions and issues driving research in vascular biology can be found in a monograph by Schwartz and Mecham (1995).

2. Expression profiling of the developing mouse aorta

To better understand the functional properties of vessel wall cells during differentiation and maturation, gene expression profiling of developing mouse aorta was done using oligonucleotide microarray technology (MU74Av2 chip from Affymetrix, Santa Clara, CA). The entire aorta was removed starting at the most distal aspect of the aortic arch up to, but not including, the common iliac vessels. The time points used in the study were embryonic day 12 (E12), E14, E16, E18, postnatal day 0 (P0) (representing the first 24 hours of life), P4, P7, P10, P14, P21, P30, P60, 5.5 months of life, and 6 months of life. Ten micrograms of total ribonucleic acid (RNA) from each sample of pooled aortae were used to create the target. Twenty-five micrograms of each complementary RNA were hybridized to the MU74Av2 chip (Affymetrix) and four biotinylated hybridization controls (BioB, BioC, BioD, and Cre) were included in each hybridization reaction to verify consistent hybridization efficiency.The image intensities of each scanned chip were analyzed by the GeneChip Analysis Suite software (Affymetrix) and were scaled to an average of 1,500 U. Probe sets were removed from the analysis if the difference between their maximum and minimum raw average difference values over the time series was <300. The normality of the chips themselves was determined by reassessing the CEL files produced by the

Microarray Suite software (Affymetrix) with DNA-Chip Analyzer (dChip) software package as described by Li and Wong (2001).

3. The vascular ECM

The array data identify several major patterns for matrix gene expression. The first and most prevalent begins around day 14, shortly after mesenchymal cells recruited to the vessel wall organize into layers that closely approximate the number that will be found in the mature tissue. This expression pattern consists of a major increase in matrix protein expression at embryonic day 14 followed by a steady rise through the first 7 to 14 days after. birth. This is followed by a decrease in expression over 2 to 3 months to low levels that persist in the adult (Fig. 1). Most of the structural matrix proteins follow this pattern. The second most prevalent pattern was one of consistent expression throughout the time series and was typical of basement membrane components, fibronectin, most integrins, and some matrix metalloproteinases. The third pattern consists of high expression levels in the embryonic/fetal period followed by decreased expression postnatally. The final and least populated pattern was low expression throughout development with an increase in the adult period.

Our expression data are in agreement with the appearance of structural matrix proteins in the vessel wall as assessed by ultrastructural studies (Nakamura, 1988). The electron micrographs in Fig. 2 compare the vessel wall of the developing mouse aorta at E12, E14, and E18. At day 12, there are few discernable collagen or elastin fibers in the extracellular space. By

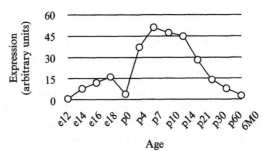

Fig. 1. Typical expression pattern for structural matrix proteins in developing mouse aorta. Expression of structural matrix proteins begins around E14, shortly after mesenchymal cells recruited to the vessel wall organize into layers that closely approximate the number that will be found in the mature tissue. This expression pattern consists of a major increase in matrix protein expression at E14 followed by a steady rise through the first 7 to 14 days after birth. This is followed by a decrease in expression over 2 to 3 months to low levels that persist in the adult. Many matrix proteins show a temporary decrease at day 0 while some show a temporary increase.

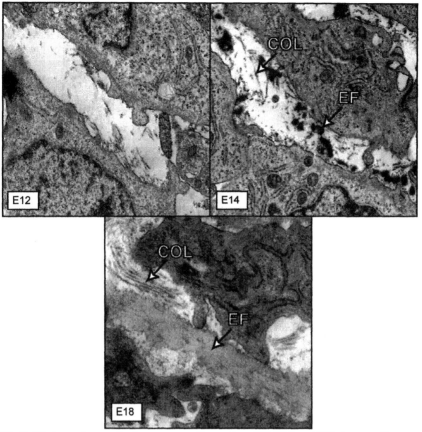

Fig. 2. Electron micrograph of mouse ascending aorta at E12, E14, and E18. Sparse collagen or elastic fibers (EF) are evident in the extracellular space between two smooth muscle cell layers at E12. At E14 there is a major increase in structural matrix protein expression, which is evident as small patches of EF and individual collagen fibers (COL). By E18, the small elastic bundles have coalesced into larger fibers that define the elastic lamellae. Bundles of mature collagen fibers are also evident. These micrographs illustrate the remarkable quantity of ECM that is deposited into the vessel wall in a relatively short period of development. From Kelleher et al. (2004), used with permission.

day 14, however, collagen fibers and small patches of elastin are beginning to form in the extracellular space. By day 18, the elastic lamellae and mature collagen fibers are clearly evident. These micrographs illustrate the tremendous rate of structural matrix protein synthesis that begins around E14. The following sections will summarize the expression patterns of several of the most abundant vascular ECM components. Further information on the expression patterns of ECM proteins can be found in Kelleher et al. (2004).

3.1. Collagens

Collagens are ubiquitous ECM proteins that impart a structural framework for tissues (Mecham, 1998). In all, 17 different collagens were identified in the developing mouse aorta by microarray analysis, with collagens I, III, IV, V, and VI having the highest expression levels. Present in lesser amounts were collagens VII, VIII, IX, X, XI, XIV, XV, XVIII, and XIX. Collagens II, XII, XIII, and XVII had low expression levels. Collagens type I, II, III, and V are fibril-forming collagens that assemble into striated fibers of varying diameter and are usually the most abundant collagens in tissues. Type VI, also a fibrillar collagen, forms a beaded filament. Collagens IV, VIII, and X are members of the network-forming collagen family and create "basket weave-like" structures through associations between their helical and non-helical domains. Type IV collagen is the major structural protein of basement membranes. Collagens IX, XIV, and XIX are FACIT collagens that attach to the surface of fibril-forming collagens but do not form fibers themselves. Collagen XIII is a collagen with a transmembrane domain that resides in adhesive structures of cells and has been implicated in cell adhesion. Collagens XV and XVIII are closely related non-fibrillar collagens that are associated with basement membranes. Type XV is thought to help anchor cells to the basement membrane and the C-terminal fragment of type XVIII collagen, called endostatin, is a potent inhibitor of angiogenesis and endothelial cell migration.

Based on the array data, the major fibrillar collagens in the aorta (types I, III, and VI) show an expression pattern typical of the matrix expression pattern described Fig. 1. A major increase in expression beginning at E14 was followed by high expression through P10. Expression then decreased relatively rapidly over several weeks and continued to fall gradually into the adult period. Collagens XV and XVIII were the only collagens to show increased expression in adult animals.

3.2. Elastic fiber proteins

The elastic fiber is a multicomponent structure whose main protein is elastin (Mecham and Davis, 1994). In contrast to the genetic diversity evident in the collagen gene family, elastin is encoded by only one gene. In the extracellular space, lysine residues within tropoelastin are specifically modified to form covalent crosslinks by one or more lysyl oxidase, a multi-gene family consisting of five members. These enzymes are also responsible for crosslinking collagen molecules. Similar to what was found for type I collagen, elastin expression in the mouse aorta begins in the embryonic period and increases (with a dip at P0) until P14, then decreases rapidly to low levels in the adult. Expression of lysyl oxidase (LOX) showed a similar pattern to elastin and collagen whereas expression of lysyl oxidase-like

(LOX-L1) protein was low and unchanging. This is consistent with data from animal studies showing that animals deficient in LOX die of ruptured arterial aneurysms shortly after birth (Hornstra et al., 2003) whereas LOX-L1 deficient mice have normal vasculature

The second major structural component of elastic fibers is the microfibril. The structural building blocks of these long linear fibers are the fibrillin molecules (Handford et al., 2000). Several microfibril-associated proteins have also been described, but their importance to microfibril structure and functions is not yet clear (Kielty et al., 2002). The human genome contains three fibrillins, but fibrillin-3 appears to have been inactivated in the mouse genome due to chromosome rearrangements (Corson et al., 2004). In the developing mouse aorta, fibrillin-1 has an expression pattern similar to elastin, except peak expression occurs at P0. Expression of fibrillin-2, on the other hand, is markedly different from fibrillin-1 and elastin. Fibrillin-2 expression is highest in the early embryonic period then decreases almost linearly throughout maturation. This suggests that fibrillin-1 is the major fibrillin in aortic microfibrils with fibrillin-2 playing a minor role.

3.3. Fibronectin

Numerous studies have illustrated the importance of fibronectin to vessel formation, particularly in the early embryonic periods (Glukhova and Koteliansky, 1995; Francis et al., 2002). In the embryonic chicken, the early vasculature is rich in fibronectin but relatively devoid of basement membrane or structural matrix proteins (Risau and Flamme, 1995). Our expression profile data suggest the same is true in the developing mouse aorta where fibronectin expression is high and relatively constant throughout development and into the adult period. Fibronectin plays an important role in facilitating cell movement during early migratory events in cell wall formation. Its continued expression after vessel maturation suggests an ongoing role in vessel homeostasis.

3.4. Basement membrane components

Along with type IV collagen, perlecan, and entactin/nidogen, the laminins are the major structural elements of the basement membrane (also referred to as basal lamina) (Ekblom and Timpl, 1996). The laminins are modular proteins with domains that interact with both cells and ECM and constitute a family of glycoproteins that affect cell proliferation, migration, and differentiation. Fifteen different laminins have been identified, each containing an α, β, and γ chain. In the late embryonic and fetal periods, our mouse expression data suggest that the predominant aortic laminin chains are $\alpha4$, $\beta1$, and $\gamma1$, which correspond to laminin-8. Laminin $\alpha1$ expression is highest at our first time point (E12) then drops slightly between E12 to E14, consistent with previous observations that laminin-1 ($\alpha1\beta1\gamma1$) is expressed earliest during embryogenesis (Smyth et al., 1999; Li et al., 2003). In the postnatal period, there are marked increases in expression of the laminin $\alpha5$ and $\beta2$ chains. Thus, after birth, laminin-9 and -11, as well as laminin-8 contribute to basement membrane structure. These findings are in agreement with immunolocalization studies showing that laminin-8 is widely distributed in vascular tissues (Petajaniemi et al., 2002) and that laminin $\alpha5$ appears during the postnatal period (Sorokin et al., 1997). Laminin $\alpha3$ shows constant, but low, expression over the entire data series, whereas expression of laminin $\alpha2$ is intermediate between the low and high expressers, suggesting the presence of some laminin-2 or -4.

Type IV collagen and perlecan show constant expression through vascular development and maturity. One interesting change evident in the array data, however, is a dramatic increase in expression of the collagen IV $\alpha5$ chain and a decrease in the $\alpha3$ chain in the postnatal period.

Entactin, also referred to as nidogen, is a highly conserved protein in the vertebrate basement membrane that bridges laminin and type IV collagen networks (Chung et al., 1977; Carlin et al., 1981; Timpl et al., 1983). There are two members of the entactin family in mammals (entactin-1 and -2, or nidogen-1 and -2). In the developing mouse aorta, entactin-1 shows a sharp increase in expression at E18 and remains high until P7, when it drops to levels that persist through the adult stages. Expression of entactin-2, in contrast, shows a sharp increase at E14 then decreases gradually until P21 when stable expression is obtained at levels lower than those observed during the embryonic period.

3.5. Proteoglycans

Versican is the largest of the vascular glycoproteins and is known to influence cell adhesion, proliferation, and migration (Iozzo and Murdoch, 1996; Wight, 2002). Versican has been localized to both the medial and endothelial layers of human aortas by *in situ* hybridization and Western blotting (Yao et al., 1994). In the developing mouse aorta, versican messenger RNA expression from E12 to birth trends downward slightly from moderate levels, rises sharply to peak at P0, and then falls sharply by P4 to relatively low levels that are maintained through adulthood. It is hypothesized that versican and the hyaluronan matrix affect cell adhesion and shape and by this mechanism affect migration and proliferation of vascular SMCs.

The small leucine-rich proteoglycans (SLRPs) are a family of secreted proteoglycans that can bind ECM molecules including collagen, fibronectin, and fibrillin-containing microfibrils (Iozzo and Murdoch, 1996; Trask et al., 2000). The SLRP family includes decorin, biglycan, fibromodulin, osteoglycin, and lumican. There is evidence that these SLRPs regulate cell proliferation or differentiation through regulation of growth factor activity

(Riquelme et al., 2001). Biglycan localizes to all layers of the human aorta by immunohistochemical staining, whereas decorin is found only in the adventitia (Theocharis and Karamanos, 2002). The expression profile of decorin in the mouse aorta closely parallels that of type I collagen in embryonic time points but peaks at P0. Decorin expression decreases somewhat in the postnatal time points but remains constitutively expressed at a moderate level. Biglycan, on the other hand, shows increasing expression over the embryonic time points to peak at P7 at levels similar to $coll\alpha1$. Expression levels fall over the first postnatal month, but rise again as the animal enters adulthood (P5.5 to P6 months). Expression of lumican, a class II SLRP, occurs from E12 to P6 months at low to moderate levels. There is a small peak in expression levels at P0.

4. The functional context of SMC differentiation

A major challenge in characterizing the differentiation stages of the vascular SMC is defining cell type-specific markers that characterize the differentiated phenotype. The SMCs are highly plastic cells that are capable of alterations in their phenotype in response to change in local environmental cues. Thus, to better understand the repertoire of genes expressed by vascular wall cells at each stage of development, hierarchical clustering was used to identify the four major groups (i.e., expression patterns) existing in the dataset (Fig. 3). With the use of the Gene Ontology Database (www.geneontology.org), the ontology annotations for each probe set were collected and organized into groups based on three ontologies: Molecular Function, Biological Process, and Cellular Components. The definitions of the ontologies are as follows: Molecular Function—the tasks performed by individual gene products; Biological Process—broad biological goals that are accomplished by ordered assemblies of molecular functions; Cellular Component—subcellular structures, locations, and macromolecular complexes. We report here the 10 most significant annotations within each group as well as genes of consequence to our interest in ECM and vascular development. The genes include ECM genes, matrix metalloproteinases (MMPs) and related genes, growth factors, signal transduction genes, genes identified as SMC markers and endothelial cell markers.

4.1. Group A: Genes upregulated in embryogenesis

Group A is the largest functional group within the dataset with 4,856 probe sets. The general pattern of expression demonstrates genes that are above baseline expression during the embryonic phases of development and go below baseline during the adult period. In the ontology of Biological

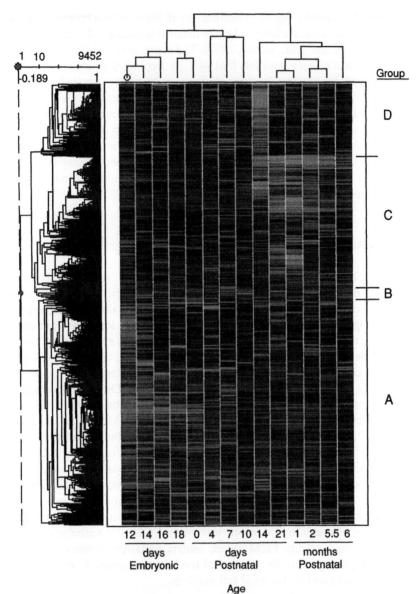

Fig. 3. Hierarchical clustering of 9,452 probe sets that changed by at least 300 points in their average difference value over the course of the time series. Numbers at the bottom indicate mouse developmental age and letters on the right side indicate functional groups derived from hierarchical clustering of the probe sets. The dendrigram at the top shows that the time points cluster into two large groups. The first cluster includes the "early" time points E12 to P4 and the second cluster consists of the "late" time points P7 to P6 months. Within the "early" cluster, the time points within the respective sub-clusters of E12 and E14 and e16 and e18 show the highest degree of similarity. P0 shows the least amount of similarity to any time point within the "early" cluster. In the "late" cluster, P60 and P6 months are the two time points that show the highest amount of similarity. Time points P7, P10, and P14 demonstrate little likeness to other time points within the series. (See Color Insert.)

Processes (**GO annotation**: # of probe sets), the largest groups were genes involved in the regulation of transcription (**6355**: 330), development (**7275**: 127), transport (**6810**: 113), protein amino acid phosphorylation (**6468**: 112), and cell adhesion (**7155**: 89) (Fig. 4A). The latter five categories within the ontology focus upon signal transduction and cell signaling, cell cycle, and cell growth. Genes of interest in this group are listed in Table 1.

Several different families of transcription factors were represented within group A. The Sox (Sry-related HMG box) genes had 14 probe sets present, representing 11 members of the Sox gene family (Sox-1, -2, -3, -4, -6, -7, -10, -11, -15, -17). Interestingly, only Sox-13 has been described to be expressed in murine embryonic arterial development (Roose et al., 1998). Regulation of vascular genes may be a new role for other members of the Sox gene family. Hox genes of the homeobox gene family are also prominent within

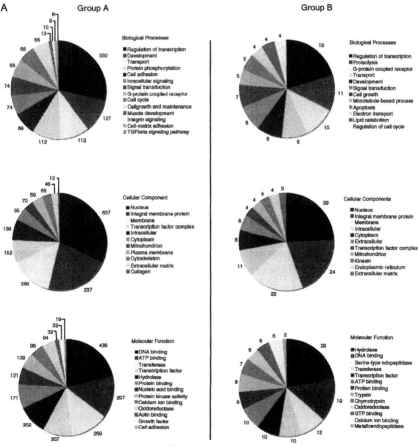

Fig. 4. (*continued*)

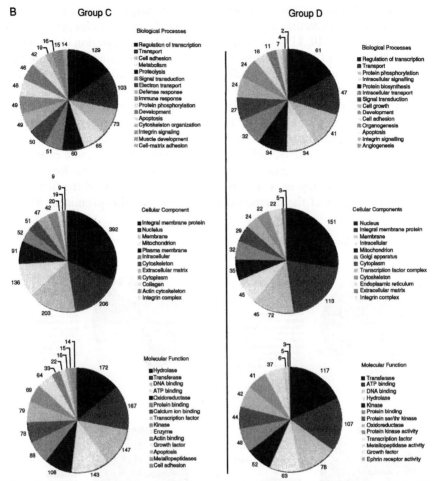

Fig. 4. Functional associations of gene clusters associated with groups A and B (panel A) and C and D (panel B). The Gene Ontology Database (www.geneontology.org) was used to organize the ontology annotations for each group into three ontologies: Molecular Function, Biological Process, and Cellular Components. The numbers indicate how many probe sets fall into each subcategory. (See Color Insert.)

the group (22 probe sets representing 21 genes). Many Hox genes have been described to be present during vascular development (reviewed in Gorski and Walsh, 2000). Other major families of transcription factors include the zinc finger protein genes (13 probe sets for 10 genes), the forkhead box gene family (9 probe sets for 9 genes), and hypoxia-inducible factor 1α(HIF-1α), a transcription factor related to angiogenesis (Forsythe et al., 1996).

Genes for SMAD proteins 1, 2, 4, 5, and 7, which function as downstream signal transduction molecules for the transforming growth factor (TGF)-ß

Table 1
Genes of interest: Group A

Hypertension Genes

102737_at	U35233	endothelin 1
103550_at	U32329	endothelin receptor type B
100766_at	U32330	endothelin 3
103593_at	K02781	natriuretic peptide precursor type A
104184_at	D16497	natriuretic peptide precursor type B
160900_at	AI847897	42 kD cGMP-dependent protein kinase anchoring protein
162359_r_at	AV220336	42 kD cGMP-dependent protein kinase anchoring protein
94934_at	U04828	angiotensin II receptor, type 2
96154_at	AA600645	renin binding protein
98480_s_at	M32352	renin 1 structural
160986_r_at	AA690434	RIKEN cDNA 2010305L05 gene (angiotensin I-converting enzyme ACE 2)
98727_at	L41933	guanylate cyclase 2e

ECM Elastic Fiber Genes

101090_at	L29454	fibrillin-1
103623_at	L39790	fibrillin-2
101095_at	L23769	microfibrillar-associated protein 2 (MAGP-1)
160095_at	D10837	lysyl oxidase
161177_f_at	AV232133	lysyl oxidase
100928_at	X75285	fibulin 2
100019_at	D45889	chondroitin sulfate proteoglycan 2 (Versican)
160133_at	L40459	latent transforming growth factor beta binding protein 3
100308_at	X66976	procollagen, type VIII, alpha 1

Notch Signal Transduction

162204_r_at	AV374287	Notch gene homolog 1 (*Drosophila*)
92652_at	AF030001	advanced glycosylation end product–specific receptor (Notch 4)
92956_at	X74760	Notch gene homolog 3 (*Drosophila*)
97497_at	Z11886	Notch gene homolog 1 (*Drosophila*)
101975_at	Z12171	delta-like 1 homolog (*Drosophila*)
92931_at	X80903	delta-like 1 (*Drosophila*)

Smooth Muscle Cell Marker Genes

101028_i_at	M15501	actin, alpha, cardiac
101029_f_at	M15501	actin, alpha, cardiac
100879_at	AF093775	actinin alpha 3
162164_f_at	AV359510	actinin alpha 3
94004_at	Z19543	calponin 2
160150_f_at	AW125626	calponin 3, acidic
92541_at	X12973	myosin light chain, alkali, fast skeletal muscle
92881_at	U77943	myosin light chain, phosphorylatable, fast skeletal muscle
100403_at	AA839903	myosin light chain, regulatory A
160487_at	M19436	myosin light chain, alkali, cardiac atria
98616_f_at	AJ223362	myosin heavy chain, cardiac muscle, fetal
101071_at	M76599	myosin heavy chain, cardiac muscle, adult
160185_at	AB031291	transgelin 3

Insulin Growth Factor

100566_at	L12447	insulin-like growth factor binding protein 5
103896_f_at	X81579	insulin-like growth factor binding protein 1

(*continued*)

Table 1 (*continued*)

94222_at	AI838737	insulin-like growth factor binding protein 4
95082_at	AI842277	insulin-like growth factor binding protein 3
95083_at	X81581	insulin-like growth factor binding protein 3
95117_at	U04710	insulin-like growth factor 2 receptor
95545_at	X04480	insulin-like growth factor 1
95546_g_at	X04480	insulin-like growth factor 1
98623_g_at	X71922	insulin-like growth factor 2
98627_at	X81580	insulin-like growth factor binding protein 2

TGF-β and TGF-β Associated Genes

93300_at	X57413	transforming growth factor, beta 2
101918_at	AJ009862	transforming growth factor, beta 1
101177_at	Z31663	activin A receptor, type 1B
93903_at	M84120	activin receptor 2B
98841_at	M65287	activin receptor 2A
92701_at	AA518586	bone morphogenetic protein 1
93243_at	X56906	bone morphogenetic protein 7
93456_r_at	L47480	bone morphogenetic protein 4

Cadherins

100006_at	D21253	cadherin 11
100409_at	X06340	cadherin 3
101650_at	D86917	protocadherin α6
101769_at	AB008178	protocadherin α14
101771_r_at	AB008180	protocadherin α13
101773_r_at	AB008183	protocadherin α10
102852_at	M31131	cadherin 2
160610_at	D86916	protocadherin α4
162097_r_at	AV322859	cadherin 3
92364_at	AB028499	cadherin EGF LAG seven-pass G-type receptor 2
94641_at	U69137	cadherin 10
97756_s_at	AB008181	protocadherin alpha 5
98140_at	X60961	cadherin 1
104083_at	AI853217	expressed sequence AA408225 (Cadherin 5)

Catenins

160430_at	M90364	catenin β
92227_s_at	D25281	catenin α2
92228_at	D25282	catenin α2
93364_at	X59990	catenin α1
98152_at	AW122407	catenin src
99856_r_at	AI852919	catenin (cadherin-associated protein), delta 2 (neural plakophilin-related arm-repeat protein)

Collagens

100308_at	X66976	procollagen, type VIII, alpha 1
100481_at	D38162	procollagen, type XI, alpha 1
100897_f_at	AF100956	procollagen, type XI, alpha 2
101039_at	X04647	procollagen, type IV, alpha 2
101080_at	AB009993	procollagen, type V, alpha 1
101093_at	M15832	procollagen, type IV, alpha 1
101110_at	AF064749	procollagen, type VI, alpha 3
101130_at	X58251	procollagen, type I, alpha 2

Table 1 *(continued)*

101881_g_at	L22545	procollagen, type XVIII, alpha 1
101882_s_at	U03715	procollagen, type XVIII, alpha 1
102070_at	AW212495	procollagen, type IX, alpha 3
102261_f_at	U30292	procollagen, type XIII, alpha 1
102990_at	AA655199	procollagen, type III, alpha 1
102991_s_at	AF100956	procollagen, type XI, alpha 2
103371_at	AF100956	procollagen, type XI, alpha 2

Connexins

104016_at	M91236	gap junction membrane channel protein beta 5
104635_r_at	X63100	gap junction membrane channel protein alpha 7
101778_at	X61675	gap junction membrane channel protein alpha 5

Ephrins and Ephrin Receptors

100289_at	U92885	ephrin A3
102869_at	U14941	ephrin A2
103007_at	U90662	ephrin A1
103692_at	AW121468	ephrin B3
160857_at	U30244	ephrin B2
94160_at	AF025288	ephrin B3
98407_at	U07602	ephrin B1
102682_at	U72207	Eph receptor A8
104673_at	X65138	Eph receptor A4
92906_at	X79083	Eph receptor A7
93469_at	Z49086	Eph receptor B3
95298_at	M68513	Eph receptor A3

FGF and FGF Receptors

162253_i_at	AV336395	fibroblast growth factor receptor 3
93090_at	M23362	fibroblast growth factor receptor 2
102574_at	U66203	fibroblast growth factor 11
161273_f_at	AV322895	fibroblast growth factor inducible 14
95316_at	AB004639	fibroblast growth factor 18
97124_at	U42384	fibroblast growth factor inducible 15
97421_at	U42385	fibroblast growth factor inducible 16
98826_at	X14849	fibroblast growth factor 4
99176_at	AI843393	fibroblast growth factor (acidic) intracellular binding protein

Fibronectin

92852_at	M18194	fibronectin 1

Endothelial Cell Markers

98452_at	D88689	FMS-like tyrosine kinase 1
98453_at	D88690	FMS-like tyrosine kinase 1
95295_s_at	X59398	FMS-like tyrosine kinase 3
95296_r_at	M64689	FMS-like tyrosine kinase 3
104265_at	X70842	kinase insert domain protein receptor (Flk-1/Kdr)
95016_at	D50086	neuropilin
161184_f_at	AV235418	tyrosine kinase receptor 1 (Tie-1)
99936_at	X80764	tyrosine kinase receptor 1 (Tie-1)
102720_at	X71426	endothelial-specific receptor tyrosine kinase (Tie-2)
92210_at	AF004326	angiopoietin 2
96118_at	AF110520	H2-K region expressed gene 2

(continued)

Table 1 (*continued*)

Integrins

100751_at	AF011379	a disintegrin and metalloprotease domain 10
100906_at	M68903	integrin beta 7
100991_at	AI852849	integrin beta 1 binding protein 1
102655_at	X53176	integrin alpha 4
102832_at	U22056	a disintegrin and metalloproteinase domain 1a (fertilin alpha)
103554_at	AA726223	a disintegrin and metalloproteinase domain 19 (meltrin beta)
161196_r_at	AV254135	a disintegrin and metalloprotease domain 3 (cyritestin)
162388_r_at	AV256143	a disintegrin and metalloprotease domain 5
92414_at	D50411	a disintegrin and metalloproteinase domain 12 (meltrin alpha)
94171_at	AF051367	integrin beta 2-like
95292_at	AA189389	integrin alpha 4
95511_at	X69902	integrin alpha 6
98834_at	X75427	integrin alpha 2

Laminins

100428_at	U43327	laminin, gamma 2
101948_at	X05212	laminin B1 subunit 1
103729_at	M36775	laminin, alpha 1
162263_f_at	AV357656	laminin B1 subunit 1
96016_at	AW045665	laminin receptor 1 (67kD, ribosomal protein SA)
97790_s_at	X84014	laminin, alpha 3

Myocyte Enhancement Factor

93417_at	D50311	myocyte enhancer factor 2B

MMP

100016_at	Z12604	matrix metalloproteinase 11
160118_at	AF022432	matrix metalloproteinase 14 (membrane-inserted)
161509_at	AV145762	matrix metalloproteinase 2
162318_r_at	AV069212	matrix metalloproteinase 7
98280_at	AB021228	matrix metalloproteinase 16
161219_r_at	AV250023	Cluster Incl AV250023:AV250023 (MMP 13)

PDGF and PDGF receptors

95079_at	M57683	platelet-derived growth factor receptor, alpha polypeptide
160332_at	AI840738	platelet-derived growth factor receptor, alpha polypeptide

TIMP

101464_at	V00755	tissue inhibitor of metalloproteinase

VEGF

94712_at	U73620	vascular endothelial growth factor C

WNT/Frizzled Signal Transduction

101142_at	U43319	frizzled homolog 6 (*Drosophila*)
101143_at	U43320	frizzled homolog 7 (*Drosophila*)
102044_at	AF100777	WNT1 inducible signaling pathway protein 1
104672_at	U68058	frizzled-related protein
161040_at	AF054623	frizzled homolog 1, (*Drosophila*)
162011_f_at	AV246963	Wnt1 responsive Cdc42 homolog
92469_at	AF117709	secreted frizzled-related sequence protein 4

Quaking

95096_at	AI846695	quaking

family of proteins, were represented within the group. SMAD 1 and SMAD 5 have been implicated in normal blood vessel formation in mutagenic mice. SMAD 1 $-/-$ mice fail to form a normal chorion-allantoic circulation (Lechleider et al., 2001; Tremblay et al., 2001). SMAD 5 $-/-$ mice suffer a gestational death between E10.5 and E11.5 due to defects in angiogenesis that include a decrease in the number of vascular SMCs (Yang et al., 1999). Also present in group A were genes for MyoD 1, myogenin, myogenic factor 6, and myocyte enhancer factor (MEF) 2B. Collectively, these transcription factors play a regulatory role in myoblast differentiation but are not thought to be specific for vascular SMC differentiation (reviewed in Owens, 1995; Owens et al., 2004).

Within the category of Development (**7275**) are many of the signal transduction pathways that are ubiquitous in mammalian development. The WNT/Frizzled signal transduction pathway represents two of the most prominent gene families within the group A dataset. Seven WNT genes (WNT: 1, 3, 5B, 7B, 10a, 10b, 11), 4 frizzled genes (frizzled 4, 6, 7, and frizzled-related protein Fzd b), and dishevelled 2 are present. Other signaling pathway genes within group A are notch-1 and notch-3 and ephrin B1, B2, and B3. Both notch and ephrin have been shown to play important roles in vascular development (reviewed in Rossant and Lorraine, 2002).

In the Transport category, two gene families are prominent. The solute carrier family (26 probe sets representing 15 genes) is a superfamily of genes that function as carrier proteins for cellular transport. The other major gene family defined in the transport category is the adenosine triphosphate (ATP)-binding cassette family (9 genes).

The Protein amino acid phosphorylation (**6468**) category has a multitude of genes that encode for receptors or molecules that play a role in intracellular signaling. Of the receptor genes, many have been directly implicated in early vascular development and in endothelial cell differentiation. The genes for the vascular endothelial growth factor (VEGF) receptors, FMS-like tyrosine kinase 1 (Flt-1), FMS-like tyrosine kinase 3 (Flt-3), and fetal liver kinase/kinase insert domain protein receptor (Flk-1/Kdr), and the angiopoietin receptors, tyrosine kinase receptor 1 (Tie-1) and endothelial-specific receptor tyrosine kinase (Tie-2), are present within this category for group A. Their roles in vascular development have been well described (reviewed in Nguyen and D'Amore, 2001). Genes for fibroblast growth factor receptor 2 and 3 and the genes for the ephrin receptors A3, A4, A7, A8, and B3 are all present. Also represented are the receptors for the TGF-β family activin receptor 2a, activin receptor 2b, and bone morphogenic protein receptor 2. The major intracellular signaling molecules that are within this category are the mitogen-activated protein kinase (MAPK) gene family (16 probe sets representing 13 genes). Members of the MAPK family are involved in the intracellular signaling in all cell types.

Many of the genes identified within the Cell adhesion (7155) category are well described in their involvement in vascular development. They include neuropilin, vascular cell adhesion molecule, thrombospondin 1 and 2, integrin family members (α2, α4, α6, ß2-like, ß7), and fibronectin (reviewed in Nguyen and D'Amore, 2001). The genes for the cell-to-cell adhesion molecules, the cadherins (1, 2, 3, 10, 11), and the related genes, protocadherins (α4, α5, α6, α10, α13) and catenins (α1, α2, src, delta2,ß), are also prominent within this category. Genes for the basement membrane protein family laminin (α1, α3, ß1 subunit 1, and α2) and genes encoding the extracellular matrix proteins versican (chondroitin sulfate proteoglycan 2), neurocan (chondroitin sulfate proteoglycan 3), nidogen-1 and -2 are identified with this category for Group A genes. The ECM family with the largest representation of genes for this category is collagen (9 genes: IVα1, VIIα1, VIIIα1, IXα1, IXα2, IXα3, XIα1, XIIα1, XIIIα1).

Examination of the cell-matrix adhesion (7160: 9) category revealed the aforementioned integrins and nidogens. As SMCs are the major component of the vascular wall, there are 13 genes of the muscle development (7517: 13) category in group A. The genes encoding the isoforms of the cytoskeletal proteins myosin heavy chain (polypeptide 6, cardiac muscle, adult [Myh6]; polypeptide 8, skeletal muscle, perinatal [Myh8]; polypeptide 3, skeletal muscle, embryonic [Myh3]), myosin light chain (2a, alkali, cardiac atria [Myla]; alkali, fast skeletal muscle [Mylf]; phophorphorylatable, fast skeletal muscle [Mylpf]), myosin-binding protein H [Mybph], troponin C (fast skeletal [Tncs], cardiac/slow skeletal [Tncc], and troponin T2 [Tnnt2]. Other cytoskeletal proteins include α-actin (cardiac isoform), ß-actin, and the catenins (α1, α2, ß, delta-2). The gene for angiotensin II receptor type 2 is present in the categories for integral membrane proteins (16021) and membrane (16020). For the integral plasma membrane protein (5887) category the most prevalent genes are related to potassium channels (14 genes) and the solute carrier gene family (12 genes). Genes for MMP-2, -7, -11, -14, -16 and tissue inhibitor of metalloproteinase-1 are present in group A for the extracellular matrix (5578) category.

The overall pattern of expression within group A is that of genes that go from an above baseline level of expression during the embryonic phases of our time series to below baseline during the latter phases. The primary emphasis of the genes in group A is that of gene regulation. This is emphasized by the largest categories within the ontologies being related to the regulation of transcription (6355), being physically present in the nucleus (5634) or transcription factor complexes (5667), and by the emphasis on deoxyribonucleic acid (DNA) binding (3677). A large number of genes are also devoted to developing the machinery for inter-cellular and intra-cellular signaling. This is represented by multiple categories, including transport (6810), protein amino acid phosphorylation (6468), intracellular signaling cascade (7242), integral membrane protein (16021), ATP binding (5524), and protein kinase activity (**4672**). Many signaling pathways essential to normal blood vessel formation are a part of Group A.

While the genes of group A in the categories related to muscle development [muscle development (**7517**), myogenesis (**7519**), cytoskeleton (**5856**)] did not represent genes specific to the vascular SMC, they may represent the genes that can be found in vascular SMC precursors; thus, the genes of group A may play an important role in defining the future SMCs of the aorta. Other significant groups of molecular categories within group A were composed of members of the ECM (**5578**), collagens (**5581**), and microfibrils (**1527**). Their presence in group A represents the early part of the elaboration of the ECM and components of the elastic fiber which are essential to normal blood vessel formation. In summary, group A genes compose many of the building blocks that will allow for differentiation and development of the various components of the aorta.

4.2. Group B: Genes with constant expression except for variation at P7 to P14

Group B is made up of a constellation of 268 genes. The global expression profile reflects genes that remain near their baseline expression value for a majority of the time points examined. The greatest variance in the expression profiles occurs in between days 7 to 14 with some of the genes peaking at day 10. Gene ontology analysis of group B shows some differences with that of group A.

The genes of group B demonstrate a variety of themes (see Table 2 for a listing of genes of interest). Two of the 10 largest categories involve gene regulation. The largest category in group B is regulation of transcription, DNA-dependent (**6355**: 19), which includes the genes for notch-2, cut-like 1 (Cutl1), forkheadbox C2, and homeo-box msh-like 1 (Msx1) (Fig. 4A). Proteolysis is another major category in group B, and it includes genes for elastase-2, MMP-12, cathepsin-B, and adam-17—all, genes capable of degrading extracellular molecules. The other major gene family within this category, kallikreins (kallikrein-5, -6, and -9), has been linked to vasodilation and hypotension in mutant mouse models of overexpression (Wang et al., 1994; Ma et al., 1995; Garbers and Dubois, 1999). Other major categories contain genes responsible for cell signaling (G protein–coupled receptor protein signaling pathway). Tissue plasminogen activator (Plat) is also present in group B. It has been shown to be present in endothelial cells, related to endothelial cell migration, and implicated as possibly playing a role in angiogenesis (reviewed in Pepper, 2001).

Due to their global pattern of expression and the descriptions provided by the three ontologies, the genes of group B appear to be serving as genes

Table 2
Genes of interest: Group B

Notch-Signal Transduction		
101342_at	D32210	notch gene homolog 2 (*Drosophila*)
Smooth Muscle Cell Marker		
101072_at	AW124250	myosin heavy chain, cardiac muscle, adult
FGF and FGF receptors		
100884_at	U04204	fibroblast growth factor regulated protein
Integrins		
160241_at	AB021709	a disintegrin and metalloproteinase domain 17
MMP		
95339_r_at	M82831	matrix metalloproteinase 12
WNT/Frizzled Signal Transduction		
93459_s_at	AW122897	frizzled homolog 4 (*Drosophila*)

functioning in the growth, differentiation, and maintenance of the developing aorta. Furthermore, the enzymatic categories from group B contain genes important to the remodeling of the ECM, which is an essential aspect of growth and differentiation. Expression of genes that play a role in anticoagulation (protein C and t-PA) are present, which is important in the maintenance of the primary role of the aorta as a conduit for vascular flow. In addition, the categories of electron transport (**6118**), lipid catabolism (**16042**), mitochondrion (**5739**), and calcium ion binding (**5509**) are consistent with the theme of maintenance. These categories contain genes focused upon cellular metabolism and energy production. Lastly, the presence of several kallikrein genes indicates the presence of genes that may function in the normal physiology of a blood vessel. Genes with near baseline levels of expression may be key to the normal existence of the aorta to avoid the abnormal physiology and pathologic states associated with under- or over-expression of any one of a plethora of genes.

4.3. Group C: Genes with low expression in the embryo but increasing expression in postnatal stages

Group C consists of 2,608 probe sets. The pattern of expression for this group of genes demonstrates a below baseline level of expression in the embryonic time points to above baseline expression in the postnatal stages. Gene ontology findings in this group reveal areas of genetic emphasis that are different from the other major functional groups.

As with the other groups in our dataset, genes that are involved in the regulation of gene transcription are the most numerous in the Biological

Processes ontology for Group C (see Table 3 for a listing of genes of interest in this group). Regulation of transcription, DNA-dependent (**6355**: 129) is the largest category (Fig. 4B). In this category are many genes for transcription factors that are critical to SMC differentiation and blood vessel maturation. Prominent within this category are four probe sets for genes within the MEF2 family of MADS box transcription factors. There are also three probe sets for MEF2C and one for MEF2A. MEF2C plays an important role in vascular maturation and SMC differentiation. MEF2C mutant mouse embryos possess endothelial cells that do not form normal vascular plexi. In addition, their myocytes fail to undergo SMC differentiation and the embryos do not have appropriate vascular remodeling (Lin et al., 1998). Another DNA regulatory molecule in group C related to vascular remodeling is the gene for the basic helix-loop-helix (bHLH)-PAS domain transcription factor endothelial PAS domain protein-1 (EPAS1). In EPAS1 mutant mice, blood vessel remodeling fails due to improper fusion of blood vessels and the inability to form large vascular structures (Peng et al., 2000).

The kruppel-like factor (KLF) gene family of transcription factors is prominent within this group, with five probe sets that represent four genes for kruppel-like factor-2 (lung kruppel-like factor/LKLF), kruppel-like factor-4 (gut) (KLF4), kruppel-like factor-9 (KLF9/BTEB1), and kruppel-like factor-13 (KLF13/BTEB3). The LKLF has been shown to be critical to blood vessel remodeling and stabilization (Kuo et al., 1997). The KLF4 and KLF13/BTEB3 have been implicated in SMC differentiation through the demonstration of their ability to bind to different sites within the SM22-α promoter and to mediate activation of the promoter (King et al., 2003; Martin et al., 2003). The KLF9/BTEB-1 does not have a demonstrated role in vascular development.

TGF-ß inducible early growth response (Tieg) gene is present within group C. It has been shown to play a role in the augmentation of TGF-β/SMAD signaling (Johnsen et al., 2002). Also present is early growth response-1 (Egr-1) gene. Tissue culture models suggest that Egr-1 may be linked to the activation of SMC proliferation and the mediation of angiotensin II signaling of SMC proliferation (Lin et al., 1998; Santiago et al., 1999).

Peroxisome proliferator activated receptor (PPAR)-α and γ are present within the regulation of transcription, DNA-dependent (**6355**) category for group C. The PPAR-γ is a nuclear receptor that is related to smooth muscle cell differentiation by the ability of its ligands to induce the expression of SMC α-actin and SMC myosin heavy chain (Abe et al., 2003). PPAR-γ may play a role in maintaining the differentiated state of an SMC by its demonstrated role of inhibiting SMC migration and proliferation in tissue culture and murine injury models (Xin et al., 1999; Goetze et al., 1999, 2001; Abe et al., 2003).

There are other transcription factors present in group C that play a role in the inhibition of SMC proliferation and in the maintenance of a

Table 3
Genes of interest: Group C

Hypertension-Related Genes

92532_at	D49730	arginine vasopressin receptor 1A
96585_at	AJ006691	arginine vasopressin receptor 2
94167_at	U53142	nitric oxide synthase 3, endothelial cell
103038_at	L36860	guanylate cyclase activator 1a (retina)
93954_at	AF020339	guanylate cyclase 1, soluble, beta 3
160927_at	J04946	angiotensin-converting enzyme
161224_f_at	AV258262	angiotensin-converting enzyme
95355_at	AA407794	DNA segment, Chr 4, Wayne State University 124, expressed (angiotensin II type I receptor–associated protein)

Elastic Fiber Genes

99517_at	AA832724	microfibrillar associated protein 5
99518_at	AW121179	microfibrillar associated protein 5
92207_at	U08210	elastin
92836_at	AA919594	elastin
103850_at	U79144	lysyl oxidase-like
94307_at	X70854	fibulin 1
93534_at	X53929	decorin
96049_at	X53928	biglycan
162347_f_at	AV166064	biglycan
103209_at	AF022889	latent TGF-β binding protein 1
92335_at	AF004874	latent TGF-β binding protein 2
97347_at	AA838868	RIKEN cDNA 2310046A13 gene (LTBP-4)

Collagens

101110_at	AF064749	procollagen, type VI, alpha 3
162459_f_at	AV010209	procollagen, type VI, alpha 1
95493_at	X66405	procollagen, type VI, alpha 1
93517_at	Z18272	procollagen, type VI, alpha 2
101881_g_at	L22545	procollagen, type XVIII, alpha 1
101882_s_at	U03715	procollagen, type XVIII, alpha 1
103709_at	AA763466	procollagen, type I, alpha 1
103828_at	AF100956	procollagen, type XI, alpha 2
104483_at	L12215	procollagen, type IX, alpha 1
160594_at	L08407	procollagen, type XVII, alpha 1
161156_r_at	AV230631	procollagen, type I, alpha 2
161984_f_at	AV234303	procollagen, type III, alpha 1
162459_f_at	AV010209	procollagen, type VI, alpha 1
162483_f_at	AV112006	procollagen, type XV
92313_at	AI844066	procollagen, type XII, alpha 1
92314_at	U25652	procollagen, type XII, alpha 1
92567_at	L02918	procollagen, type V, alpha 2
93220_at	Z35168	procollagen, type IV, alpha 5
93383_at	U32107	procollagen, type VII, alpha 1
93517_at	Z18272	procollagen, type VI, alpha 2
94305_at	U03419	procollagen, type I, alpha 1
95158_at	AF100956	procollagen, type XI, alpha 2
95410_s_at	M63709	procollagen, type II, alpha 1
95493_at	X66405	procollagen, type VI, alpha 1
98027_at	Z22923	procollagen, type IX, alpha 2

Table 3 (*continued*)

98331_at	X52046	procollagen, type III, alpha 1
98333_at	AF100956	procollagen, type XI, alpha 2
99476_at	AJ131395	procollagen, type XIV, alpha 1

Smooth Muscle Cell Marker

93100_at	X13297	actin, alpha 2, smooth muscle, aorta
100381_at	M12347	actin, alpha 1, skeletal muscle
96343_at	AI836968	alpha actinin 4
104578_f_at	AI195392	RIKEN cDNA 3110023F10 gene (alpha actinin 1)
104579_r_at	AI195392	RIKEN cDNA 3110023F10 gene (alpha actinin 1)
92280_at	AA867778	RIKEN cDNA 3110023F10 gene (alpha actinin 1)
99942_s_at	U28932	calponin 1
93050_at	M91602	myosin light chain, phosphorylatable, cardiac ventricles
97990_at	D85923	myosin heavy chain 11, smooth muscle
98559_at	AJ010305	smoothelin
93541_at	Z68618	transgelin
160162_at	AI852545	transgelin 2
93266_at	U04541	tropomyosin 3, gamma
100605_at	M81086	tropomyosin 2, beta
160532_at	M22479	tropomyosin 1, alpha
94964_at	L18880	vinculin
94963_at	AI462105	expressed sequence AI462105 (Vinculin)

Insulin-Like Growth Factor and Receptors

102224_at	AF056187	insulin-like growth factor I receptor
103904_at	X81584	insulin-like growth factor binding protein 6
160527_at	AB012886	insulin-like growth factor binding protein 7

TGF-β and TGF-β receptors

102751_at	M32745	transforming growth factor, beta 3
100448_at	Z31664	activin A receptor, type II-like 1
100449_g_at	Z31664	activin A receptor, type II-like 1
100450_r_at	L48015	activin A receptor, type II-like 1
92372_at	X80992	bone morphogenetic protein 6
93455_s_at	X56848	bone morphogenetic protein 4
161028_at	AI850533	bone morphogenetic protein 6
92767_at	D16250	bone morphogenetic protein receptor, type1A
97725_at	Z23143	bone morphogenetic protein receptor, type1B
100134_at	X77952	endoglin

Cadherins

101730_at	D82029	cadherin 6
101772_r_at	AB008182	protocadherin alpha 11
101796_at	AB011255	protocadherin 7
102280_at	AB006758	protocadherin 7
104743_at	AB022100	cadherin 13

Connexins

100064_f_at	M63801	gap junction membrane channel protein alpha 1
100065_r_at	M63801	gap junction membrane channel protein alpha 1
101286_at	M91243	gap junction membrane channel protein alpha 8

Ephrins and Receptors

162371_r_at	AV240015	Eph receptor B6

(*continued*)

Table 3 (*continued*)

FGF and FGF Receptors

100494_at	M30641	fibroblast growth factor 1
93294_at	M70642	connective tissue growth factor

Fibronectin and Fibronectin Receptor

100123_f_at	X15202	integrin beta 1 (fibronectin receptor beta)
100124_r_at	X15202	integrin beta 1 (fibronectin receptor beta)

Endothelial Marker

102929_s_at	L23636	FMS-like tyrosine kinase 3 ligand
160358_at	AI847784	expressed sequence AU040960 (CD-34)
96119_s_at	AA797604	angiopoietin-like 4
102114_f_at	AI326963	angiopoietin-like 4

Integrins

100123_f_at	X15202	integrin beta 1 (fibronectin receptor beta)
100124_r_at	X15202	integrin beta 1 (fibronectin receptor beta)
100601_at	AF022110	integrin beta 5
100989_at	AJ001373	integrin beta 1 binding protein 1
101499_at	U94479	integrin linked kinase
101626_at	AF013107	a disintegrin and metalloprotease domain 7
102353_at	M31039	integrin beta 2
103305_at	L04678	integrin beta 4
103588_at	AF006196	a disintegrin and metalloproteinase domain 15 (metargidin)
103611_at	AB012693	integrin-associated protein
104211_at	D13867	integrin alpha 3
104308_at	AI035495	integrin alpha X
161497_f_at	AV093331	integrin alpha 7
161786_f_at	AV363187	integrin beta 5
162218_f_at	AV106844	integrin beta 5
93061_at	L23423	integrin alpha 7
93920_at	AB009676	a disintegrin and metalloprotease domain 11
96738_at	U41765	a disintegrin and metalloproteinase domain 9 (meltrin gamma)

Laminins

101359_at	U43541	laminin, beta 2
104587_at	U69176	laminin, alpha 4
161702_f_at	AV236263	laminin, alpha 5
161706_f_at	AV244043	laminin, gamma 1
92366_at	U12147	laminin, alpha 2
99931_at	U37501	laminin, alpha 5

Myocyte Enhancer Factor

104590_at	L13171	MEF2C
104592_i_at	AI595996	MEF2C
93852_at	AW045443	MEF2A

MMP

102037_at	AI844269	matrix metalloproteinase 14 (membrane-inserted)
92461_at	AB021224	matrix metalloproteinase 17
95338_s_at	M82831	matrix metalloproteinase 12
98833_at	X66402	matrix metalloproteinase 3

Table 3 (*continued*)

TIMP		
93507_at	X62622	tissuc inhibitor of metalloproteinase 2
160519_at	U26437	tissue inhibitor of metalloproteinase 3
VEGF		
103001_at	U43836	vascular endothelial growth factor B
WNT/Frizzled		
93503_at	U88567	secreted frizzled-related sequence protein 2
93681_at	AW123618	frizzled homolog 2 (*Drosophila*)
94704_at	AF100778	WNT1 inducible signaling pathway protein 2

differentiated phenotype within the SMC. The genes for GATA binding protein-6 (GATA6) and mesenchyme homeobox-2 (Meox2/Gax) have been linked to the inhibition of SMC proliferation (Gorski and Walsh, 2000). Another important regulator gene in this group is paired related homeobox (PRX-1). Various studies have demonstrated that it is an important gene regulation protein for normal vascular remodeling (Hautmann et al., 1997; Bergwerff et al., 1998, 2000).

Other genes in group C with links to blood vessel development are homeobox A3 (Hoxa3) and Sox-18 (Gorski and Walsh, 2000; Downes and Koopman, 2001). Other genes linked to developmental processes that have not been described in vascular development include Hox B6, Hox A7, aristaless 3, aristaless 4, distal-less homeobox 3, four CCAAT/enhancer binding proteins (C/EBP) (alpha, beta, gamma, and delta), and members of the signal transducer and activator of transcription (STAT) gene family (STAT1, STAT5A, STAT5B, STAT6).

The second largest category within group C is the Transport (**6810**: 103) category. Transport is composed of genes that encode for proteins that serve as receptors and that act as intracellular transport proteins. The two most prominent families of genes within the category are the ATP-binding cassette transporters and the solute carrier transporter superfamily with 7 and 22 probe sets present, respectively. The two gene families encode proteins that are involved in the influx and efflux of a wide variety of molecules to and from a cell.

The third largest category within group C is Cell Adhesion (**7155**: 73). In this category, the collagens are the largest family of genes. Eleven collagen types are represented: Iα1, Iα2, IIα1, IIIα1, IVα2, IVα5, Vα1, Vα2, VIα1, VIα2, VIα3, XIVα1, XIXα1, XV, XVIIα1, XVIIIα1. The collagens with demonstrated vascular phenotypes in knockout mouse models are the fibrillar collagens: collagen Iα1 (embryonic vessel rupture), collagen IIIα1 (vessel rupture in adults), and collagen VIα1 (altered capillary lumen size) (reviewed in Kelleher et al., 2004). The phenotypes observed indicate that these collagens play a prominent role in vascular wall development. Another prominent family of genes in the group is the integrins with nine probe sets representing five genes. Integrins αV, α7 and ß2, ß1 and ß5 are present. Integrins play the important role of mediating cell adhesion to the ECM. The involvement of the integrins in blood vessel development is well documented (Hynes, 2002; Ruegg and Mariotti, 2003). The genes for the laminin chains found in group C are α2, α4, α5, ß2, and γ1. Expression of this set of laminin chains may lead to the production of the proteins laminin-4, -9, and -11. The only laminin chain knockout mouse with a vascular phenotype is the laminin α4 chain. Laminin α4 null mice have deficient structural integrity of the basement membrane in microvessels leading to hemorrhage and microvascular degeneration (Thyboll et al., 2002).

Molecules that play an important role in leukocyte adhesion (ICAM-1, ICAM-5, VCAM-1, P-selectin) are found in group C as part of the Cell Adhesion ontology. None of the molecules produced by these genes have a described role in vascular development, but they have been implicated in pathologic processes within blood vessels. Other important genes within the cell adhesion ontology with a described vascular function include CD36, endoglin, tenascin C, p-selectin, vinculin, and vitronectin. CD36 is an integral membrane glycoprotein. It plays the important role of mediating thrombospondin-1–initiated apoptosis in endothelial cells and thrombospondin-1 inhibition of angiogenesis (Dawson et al., 1997; Jimenez et al., 2000; Febbraio et al., 2001). Endoglin is a TGF-ß–binding protein that is located on the surface of endothelial cells. The endoglin null mouse has defective vascular development that manifests through poor vascular SMC development and arrested endothelial tube remodeling (Li et al., 1999). Tenascin-C is an ECM glycoprotein that has been linked to cell migration, proliferation, and apoptosis. Germane to vascular biology it has been implicated in playing a role in tumor angiogenesis, pulmonary hypertension in a rat monocrotaline model, and SMC proliferation (Jones and Rabinovitch, 1996; Tanaka et al., 2004). Vinculin is a cytoskeletal protein found in focal adhesion plaques. The vinculin knockout mouse has an embryonic lethal phenotype by E9.5 due to abnormal cardiac development (Xu et al., 1998). Interestingly, vinculin knockout fibroblast had increased cell migration in comparison to wild type cells due to their poor ability to adhere to the ECM (Xu et al., 1998). This may indicate an important role for vinculin in maintaining stability in the components of the blood vessel wall. Vitronectin is a glycoprotein that is localized to the ECM. The vitronectin null mouse has no vascular phenotype. Vitronectin has been implicated in SMC migration in a rat carotid injury model (Dufourcq et al., 2002). Other genes with the Cell Adhesion ontology do not have a described role in vascular development, disease, or physiology.

Proteolysis and peptidolysis (**6508**: 60) is the next category with a significant number of genes related to vascular development or physiology. The largest family of genes expressed within this category is the lysosomal proteolytic cathepsins with eight cathepsin genes represented. Cathepsins L, D, and B are the only cathepsins expressed in group C that are related to vascular biology. Cathepsin L has the ability to generate endostatin from collagen XVIII (Felbor et al., 2000) and cathepsin D has been implicated in the creation of angiostatin (Morikawa et al., 2000). Cathepsin B may play a role in angiogenesis in cancer and arthritis. It has been suggested that cathepsins L, D, and B all play a role in aortic aneurysm formation (Gacko and Chyczewski, 1997; Gacko and Glowinski, 1998). Another proteolytic enzyme family in group C is the ADAM family with three genes within the category. ADAMs may play a role in cell adhesion to the ECM and in proteolysis. ADAM-7 and -11 are expressed within group C, but they have not been described as having a role in vascular development. ADAM-15 is also present in the group for this category. ADAM-15 has high expression levels in vascular cells, but the ADAM-15 null mouse has no phenotypic abnormalities. The ADAM-15 knockout does have decreased neovascularization when compared to WT in a mouse model for retinopathy of prematurity (Horiuchi et al., 2003).

Angiotensin-converting enzyme (ACE) is a gene in group C that plays an important role in vascular development. ACE is a part of the renin-angiotensin system that plays a prominent role in the modulation of blood pressure, and is responsible for the conversion of angiotensin I to angiotensin II and is expressed in many vascular beds during development. Mice lacking ACE demonstrate the phenotype of hypercellularity, cellular disorganization, and thickening in the arterial wall and luminal narrowing (Krege et al., 1995; Hilgers et al., 1997). The findings were demonstrated in the renal vasculature. Further analysis must be performed to determine if this applies to all vascular beds. If so, ACE and the renin-angiotensin system may play an important role in normal developmental formation and organization of the vascular wall.

The MMPs are a family of genes whose protein products are proteases with the capability of modifying a variety of substrates including proteins within the ECM. Within the proteolysis and peptidolysis category for group C are MMP-3, -12, and -17. The MMP-3 has not been described as having a direct and essential role in vascular development since the MMP-3 knockout mouse has no developmental vascular phenotype. However, MMP-3 has been shown to act upon many substrates within the matrix that are important to vascular development. In addition, MMP-3 is expressed by endothelial cells, has the ability to release FGF and TGF-ß from the ECM, and may function to cleave protein substrates to form angiostatin (Dollery et al., 1995; Lijnen et al., 1998; Mudgett et al., 1998; McCawley and Matrisian, 2001). The MMP-12 likewise has no identified developmental function; it is present within vascular SMCs and macrophages (Shapiro et al., 1992; Wu et al., 2003) and its expression is elevated in abdominal aortic aneurysms and in atherosclerotic lesions. Similar to MMP-3, it possesses the ability to generate angiostatin (Cornelius et al., 1998; McCawley and Matrisian, 2001; Wu et al., 2003). MMP-17 has no known function related to blood vessel development or function.

The plasminogen gene is present within group C. The plasminogen gene product is the precursor of the enzyme plasmin that is responsible for a variety of functions that include the degradation of fibrin, remodeling of ECM proteins, and activation of other proteinases (Carmeliet et al., 1997). In addition, plasmin has the capability of liberating growth factors, cytokines, and angiostatin from the matrix (Carmeliet et al., 1997; O'Reilly et al., 1997). The plasminogen knockout does not have impaired vascular development, but does show increased fibrin deposition (Ploplis et al., 1995). However, in a mouse vascular injury model the plasminogen knockout mouse has impaired neointima formation due to altered SMC migration (Carmeliet et al., 1997). Other genes within the proteolysis and peptidolysis category do not have a demonstrated role in vascular development.

Signal Transduction (7165: 51) is another major category within the Biological Processes ontology for group C. Members of the chemokine ligand gene family (five genes) and the RGS gene family (five genes) are the largest gene families represented in the category for group C. The family of genes that encode for the proteins known as regulators of G-protein signaling (RGS) are a group of molecules that accelerate GTPase-activity intrinsic to G-proteins and play a role in control of G-protein signaling (Ishii and Kurachi, 2003). RGS-2, -4, -5, -7, and -10 are represented in group C. The RGS-2 and -5 have demonstrated functions in blood vessel physiology. RGS-2 has been shown to play a role in vascular SMC relaxation induced by nitrovasodilators and cGMP; furthermore, RGS-2 knockout mice develop marked levels of hypertension (Heximer et al., 2003; Tang et al., 2003). RGS-5 has been identified as a marker for pericytes and vascular SMCs during development, and it has been shown to have high expression levels within the aorta when compared to the vena cava (Adams et al., 2000; Bondjers et al., 2003; Cho et al., 2003). In addition, RGS-5 may play an important role in vascular SMC proliferation (Cho et al., 2003). RGS-4 is differentially expressed in the heart and aorta and may have a small role in endothelin signaling. To date, the RGS-7 and -10 have no defined function in relation to vascular development.

The chemokine ligand genes do not have a function that has been related specifically to vascular development. FGF-1 is present within this category for group C. The FGF-1 is a member of the heparin-binding FGF family that has been shown to act directly on vascular cells to induce endothelial cell growth and angiogenesis. These characteristics were demonstrated in an *in vivo* model of arterial gene transfection with the recombinant FGF-1 gene. In the model, local intimal hyperplasia and angiogenesis occurred in the form of capillary formation (Nabel et al., 1993). The genes for leptin and the leptin receptor are present within this functional category. Leptin is an endocrine hormone that regulates adipose tissue mass and the leptin receptor is expressed on endothelial cells. Leptin has been shown to induce angiogenesis in *in vitro* and *in ovo* studies (Bouloumie et al., 1998). Of note, EPAS-1, laminin-γ-1, and members of the Stat gene family (Stat-1, -5a, -5b, and -6) are present within this category, and they have been previously discussed.

Protein amino acid phosphorylation (**6468**: 46) is another category within the Biological Processes ontology for discussion within group C. The category contains the genes for many ubiquitous proteins including members of the MAP kinase gene family, Janus kinase, and serine/threonine kinases. The gene for activin receptor-like kinase 1 (ALK1/Acvrl1) is present within group C. ALK1 is a receptor on endothelial cells that mediates signaling for TGF-ß1. Hereditary hemorrhagic telangiectasia II (HHT-II) in humans has been linked to ALK1. Separate groups have produced ALK-1 knockout mice that demonstrate vascular defects and to an extent recapitulate HHT-II. The ALK1 knockout mouse is embryonic lethal between E10.5 and E11.5. The embryos have defective blood vessel formation with the development of arteriovenous malformations, excessive fusion of capillary plexes, hyperdilation of arteries, and defective recruitment of vascular SMCs (Oh et al., 2000; Urness et al., 2000). Genes for bone morphogenetic protein receptor type 1a (BMPR1a/ALK-3) and bone morphogenetic protein receptor type 1b (BMPR1b/ALK-6) are other members of the TGF-β family present within group C. Mice deficient in these two genes do not have a vascular phenotype, although both are expressed in developing blood vessels (Agrotis et al., 1996; Goumans and Mummery, 2000).

Insulin-like growth factor-I receptor (IGF-1R) is present within Protein amino acid phosphorylation category for group C. The receptor is a receptor tyrosine kinase with two ligands, IGF-I and IGF-II. The IGF-1R is expressed in the pulmonary endothelium during lung development and to a much lesser extent within SMCs of the developing pulmonary vasculature and airways (Han et al., 2003). In fetal lung explant cultures, neutralizing antibodies against IGF-1R demonstrated reduced endothelial cell numbers and endothelial tube growth; thus, IGF-1R appears to play a role in the vasculogenesis of the developing pulmonary vasculature (Han et al., 2003). No role has been identified for IGF-1R in other vascular beds nor in aortic development.

The other gene of note in the protein amino acid phosphorylation category for Group C is platelet derived growth factor (PDGF)-ß. PDGF-β is a receptor tyrosine kinase that facilitates the signaling for the mitogen PDGF-B. PDGF-B and PDGFR-β are thought to play a role in the recruitment of SMCs to the developing endothelial tubes. Arteries from the PDGFR-β $-/-$ mouse demonstrate dilation, microaneurysm formation in vascular beds, and a decreased number of vascular smooth muscle cells and pericytes and in developing blood vessels when compared to wild type mice.

PDGFR-β $-/-$ mice die from hemorrhage at or just before birth, which is thought to be due to the lack of proper vascular SMC recruitment (Soriano, 1994; Hellstrom et al., 1999).

Many of the genes that produce the cytoskeletal proteins present within vascular SMCs can be found in the categories of Cytoskeleton organization and biogenesis (**7010**: 19) and Muscle development (**7517**: 15). Many of the genes within the two categories can be found in a variety of cell types. Smooth muscle MyHC, smooth muscle α-actin, and transgelin/SM22-α are genes present in one or both categories. While these genes play a role in the contractile apparatus of smooth muscle cells, these three genes and other contractile proteins have been shown to be useful markers for studying vascular SMC differentiation (reviewed in Owens, 1995; Owens et al., 2004). Along the time line of vascular smooth muscle maturation, SMC α-actin has been identified as the earliest specific marker for a vascular SMC. In that continuum, transgelin is an early to intermediate marker, and the appearance of smooth muscle MyHC represents a fully differentiated vascular SMC. The α-tropomyosin and desmin are present in the Muscle development set of genes. Tropomyosin has multiple isoforms that are present in muscle and in non-muscle cells. Smooth muscle α-tropomyosin is a splice variant of α-tropomyosin gene particular to SMCs (Owens, 1995). In adult animals, the expression of α-tropomyosin is limited to SMCs; therefore, it is a good marker for a differentiated state in SMCs. Desmin is an intermediate filament protein present in a variety of muscle cell types. In prior work, Desmin has not been shown to be a good marker for SMC lineage, but it is good for assessing the relative state of differentiation and maturation (Owens, 1995). The presence of early and late vascular SMC markers in group C indicates that genes involved in SMC differentiation will have an expression pattern consistent with those of group C.

Two subsets of genes within group C that are of interest to vascular biology are Cell Matrix Adhesion (**7160**: 14) and Angiogenesis (**1525**: 6). The Cell Matrix Adhesion subset contains members of the integrin family of genes that have been previously discussed. Another protein in his set, nephronectin, is an integrin binding ECM protein that is important in kidney development. There is no known role for nephronectin in vascular development. Ras homolog gene family member A2 (Rho A) is another gene present within this subset. Rho A is a small GTPase that effects many cellular signaling pathways with many different functional outcomes. Recently Rho A has been implicated in mediating tropoelastin signaling to induce actin polymerization and a mature phenotype within vascular SMCs (Karnik et al., 2003). Connective tissue growth factor (CTGF) is a regulatory factor involved in angiogenesis, ECM production, and fibrosis. CTGF is present in both the Cell Matrix Adhesion and Angiogenesis categories within group C. CTGF was described as a mitogenic molecule produced in culture by human umbilical vein endothelial cells (Bradham et al., 1991). During murine embryonic development (E12 to E18), CTGF is predominantly expressed in the intima and sparsely expressed in the media of the developing aorta (Friedrichsen et al., 2003). In cell culture of fibroblasts and mesangial cells, CTGF has been shown to increase the production of ECM proteins including fibronectin, laminin, integrin 5, collagen I, and collagen IV (Frazier et al., 1996). CTGF also plays a role in inducing apoptosis in mesangial and SMCs (Hishikawa et al., 1999, 2000). Other genes in the Angiogenesis subset are endoglin, Epas-1, procollagen type XVIIIα-1, and angiogenin. Endoglin and Epas-1 have been discussed in earlier sections. Angiogenin is a ribonuclease that is a potent pro-angiogenic factor. It had been shown to participate in many facets of angiogenesis including the ability to bind to endothelial cells, to induce the activation of basement membrane proteases, to mediate cell adhesion, to induce the construction of tubular structures by endothelial cells, and to stimulate cell proliferation (reviewed in Strydom, 1998). Collagen type XVIII is a basement membrane collagen. Cleavage of the c-terminus produces the potent anti-angiogenic factor endostatin (Muragaki et al., 1995; O'Reilly et al., 1997). A role for endostatin in vascular development has not been identified, but it may serve to limit vascular proliferation.

Critical to the development of the vascular wall is the assembly of the ECM. The GO cell compartment ontology divides the gene for these important proteins into the following categories: Extracellular matrix (**5578**: 47), Collagen (**5581**: 20), and Basement membrane (**5604**: 9). These categories are of great prominence to group C and indicate that the gene expression pattern and genes in group C are largely responsible for assembling, modifying, and maintaining the elements of the vascular wall. The Extracellular matrix subset consists of genes that produce proteins that are a part of the elastic fiber: elastin, fibulin-1, latent TGF-ß–binding protein-1 (LTBP-1), LTBP-2, and microfibillar associated protein-5 (MAGP-2). Of the genes that are a part of the elastic fiber represented in group C, elastin and fibulin-1 have been shown to be critical for normal blood vessel formation. The elastin gene has been associated with human inheritable disease states Williams' Syndrome and supravalvular aortic stenosis (SVAS), which results in congenital narrowing of large arteries (Ewart et al., 1993a,b; Olson et al., 1995). The elastin knockout mouse mirrors the human disease states. Mice null for elastin die during the early perinatal period. Their blood vessels have areas of narrowing within large vessels and SVAS (Li et al., 1998a,b). Closer analysis demonstrates disorganization among SMCs in their spatial orientation in the vascular wall.

Proteoglycans have representation within the extracellular matrix subset. Large proteoglycans present in group C include aggrecan-1 and the heparan sulfate proteoglycan perlecan. Small leucine-rich proteoglycans present are biglycan, lumican, and decorin. Biglycan and lumican localizes to all layers of the aorta by immunohistochemical staining, and decorin is found only in the adventitia (Onda et al., 2002; Theocharis and Karamanos, 2002).

Biglycan, lumican, and decorin bind to and regulate collagen fibrillogenesis and all have the capacity to bind and sequester growth factors into the ECM.

Multiple matricellular genes are present within the ECM subset including secreted acidic cysteine rich glycoprotein (SPARC), SPARC-like-1 (SPARCL-1), tenascin C, CTGF (previously discussed), and TGF-ß–induced (also known as BIG-H3). Tenascin C has been shown to have diverse biological effects including stimulation and inhibition of cellular proliferation and the inhibition of cell attachment. There are no described vascular diseases or knockout mouse models with vascular phenotypes related to tenascin C. SPARC is a matricellular protein with the capability to bind many other matrix molecules, PDGF, and VEGF. SPARC has a variety of functions including the modification of growth factor activity, modulation of MMP expression, and effects on cell shape and adhesion (Brekken and Sage, 2001). While the SPARC null mouse shows a variety of defects, no vascular abnormalities were observed.

The Extracellular matrix subset MMP-3, -12, -17, and Adam-15 all have the capability of modifying ECM proteins. These proteins have been previously discussed. The MMP regulatory molecules tissue inhibitor of metalloproteinase (TIMP)-2 and -3 are also present in this subset of genes. TIMP-2 and -3 are inhibitors of MMPs. They have both been shown to inhibit cell migration (endothelial cells and SMCs), to have antiproliferative effects (TIMP-2), and to induce apoptosis (TIMP-3) (Cheng et al., 1998; Shi et al., 1999; Spurbeck et al., 2002). The role of TIMPs in aortic development has not been studied. It is possible that they may serve to stabilize the vascular wall by helping to maintain a non-proliferative and a differentiated state for vascular wall cells.

The Growth Factor subset of genes for the Molecular Function ontology of group C demonstrates genes that may be important modulators of vascular development. The genes present include bone morphogenetic proteins-4 (BMP-4) and -6 (BMP-6), platelet-derived growth factor-α (PDGF-α), TGF-ß3, and VEGFb. PDGF-A is a mitogenic factor that interacts with PDGF receptors. PDGF-A is thought to be important for the development of mesenchymal structures. The PDGF-A knockout mouse failed to demonstrate an aortic phenotype, although altered lung myoblast development and altered lung elastogenesis were observed (Bostrom et al., 1996).

4.4. Group D: Genes that decrease at birth

Group D consists of 1,524 probe sets. The global pattern of expression is such that the genes remain at or near baseline with a majority having a dramatic decrease in expression at the time of birth. In the Biological Processes ontology for group D (see Table 4 for genes of interest in this group) genes representing proteins involved in gene regulation are the most prevalent (Fig. 4B). For group D, the Regulation of transcription,

Table 4
Genes of interest: Group D

Hypertension-Related Genes

| 102738_s_at | U07982 | endothelin 1 |

Elastic Fiber Genes

| 94308_at | X70853 | fibulin 1 |
| 94309_g_at | X70853 | fibulin 1 |

IGF and IGF-Related Genes

| 101571_g_at | X76066 | insulin-like growth factor binding protein 4 |
| 97987_at | U66900 | insulin-like growth factor binding protein, acid labile subunit |

Smooth Muscle Cell Marker

| 96980_at | U34303 | myosin heavy chain 10, non-muscle |

TGF-β and TGF-β Receptors

93460_at	L15436	activin A receptor, type 1
95557_at	L24755	bone morphogenetic protein 1
99393_at	L41145	bone morphogenetic protein 5
102637_at	AF039601	transforming growth factor, beta receptor III
92427_at	D25540	transforming growth factor, beta receptor I

Cadherins

92363_at	AI586083	cadherin EGF LAG seven-pass G-type receptor 2
93515_at	AF016271	cadherin 16
94449_at	AI854522	protocadherin 13

Catenins

| 94174_at | AF006071 | catenin alpha-like 1 |
| 98151_s_at | Z17804 | catenin src |

Collagens

99637_at	AF011450	procollagen, type XV
99638_at	D17546	procollagen, type XV
99842_at	AB000636	procollagen, type XIX, alpha 1
99953_at	AF100956	procollagen, type XI, alpha 2

Connexins

| 104729_at | X57971 | gap junction membrane channel protein alpha 4 |

Ephrins and Ephrin Receptors

98446_s_at	U06834	Eph receptor B4
102871_at	L77867	Eph receptor B6
98771_at	L25890	Eph receptor B2

FGF and FGF Receptors

93091_s_at	M63503	fibroblast growth factor receptor 2
97509_f_at	U22324	fibroblast growth factor receptor 1
93309_at	U42386	fibroblast growth factor inducible 14

Endothelial Cell Marker

104416_at	L07296	FMS-like tyrosine kinase 4
97830_at	L06039	platelet/endothelial cell adhesion molecule
97773_at	AI173145	expressed sequence AU040960m (CD34)

(*continued*)

Table 4 *(continued)*

Integrins		
103039_at	X79003	integrin alpha 5 (fibronectin receptor alpha)
100990_g_at	AJ001373	integrin beta 1 binding protein 1
103039_at	X79003	integrin alpha 5 (fibronectin receptor alpha)
92357_at	AB009673	a disintegrin and metalloprotease domain 23
94764_at	M60778	integrin alpha L
94826_at	Y11460	integrin beta 4 binding protein
98366_at	U14135	integrin alpha V
Laminins		
97750_at	X06406	laminin receptor 1 (67kD, ribosomal protein SA)
Myocyte Enhancer Factor		
104591_g_at	L13171	myocyte enhancer factor 2C
MMP		
93612_at	D86332	matrix metalloproteinase 15
VEGF		
103520_at	M95200	vascular endothelial growth factor
WNT/Frizzled		
95771_i_at	U43317	frizzled homolog 4 (*Drosophila*)
96747_at	AW121294	Wnt1 responsive Cdc42 homolog
97997_at	U88566	secreted frizzled-related sequence protein 1
98348_at	U43205	frizzled homolog 3 (*Drosophila*)
99415_at	U43321	frizzled homolog 8 (*Drosophila*)
99844_at	Y17709	frizzled homolog 9 (*Drosophila*)

DNA-dependent (**6355**: 61) category contains gene regulatory molecules with multiple functions. Many play a role in repressing cell proliferation to maintain a differentiated state. The genes present include MAD homolog 6 (Madh6), which encodes the protein SMAD 6, a repressor of BMP signaling. Madh 6 −/− mice have cardiovascular abnormalities. Newborns possess a stenotic ascending aorta; moreover, the adults are plagued by defective aortic contractility, ossification of the aorta, and hypertension in comparison to WT mice (Galvin et al., 2000).

Another group of inhibitory genes linked to vascular development are the retinoic acid receptors. Four genes are present from the two families of genes. They are retinoic acid receptor (RAR)-α, -γ, retinoid X receptor (RXR)-α, and RXR-β. These receptors have been demonstrated in cultured SMC to mediate growth inhibition (Miano et al., 1996). Another repressor present in group D is the Ngfi-A binding protein 2 (Nab2). Nab2 is a transcriptional repressor gene that is capable of down-regulating Egr-1 and -2. In cultured vascular SMCs, Nab2 has been shown to inhibit Egr-1–dependent gene expression in the presence of a pharmacologic agonist; moreover, the level of expression in a murine *in vivo* model of arterial injury demonstrated an increase in Nab2 transcription (Silverman et al., 1999).

In group D, other genes that encode for repressors of transcription include CpG binding protein (Cgbp-pending), Est2 repressor factor (Erf), hairy and enhancer of split-1 (Hes1), kruppel-like factor 3 (basic) (Klf3), max-binding protein (Mnt), upstream transcription factor 1 (Usf1), and upstream transcription factor 2 (Usf2). These genes have not been linked to vascular development or to cellular elements of blood vessels. Continuing with the theme of genetic stability, the transformation-related protein 53 (Trp53) that encodes the protein p53 is present in group D. p53 is a ubiquitous protein that regulates cell responses to DNA damage in order to maintain genetic stability (reviewed in Xu and el-Gewely, 2001).

Ets variant gene 6 (or TEL) (Etv6) and aryl hydrocarbon receptor nuclear translocator (Arnt) are genes of group D that are related to angiogenesis based upon defects seen in mutant mice. The Etv6 $-/-$ mouse is embryonic lethal with defective yolk-sac angiogenesis (Wang et al., 1997). The Arnt $-/-$ mouse displays normal vasculogenesis but has failed angiogenesis in the yolk sac and branchial arches (Maltepe et al., 1997). Another gene from group D that provokes abnormal blood vessel formation in mutant mice is the gene for forkhead box F1a (Foxf1a). Heterozygote mice for the Foxf1a have defective lung vasculogenesis and decreased expression of VEGF-A, Flk-1, and BMP4 that are important for normal blood vessel formation (Kalinichenko et al., 2003).

Within group D there are probe sets for several genes linked to myoblast differentiation. They include Mef2c, nuclear receptor coactivator 2 (Ncoa2), Six4, and Six5 (Esteve and Bovolenta, 1999; Chen et al., 2001; Kirby et al., 2001; Ozaki et al., 2001). Sox-18, a transcription factor expressed in endothelial cells during development (Downes and Koopman, 2001), is also present. Other genes that have been identified in development that are present in group D include: Hoxb4, Hoxc5, Sox-17, Stat4, and runt-related transcription factor 2 (Runx2). The aforementioned transcription factors have not been linked to vascular development.

The second largest subset of genes within the biological processes ontology for group D is Transport (**6810**: 47). The genes within this group are expressed in many cell types and organ systems but none are specific to vascular development. The most prominent member of the transport subset is the solute carrier family (11 genes) of genes that function to facilitate the movement of many different types of macromolecules.

Protein amino acid phosphorylation (**6468**: 41) is the third largest subset of genes for the biological processes of ontology in group D. The largest family within the subset is made up of genes related to MAPK signaling with seven genes. This family of genes is not specific for vascular development. Two receptors associated with the TGF-β family are present within this gene subset. Activin A receptor type 1 (also known as ALK-2) is a receptor that can be found on a variety of cell types including vascular SMCs (Agrotis et al., 1996). ALK-2 is capable of binding TGF-β, activin, and BMPs, but it binds preferentially to the 60A subgroup of BMPs: BMP-5, -6, and -7 (Macias-Silva et al., 1998). The conventional ALK-2-knockout mouse has an early embryonic lethal phenotype with defects related to gastrulation (Gu et al., 1999). A conditional knockout for ALK-2 in neural crest cells reveals a vascular specific phenotype. The ALK-2 conditional knockout mouse has cardiovascular defects. The most prominent abnormality is in the cardiac outflow tract where all mice had a persistent truncus arteriosus, a significantly enlarged heart, and ventricular septation defects (Kaartinen et al., 2004). Interestingly, simultaneous deletion of BMP-6 and -7 leads to cardiac outflow tract, valve, and septation defects (Kim et al., 2001). The other receptor for TGF-β family growth factors in group D is TGF-β receptor I (also known as ALK-5). ALK-5 is a type I TGF-β receptor that serves as a receptor for TGF-β-1, -2, and -3. It is expressed in vascular SMCs (Agrotis et al., 1996) and plays a role in early blood vessel formation. The ALK-5 knockout mouse is embryonic lethal at day 10.5 due to abnormal vascular development in the yolk sac (Larsson et al., 2001). Isolated endothelial cells from ALK-5 knockout mice have impaired fibronectin production and migration (Larsson et al., 2001).

The FGF receptors FGFR1 and FGFR2 are present in this subset of genes. Both FGFR1 and FGFR2 have multiple isoforms that arise through alternative splicing. The receptors have differential tissue expression and affinity for members of the FGF family (Ornitz et al., 1996; Arman et al., 1998; Groth and Lardelli, 2002). FGFR1 has been shown to bind to more than FGFs; it also binds to heparan sulfate proteoglycans and neural cell adhesion molecules, but binds to FGF1 and FGF2 with a higher affinity (Ornitz et al., 1996; Groth and Lardelli, 2002). Conventional homozygous knockout mice for FGFR1 and FGFR2 are both embryonic lethal. FGFR2 null mice fail to have post-implantation growth, and FGFR1 null mice are embryonic lethal between E7.5 and E9.5 with defects in the mesoderm in early gastrulation (Yamaguchi et al., 1994; Ciruna et al., 1997; Arman et al., 1998). While there are no studies that directly demonstrate a role in vascular development for FGFR1 and FGFR2, in an in vitro study of angiogenesis it was suggested that FGFR1 mediates the angiogenic effects of FGF-2 (Akimoto and Hammerman, 2003).

Epidermal growth factor receptor (EGFR) is an integral membrane receptor within the transport subset of genes. The EGFR mediates the cellular signals for multiple ligands including EGF, TGF-α, and heparin-binding EGF (HB-EGF). The EGFR knockout mice exhibit a variety of defects that are strain dependent (Miettinen et al., 1995; Sibilia and Wagner, 1995; Threadgill et al., 1995; Sibilia et al., 1998; Chen et al., 2000). In some strains, the lack of EGFR is embryonic lethal and in others there is a shortened postnatal life. The defects occur in a variety of organ systems, but there is no evidence for direct vascular involvement. A study in zebrafish using EGFR kinase inhibitors and EGFR antisense morpholino oligos demonstrated defects in the developing cardiovascular system. The defects included obstruction of the outflow tract, altered segmental vessel development, and dilated heart chambers. The EGFR may also be involved in vascular development and physiology by mediating angiotensin II-induced vascular SMC hyperplasia and hypertrophy, cardiac hypertrophy, and hypertension via transactivation by the angiotensin II type I receptor (Sambhi et al., 1992; Swaminathan et al., 1996; Eguchi et al., 1998; Kagiyama et al., 2002).

The ephrin receptor family also has representation within the Transport subset of genes in group D. The ephrin receptors and their ligands are thought to play a role in juxtacrine cell-cell contacts, cell adhesion to matrix, and cell migration (reviewed in Cheng et al., 2002). The receptors present in group D are EphB2, EphB4, and EphB6. EphB4 is expressed in venous endothelial cells and at low levels in arterial endothelium (Adams et al., 1999; Wang et al., 1998). With its ligand EphB2, EphB4 is thought to play a role in establishing spatial orientation between presumptive arteries and presumptive veins in vascular development (Wang et al., 1998). EphB2 is expressed in both endothelium and mesenchymal supporting cells, which suggests a role of EphB2 in vessel wall development through interactions between endothelium and mesenchyme (Adams et al., 1999). The EphB4 knockout mouse shares the phenotypes of its ligand EphB2 null mouse. The vascular phenotypes in the E10.5 embryonic lethal EphB4 null mouse are endothelial cell disorganization, decreased vessel remodeling, and vasculogenesis restricted to a primary plexus stage (Gerety et al., 1999). EphB2 null mice show no overt vasculars defect, but a double knockout for EphB2 and EphB3 demonstrates defects in vascular remodeling coupled with a 30% embryonic lethal phenotype at E10.5 (Adams et al., 2000). EphB6 has no known function in blood vessel development.

The fourth subset of genes is the Intracellular Signaling Cascade (**7242**: 34) category. In this category are genes related to cell signaling that are present in many cell types and organ systems and are not specific for vascular development. Within the subset is dishevelled-2 (Dvl-2), a gene that is specifically related to cardiac development. Dvl-2 is a member of the Wingless/WNT developmental pathway that is related to determining cell fate. The Dvl-2 knockout mouse has a phenotype of 50% lethality due to severe cardiac outflow tract defects that include double outlet right ventricle, transposition of the great arteries, and persistent truncus arteriosus (Hamblet et al., 2002). The abnormalities are related to abnormal cardiac neural crest development.

The categories Protein Biosynthesis (**6412**: 34), Intracellular Protein Transport (**6886**: 32), and Protein Transport (**15031**: 28) contain mostly ubiquitous genes that are not specific to vascular development. The Signal Transduction category (**7165**: 27), however, contains several signaling molecules important for vascular function, including RGS proteins and the signaling protein Wnt5a. Wnt5a is involved in a host of developmental processes, but has not been demonstrated to have a direct link to vascular development.

Within the Cell Growth and/or Maintenance (**8151**: 24) group are oncogenes, genes related to apoptosis, and cell cycle genes. EGFR and Etv6 are a part of the subset and have been previously discussed. The other gene related to vascular development within the subset is thymomo viral proto-oncogene 1 (Akt1). Akt1 is a serine/threonine kinase that has been shown to be an anti-apoptotic gene (Shiojima and Walsh, 2002). Akt1 is important to normal blood vessel physiology through its ability to phosphorylate, activate, and regulate endothelial nitric oxide synthase (Dimmeler et al., 1999). In spontaneously hypertensive rats, Akt1 was demonstrated to participate in endothelial dysfunction by inappropriate localization in endothelium leading to impaired endothelial nitric oxide synthase phosphorylation (Iaccarino et al., 2004). Akt1 has also been shown to mediate the hypertrophy and polyploidization of vascular SMCs (Hixon et al., 2000). Akt2 is also present within the gene subset. It has no direct relationship to vascular development or disease although it does play a role in skeletal muscle differentiation (Vandromme et al., 2001). Closer analysis in vascular SMCs may demonstrate a role for Akt2 in vascular SMC differentiation.

Development (**7275**: 24) is the last subset of the 10 most prominent genes for group D Biological Processes. Dvl2, Hoxb4, Hoxc5, Six4, Six5, and Wnt5a are present within the subset and have been discussed in the previous sections. Many members of the frizzled family are present in this subset, but they have not been related to a role in vascular development (Fzd1, Fzd2, Fzd6, Fzd8, and Fzd9), although Fzd2 is transiently expressed in the outflow tract, pulmonary artery, and aorta in murine development (van Gijn et al., 2001). Interestingly, the pattern of Fzd2 expression in the great vessels is consistent with the distribution of neural crest cells. Tbox2 is a transcription factor present within the subset that has been linked to cardiac development and Tbox2 specifically has been shown to have a role in outflow tract formation. Null mice for Tbox2 have defects in atrioventricular canal formation and fail to form an aorticopulmonary septum (Harrelson et al., 2004). Other genes within the subset have no demonstrated role in vascular development.

5. Conclusions

The focus of this review was to catalog the extensive set of genes that are expressed during mouse aortic development with a focus on vessel wall ECM. The structural matrix proteins, which are important for vascular strength and compliance, are produced during a relatively narrow developmental window in the mouse that begins around the last trimester of development and continues for only a few weeks after birth. Prior to this synthetic "matrix phase," the cells in the vessel wall are highly proliferative and express matrix proteins that support cell motility, establish polarity, and bind and sequester growth factors. As the cells shift out of the matrix phase, the spectrum of contractile proteins changes as the SMC prepares for its unique contractile function and general cell maintenance genes are expressed. It is clear from the genetic characterization previously outlined that the vascular SMC can exhibit a wide range of phenotypes at different stages of development. Developing a molecular snapshot of normal development will provide new avenues of investigation into vascular cell phenotypic modulation in health and disease.

References

Abe, M., Hasegawa, K., Wada, H., Morimoto, T., Yanazume, T., Kawamura, T., Hirai, M., Furukawa, Y., Kita, T. 2003. GATA-6 is involved in PPARγ-mediated activation of differentiated phenotype in human vascular smooth muscle cells. Arterioscler. Thromb. Vasc. Biol. 23, 404–410.

Adams, L.D., Geary, R.L., McManus, B., Schwartz, S.M. 2000. A comparison of aorta and vena cava medial message expression by cDNA array analysis identifies a set of 68 consistently differentially expressed genes, all in aortic media. Circ. Res. 87, 623–631.

Adams, R.H., Wilkinson, G.A., Weiss, C., Diella, F., Gale, N.W., Deutsch, U., Risau, W., Klein, R. 1999. Roles of ephrinB ligands and EphB receptors in cardiovascular development: Demarcation of arterial/venous domains, vascular morphogenesis, and sprouting angiogenesis. Genes Dev. 13, 295–306.

Agrotis, A., Samuel, M., Prapas, G., Bobik, A. 1996. Vascular smooth muscle cells express multiple type I receptors for TGF-ß, activin, and bone morphogenetic proteins. Biochem. Biophys. Res. Commun. 219, 613–618.

Akimoto, T., Hammerman, M.R. 2003. Fibroblast growth factor 2 promotes microvessel formation from mouse embryonic aorta. Am. J. Physiol. Cell. Physiol. 284, C371–C377.

Albert, E.N. 1972. Developing elastic tissue. An electron microscopic study. Am. J. Pathol. 69, 89–102.

Arman, E., Haffner-Krausz, R., Chen, Y., Heath, J.K., Lonai, P. 1998. Targeted disruption of fibroblast growth factor (FGF) receptor 2 suggests a role for FGF signaling in pregastrulation mammalian development. Proc. Natl. Acad. Sci. USA 95, 5082–5087.

Bergwerff, M., Gittenberger-de Groot, A.C., De Ruiter, M.C., van Iperen, L., Meijlink, F., Poelmann, R.E. 1998. Patterns of paired-related homeobox genes Prx1 and Prx2 suggest involvement in matrix modulation in the developing chick vascular system. Dev. Dyn. 213, 59–70.

Bergwerff, M., Gittenberger-de Groot, A.C., Wisse, L.J., De Ruiter, M.C., Wessels, A., Martin, J.F., Olson, E.N., Kern, M.J. 2000. Loss of function of the Prx1 and Prx2 homeobox genes alters architecture of the great elastic arteries and ductus arteriosus. Virchows Archiv. 436, 12–19.

Berry, C.L., Looker, T., Germain, J. 1972. The growth and development of the rat aorta. I. Morphological aspects. J. Anat. 113, 1–16.

Bondjers, C., Kalen, M., Hellstrom, M., Scheidl, S.J., Abramsson, A., Renner, O., Lindahl, P., Cho, H., Kehrl, J., Betsholtz, C. 2003. Transcription profiling of platelet-derived growth factor-B-deficient mouse embryos identifies RGS5 as a novel marker for pericytes and vascular smooth muscle cells. Am. J. Pathol. 162, 721–729.

Bostrom, H., Willetts, K., Pekny, M., Leveen, P., Lindahl, P., Hedstrand, H., Pekna, M., Hellstrom, M., Gebre-Medhin, S., Schalling, M., Nilsson, M., Kurland, S., Tornell, J., Heath, J.K., Betsholtz, C. 1996. PDGF-A signaling is a critical event in lung alveolar myofibroblast development and alveogenesis. Cell 85, 863–873.

Bouloumie, A., Drexler, H.C., Lafontan, M., Busse, R. 1998. Leptin, the product of Ob gene, promotes angiogenesis. Circ. Res. 83, 1059–1066.

Bradham, D., Igarashi, A., Potter, R., Grotendorst, G. 1991. Connective tissue growth factor: A cysteine-rich mitogen secreted by human vascular endothelial cells is related to the SRC-induced immediate early gene product CEF-10. J. Cell. Biol. 114, 1285–1294.

Brekken, R.A., Sage, E.H. 2001. SPARC, a matricellular protein: At the crossroads of cell-matrix communication. Matrix Biol. 19, 816–827.

Carlin, B., Jaffe, R., Bender, B., Chung, A.E. 1981. Entactin, a novel basal lamina-associated sulfated glycoprotein. J. Biol. Chem. 256, 5209–5214.

Carmeliet, P., Moons, L., Ploplis, V., Plow, E., Collen, D. 1997. Impaired arterial neointima formation in mice with disruption of the plasminogen gene. J. Clin. Invest. 99, 200–208.

Chen, B., Bronson, R.T., Klaman, L.D., Hampton, T.G., Wang, J.F., Green, P.J., Magnuson, T., Douglas, P.S., Morgan, J.P., Neel, B.G. 2000. Mice mutant for Egfr and Shp2 have defective cardiac semilunar valvulogenesis. Nat. Genet. 24, 296–299.

Chen, S.L., Wang, S.C., Hosking, B., Muscat, G.E. 2001. Subcellular localization of the steroid receptor coactivators (SRCs) and MEF2 in muscle and rhabdomyosarcoma cells. Mol. Endocrinol. 15, 296–299.

Cheng, L., Mantile, G., Pauly, R., Nater, C., Felici, A., Monticone, R., Bilato, C., Gluzband, Y.A., Crow, M.T., Stetler-Stevenson, W., Capogrossi, M.C. 1998. Adenovirus-mediated gene transfer of the human tissue inhibitor of metalloproteinase-2 blocks vascular smooth muscle cell invasiveness in vitro and modulates neointimal development in vivo. Circulation 98, 2195–2201.

Cheng, N., Brantley, D.M., Chen, J. 2002. The ephrins and Eph receptors in angiogenesis. Cytokine Growth Factor Rev. 13, 75–85.

Cho, H., Harrison, K., Schwartz, O., Kehrl, J.H. 2003. The aorta and heart differentially express RGS (regulators of G-protein signaling) proteins that selectively regulate sphingosine 1-phosphate, angiotensin II and endothelin-1 signaling. Biochem. J. 371, 973–980.

Chung, A.E., Freeman, I.L., Braginski, J.E. 1977. A novel extracellular membrane elaborated by a mouse embryonal carcinoma-derived cell line. Biochem. Biophys. Res. Commun. 79, 859–868.

Ciruna, B., Schwartz, L., Harpal, K., Yamaguchi, T., Rossant, J. 1997. Chimeric analysis of fibroblast growth factor receptor-1 (Fgfr1) function: A role for FGFR1 in morphogenetic movement through the primitive streak. Development 124, 2829–2841.

Cornelius, L.A., Nehring, L.C., Harding, E., Bolanowski, M., Welgus, H.G., Kobayashi, D.K., Pierce, R.A., Shapiro, S.D. 1998. Matrix metalloproteinases generate angiostatin: Effects on neovascularization. J. Immunol. 161, 6845–6852.

Corson, G.M., Charbonneau, N.L., Keene, D.R., Sakai, L.Y. 2004. Differential expression of fibrillin-3 adds to microfibril variety in human and avian, but not rodent, connective tissues. Genomics 83, 461–472.

Dawson, D.W., Pearce, S.F., Zhong, R., Silverstein, R.L., Frazier, W.A., Bouck, N.P. 1997. CD36 mediates the In vitro inhibitory effects of thrombospondin-1 on endothelial cells. J. Cell. Biol. 138, 707–717.

Dimmeler, S., Fleming, I., Fisslthaler, B., Hermann, C., Busse, R., Zeiher, A.M. 1999. Activation of nitric oxide synthase in endothelial cells by Akt-dependent phosphorylation. Nature 399, 601–605.

Dollery, C.M., McEwan, J.R., Henney, A.M. 1995. Matrix metalloproteinases and cardiovascular disease. Circ. Res. 77, 863–868.

Downes, M., Koopman, P. 2001. SOX18 and the transcriptional regulation of blood vessel development. Trends Cardiovasc. Med. 11, 318–324.

Dufourcq, P., Couffinhal, T., Alzieu, P., Daret, D., Moreau, C., Duplaa, C., Bonnet, J. 2002. Vitronectin is up-regulated after vascular injury and vitronectin blockade prevents neointima formation. Circ. Res. 53, 952–962.

Eguchi, S., Numaguchi, K., Iwasaki, H., Matsumoto, T., Yamakawa, T., Utsunomiya, H., Motley, E.D., Kawakatsu, H., Owada, K.M., Hirata, Y., Marumo, F., Inagami, T. 1998. Calcium-dependent epidermal growth factor receptor transactivation mediates the angiotensin II-induced mitogen-activated protein kinase activation in vascular smooth muscle cells. J. Biol. Chem. 273, 8890–8896.

Ekblom, P., Timpl, R. 1996. *The Laminins,* New York: Harwold Academic Publishers.

Esteve, P., Bovolenta, P. 1999. cSix4, a member of the six gene family of transcription factors, is expressed during placode and somite development. Mech. Dev. 85, 161–165.

Ewart, A.K., Morris, C.A., Atkinson, D., Jin, W., Sternes, K., Spallone, P., Stock, A.D., Leppert, M., Keating, M.T. 1993a. Hemizygosity at the elastin locus in a developmental disorder, Williams syndrome. Nat. Genet. 5, 11–16.

Ewart, A.K., Morris, C.A., Ensing, G.J., Loker, J., Moore, C., Leppert, M., Keating, M. 1993b. A human vascular disorder, supravalvular aortic stenosis, maps to chromosome 7. Proc Natl. Acad. Sci. USA 90, 3226–3230.

Febbraio, M., Hajjar, D.P., Silverstein, R.L. 2001. CD36: A class B scavenger receptor involved in angiogenesis, atherosclerosis, inflammation, and lipid metabolism. J. Clin. Invest. 108, 785–791.

Felbor, U., Dreier, L., Bryant, R.A., Ploegh, H.L., Olsen, B.R., Mothes, W. 2000. Secreted cathepsin L generates endostatin from collagen XVIII. EMBO J. 19, 1187–1194.

Forsythe, J.A., Jiang, B.H., Iyer, N.V., Agani, F., Leung, S.W., Koos, R.D., Semenza, G.L. 1996. Activation of vascular endothelial growth factor gene transcription by hypoxia-inducible factor 1. Mol. Cell Biol. 16, 4604–4613.

Francis, S.E., Goh, K.L., Hodivala-Dilke, K., Bader, B.L., Stark, M., Davidson, D., Hynes, R.O. 2002. Central roles of alpha5beta1 integrin and fibronectin in vascular development in mouse embryos and embryoid bodies. Arterioscler. Thromb. Vasc. Biol. 22, 927–933.

Frazier, K., Williams, S., Kothapalli, D., Klapper, H., Grotendorst, G.R. 1996. Stimulation of fibroblast cell growth, matrix production, and granulation tissue formation by connective tissue growth factor. J. Invest. Derm. 107, 404–411.

Frid, M.G., Moiseeva, E.P., Stenmark, K.R. 1994. Multiple phenotypically distinct smooth muscle cell populations exist in the adult and developing bovine pulmonary arterial media *in vivo.* Circ. Res. 75, 669–681.

Friedrichsen, S., Heuer, H., Christ, S., Winckler, M., Brauer, D., Bauer, K., Raivich, G. 2003. CTGF expression during mouse embryonic development. Cell Tiss. Res. 312, 175–188.

Gacko, M., Chyczewski, L. 1997. Activity and localization of cathepsin B, D and G in aortic aneurysm. Int. Surg. 82, 398–402.

Gacko, M., Glowinski, S. 1998. Cathepsin D and cathepsin L activities in aortic aneurysm wall and parietal thrombus. Clin. Chem. Lab. Med. 36, 449–452.

Galvin, K.M., Donovan, M.J., Lynch, C.A., Meyer, R.I., Paul, R.J., Lorenz, J.N., Fairchild-Huntress, V., Dixon, K.L., Dunmore, J.H., Gimbrone, M.A., Jr., Falb, D., Huszar, D. 2000. A role for smad6 in development and homeostasis of the cardiovascular system. Nat. Genet. 24, 171–174.

Garbers, D.L., Dubois, S.K. 1999. The molecular basis of hypertension. Ann. Rev. Biochem. 68, 127–155.

Gerety, S.S., Wang, H.U., Chen, Z.F., Anderson, D.J. 1999. Symmetrical mutant phenotypes of the receptor EphB4 and its specific transmembrane ligand ephrin-B2 in cardiovascular development. Molec. Cell 4, 403–414.

Gerrity, R.G., Cliff, W.J. 1975. The aortic tunica media of the developing rat. I. Quantitative stereologic and biochemical analysis. Lab. Invest. 32, 585–600.

Gittenberger-de Groot, A.C., De Ruiter, M.C., Bergwerff, M., Poelmann, R.E. 1999. Smooth muscle cell origin and its relation to heterogeneity in development and disease. Arterioscler Thromb. Vasc. Biol. 19, 1589–1594.

Glukhova, M.A., Koteliansky, V.E. 1995. Integrins, Cytoskeletal and Extracellular Matrix Proteins in Developing Smooth Muscle Cells of Human Aorta. In: *The Vascular Smooth Muscle Cell: Molecular and Biological Responses to the Extracellular Matrix* (Schwartz, Mecham, Eds.), San Diego: Academic Press, pp. 37–79.

Goetze, S., Kintscher, U., Kim, S., Meehan, W.P., Kaneshiro, K., Collins, A.R., Fleck, E., Hsueh, W.A., Law, R.E. 2001. Peroxisome proliferator-activated receptor-gamma ligands inhibit nuclear but not cytosolic extracellular signal-regulated kinase/mitogen-activated protein kinase-regulated steps in vascular smooth muscle cell migration. J. Cardiovasc. Pharmacol. 38, 909–921.

Goetze, S., Xi, X.P., Kawano, H., Gotlibowski, T., Fleck, E., Hsueh, W.A., Law, R.E. 1999. PPAR gamma-ligands inhibit migration mediated by multiple chemoattractants in vascular smooth muscle cells. J. Cardiovasc. Pharmacol. 33, 798–806.

Gorski, D.H., Walsh, K. 2000. The role of homeobox genes in vascular remodeling and angiogenesis. Circ. Res. 87, 865–872.

Goumans, M.J., Mummery, C. 2000. Functional analysis of the TGF-ß receptor/Smad pathway through gene ablation in mice. Int. J. Dev. Biol. 44, 253–265.

Groth, C., Lardelli, M. 2002. The structure and function of vertebrate fibroblast growth factor receptor 1. Int. J. Develop. Biol. 46, 393–400.

Gu, Z., Reynolds, E., Song, J., Lei, H., Feijen, A., Yu, L., He, W., MacLaughlin, D., van den Eijnden-van Raaij, J., Donahoe, P., Li, E. 1999. The type I serine/threonine kinase receptor ActRIA (ALK2) is required for gastrulation of the mouse embryo. Development 126, 2551–2561.

Hamblet, N.S., Lijam, N., Ruiz-Lozano, P., Wang, J., Yang, Y., Luo, Z., Mei, L., Chien, K.R., Sussman, D.J., Wynshaw-Boris, A. 2002. Dishevelled 2 is essential for cardiac outflow tract development, somite segmentation and neural tube closure. Development 129, 5827–5838.

Han, R.N., Post, M., Tanswell, A.K., Lye, S.J. 2003. Insulin-like growth factor-I receptor-mediated vasculogenesis/angiogenesis in human lung development. Am. J. Resp. Cell. Molec. Biol. 28, 159–169.

Handford, P.A., Downing, A.K., Reinhardt, D.P., Sakai, L.Y. 2000. Fibrillin: From domain structure to supramolecular assembly. Matrix Biol. 19, 457–470.

Harrelson, Z., Kelly, R.G., Goldin, S.N., Gibson-Brown, J.J., Bollag, R.J., Silver, L.M., Papaioannou, V.E. 2004. Tbx2 is essential for patterning the atrioventricular canal and for morphogenesis of the outflow tract during heart development. Development 131, 5041–5052.

Haust, M.D., More, R.H., Benscome, S.A., Balis, J.U. 1965. Elastogenesis in human aorta: An electron microscopic study. Exp. Mol. Pathol. 4, 508–524.

Hautmann, M.B., Thompson, M.M., Swartz, E.A., Olson, E.N., Owens, G.K. 1997. Angiotensin II-induced stimulation of smooth muscle alpha-actin expression by serum response factor and the homeodomain transcription factor MHox. Circ. Res. 81, 600–610.

Hellstrom, M., Kal n, M., Lindahl, P., Abramsson, A., Betsholtz, C. 1999. Role of PDGF-B and PDGFR-ß in recruitment of vascular smooth muscle cells and pericytes during embryonic blood vessel formation in the mouse. Development 126, 3047–3055.

Heximer, S.P., Knutsen, R.H., Sun, X., Kaltenbronn, K.M., Rhee, M.H., Peng, N., Oliveira-dos-Santos, A., Penninger, J.M., Muslin, A.J., Steinberg, T.H., Wyss, J.M., Mecham, R.P., Blumer, K.J. 2003. Hypertension and prolonged vasoconstrictor signaling in RGS2-deficient mice. J. Clin. Invest. 111, 445–452.

Hilgers, K.F., Reddi, V., Krege, J.H., Smithies, O., Gomez, R.A. 1997. Aberrant renal vascular morphology and renin expression in mutant mice lacking angiotensin-converting enzyme. Hypertension 29, 216–221.

Hishikawa, K., Nakaki, T., Fujii, T. 1999. Transforming growth factor-beta(1) induces apoptosis via connective tissue growth factor in human aortic smooth muscle cells. Eur. J. Pharmacol. 385, 287–290.

Hishikawa, K., Nakaki, T., Fujii, T. 2000. Connective tissue growth factor induces apoptosis via caspase 3 in cultured human aortic smooth muscle cells. Eur. J. Pharmacol. 392, 19–22.

Hixon, M.L., Muro-Cacho, C., Wagner, M.W., Obejero-Paz, C., Millie, E., Fujio, Y., Kureishi, Y., Hassold, T., Walsh, K., Gualberto, A. 2000. Akt1/PKB upregulation leads to vascular smooth muscle cell hypertrophy and polyploidization. J. Clin. Invest. 106, 1011–1020.

Horiuchi, K., Weskamp, G., Lum, L., Hammes, H.P., Cai, H., Brodie, T.A., Ludwig, T., Chiusaroli, R., Baron, R., Preissner, K.T., Manova, K., Blobel, C.P. 2003. Potential role for ADAM15 in pathological neovascularization in mice. Molec. Cell. Biol. 23, 5614–5624.

Hornstra, I.K., Birge, S., Starcher, B., Bailey, A.J., Mecham, R.P., Shapiro, S.D. 2003. Lysyl oxidase is required for vascular and diaphragmatic development in mice. J. Biol. Chem. 278, 14387–14393.

Hungerford, J.E., Owens, G.K., Argraves, W.S., Little, C.D. 1996. Development of the aortic vessel wall as defined by vascular smooth muscle and extracellular matrix markers. Develop. Biol. 178, 375–392.

Hynes, R.O. 2002. A reevaluation of integrins as regulators of angiogenesis. Nat. Med. 8, 918–921.

Iaccarino, G., Ciccarelli, M., Sorriento, D., Cipolletta, E., Cerullo, V., Iovino, G.L., Paudice, A., Elia, A., Santulli, G., Campanile, A., Arcucci, O., Pastore, L., Salvatore, F., Condorelli, G., Trimarco, B. 2004. AKT Participates in Endothelial Dysfunction in Hypertension. Circulation 109, 2587–2593.

Iozzo, R.V., Murdoch, A.D. 1996. Proteoglycans of the extracellular environment: Clues from the gene and protein side offer novel perspectives in molecular diversity and function. FASEB J. 10, 598–614.

Ishii, M., Kurachi, Y. 2003. Physiological actions of regulators of G-protein signaling (RGS) proteins. Life Sci. 74, 163–171.

Jimenez, B., Volpert, O.V., Crawford, S.E., Febbraio, M., Silverstein, R.L., Bouck, N. 2000. Signals leading to apoptosis-dependent inhibition of neovascularization by thrombospondin-1. Nat. Med. 6, 41–48.

Johnsen, S.A., Subramaniam, M., Monroe, D.G., Janknecht, R., Spelsberg, T.C. 2002. Modulation of transforming growth factor beta (TGF-ß)/Smad transcriptional responses through targeted degradation of TGF-ß-inducible early gene-1 by human seven in absentia homologue. J. Biol. Chem. 277, 30754–30759.

Jones, P.L., Rabinovitch, M. 1996. Tenascin-C is induced with progressive pulmonary vascular disease in rats and is functionally related to increased smooth muscle cell proliferation. Circ. Res. 79, 1131–1142.

Kaartinen, V., Dudas, M., Nagy, A., Sridurongrit, S., Lu, M.M., Epstein, J.A. 2004. Cardiac outflow tract defects in mice lacking ALK2 in neural crest cells. Development 131, 3481–3490.

Kagiyama, S., Eguchi, S., Frank, G.D., Inagami, T., Zhang, Y.C., Phillips, M.I. 2002. Angiotensin II-induced cardiac hypertrophy and hypertension are attenuated by epidermal growth factor receptor antisense. Circulation 106, 909–912.

Kalinichenko, V.V., Gusarova, G.A., Tan, Y., Wang, I.C., Major, M.L., Wang, X., Yoder, H. M., Costal, R.H. 2003. Ubiquitous expression of the forkhead box M1B transgene accelerates proliferation of distinct pulmonary cell types following lung injury. J. Biol. Chem. 278, 37888–37894.

Karnik, S.K., Brooke, B.S., Bayes-Genis, A., Sorensen, L., Wythe, J.D., Schwartz, R.S., Keating, M.T., Li, D.Y. 2003. A critical role for elastin signaling in vascular morphogenesis and disease. Development 130, 411–423.

Karrer, H.E. 1961. An electron microscope study of the aorta in young and aging mice. J. Ultrastruct. Res. 5, 1–17.

Kelleher, C.M., McLean, S.E., Mecham, R.P. 2004. Vascular extracellular matrix and aortic development. Curr. Top. Dev. Biol. 62, 153–188.

Kielty, C.M., Sherratt, M.J., Shuttleworth, C.A. 2002. Elastic fibres. J. Cell. Sci. 115, 2817–2828.

Kim, R.Y., Robertson, E.J., Solloway, M.J. 2001. Bmp6 and Bmp7 are required for cushion formation and septation in the developing mouse heart. Develop. Biol. 235, 449–466.

King, K.E., Iyemere, V.P., Weissberg, P.L., Shanahan, C.M. 2003. Kruppel-like factor 4 (KLF4/GKLF) is a target of bone morphogenetic proteins and transforming growth factor beta 1 in the regulation of vascular smooth muscle cell phenotype. J. Biol. Chem. 278, 11661–11669.

Kirby, R.J., Hamilton, G.M., Finnegan, D.J., Johnson, K.J., Jarman, A.P. 2001. Drosophila homolog of the myotonic dystrophy-associated gene, SIX5, is required for muscle and gonad development. Curr. Biol. 11, 1044–1049.

Krege, J.H., John, S.W., Langenbach, L.L., Hodgin, J.B., Hagaman, J.R., Bachman, E.S., Jennette, J.C., O'Brien, D.A., Smithies, O. 1995. Male-female differences in fertility and blood pressure in ACE-deficient mice. Nature 375, 146–148.

Kuo, C.T., Veselits, M.L., Barton, K.P., Lu, M.M., Clendenin, C., Leiden, J.M. 1997. The LKLF transcription factor is required for normal tunica media formation and blood vessel stabilization during murine embryogenesis. Genes Dev. 11, 2996–3006.

Langille, B.L. 1996. Arterial remodeling: Relation to hemodynamics. Can. J. Physiol. Pharmacol. 74, 834–841.

Larsson, J., Goumans, M.J., Sjostrand, L.J., van Rooijen, M.A., Ward, D., Leveen, P., Xu, X., ten Dijke, P., Mummery, C.L., Karlsson, S. 2001. Abnormal angiogenesis but intact hematopoietic potential in TGF-beta type I receptor-deficient mice. EMBO J. 20, 1663–1673.

Lechleider, R.J., Ryan, J.L., Garrett, L., Eng, C., Deng, C., Wynshaw-Boris, A., Roberts, A.B. 2001. Targeted mutagenesis of Smad1 reveals an essential role in chorioallantoic fusion. Dev. Biol. 240, 157–167.

Li, C., Wong, W.H. 2001. Model-based analysis of oligonucleotide arrays: Expression index computation and outlier detection. Proc. Natl. Acad. Sci. USA 98, 31–36.

Li, D.Y., Brooke, B., Davis, E.C., Mecham, R.P., Sorensen, L.K., Boak, B.B., Eichwald, E., Keating, M.T. 1998a. Elastin is an essential determinant of arterial morphogenesis. Nature 393, 276–280.

Li, D.Y., Faury, G., Taylor, D.G., Davis, E.C., Boyle, W.A., Mecham, R.P., Stenzel, P., Boak, B., Keating, M.T. 1998b. Novel arterial pathology in mice and humans hemizygous for elastin. J. Clin. Invest. 102, 1783–1777.

Li, D.Y., Sorensen, L.K., Brooke, B.S., Urness, L.D., Davis, E.C., Taylor, D.G., Boak, B.B., Wendel, D.P. 1999. Defective angiogenesis in mice lacking endoglin. Science 284, 1534–1537.

Li, S., Edgar, D., Fassler, R., Wadsworth, W., Yurchenco, P.D. 2003. The role of laminin in embryonic cell polarization and tissue organization. Dev. Cell 4, 613–624.

Lijnen, H.R., Ugwu, F., Bini, A., Collen, D. 1998. Generation of an angiostatin-like fragment from plasminogen by stromelysin-1 (MMP-3). Biochemistry 37, 4699–4702.

Lin, Q., Lu, J., Yanagisawa, H., Webb, R., Lyons, G.E., Richardson, J.A., Olson, E.N. 1998. Requirement of the MADS-box transcription factor MEF2C for vascular development. Development 125, 4565–4574.

Ma, J.X., Yang, Z., Chao, J., Chao, L. 1995. Intramuscular delivery of rat kallikrein-binding protein gene reverses hypotension in transgenic mice expressing human tissue kallikrein. J. Biol. Chem. 270, 451–455.

Macias-Silva, M., Hoodless, P.A., Tang, S.J., Buchwald, M., Wrana, J.L. 1998. Specific Activation of Smad1 Signaling Pathways by the BMP7 Type I Receptor, ALK2. J. Biol. Chem. 273, 25628–25636.

Maltepe, E., Schmidt, J.V., Baunoch, D., Bradfield, C.A., Simon, M.C. 1997. Abnormal angiogenesis and responses to glucose and oxygen deprivation in mice lacking the protein ARNT. Nature 386, 403–407.

Martin, K.M., Ellis, P.D., Metcalfe, J.C., Kemp, P.R. 2003. Selective modulation of the SM22alpha promoter by the binding of BTEB3 (basal transcription element-binding protein 3) to TGGG repeats. Biochem. J. 375, 457–463.

McCawley, L.J., Matrisian, L.M. 2001. Matrix metalloproteinases: They're not just for matrix anymore!. Curr. Opin. Cell Biol. 13, 534–540.

Mecham, R.P. 1998. Overview of extracellular matrix. *Current Protocols in Cell Biology*, New York: John Wiley & Sons Inc., pp. 10.1.1–10.1.13.

Mecham, R.P., Davis, E.C. 1994. Elastic fiber structure and assembly. In: *Extracellular Matrix Assembly and Structure* (Yurchenko, Birk, Mecham, Eds.), San Diego: Academic Press, pp. 281–314.

Miano, J.M., Topouzis, S., Majesky, M.W., Olson, E.N. 1996. Retinoid receptor expression and all-trans retinoic acid-mediated growth inhibition in vascular smooth muscle cells. Circulation 93, 1886–1895.

Miettinen, P.J., Berger, J.E., Meneses, J., Phung, Y., Pedersen, R.A., Werb, Z., Derynck, R. 1995. Epithelial immaturity and multiorgan failure in mice lacking epidermal growth factor receptor. Nature 376, 337–341.

Morikawa, W., Yamamoto, K., Ishikawa, S., Takemoto, S., Ono, M., Fukushi, J., Naito, S., Nozaki, C., Iwanaga, S., Kuwano, M. 2000. Angiostatin generation by cathepsin D secreted by human prostate carcinoma cells. J. Biol. Chem. 275, 38912–38920.

Mudgett, J.S., Hutchinson, N.I., Chartrain, N.A., Forsyth, A.J., McDonnell, J., Singer, II, Bayne, E.K., Flanagan, J., Kawka, D., Shen, C.F., Stevens, K., Chen, H., Trumbauer, M., Visco, D.M. 1998. Susceptibility of stromelysin 1-deficient mice to collagen-induced arthritis and cartilage destruction. Arthritis. Rheum. 41, 110–121.

Muragaki, Y., Timmons, S., Griffith, C.M., Oh, S.P., Fadel, B., Quertermous, T., Olsen, B.R. 1995. Mouse Col18a1 is expressed in a tissue-specific manner as three alternative variants and is localized in basement membrane zones. Proc. Natl. Acad. Sci. USA 92, 8763–8767.

Nabel, E.G., Yang, Z.Y., Plautz, G., Forough, R., Zhan, X., Haudenschild, C.C., Maciag, T., Nabel, G.J. 1993. Recombinant fibroblast growth factor-1 promotes intimal hyperplasia and angiogenesis in arteries *in vivo*. Nature 362, 844–846.

Nakamura, H. 1988. Electron microscopic study of the prenatal development of the thoracic aorta in the rat. Am. J. Anat. 181, 406–418.

Nguyen, L.L., D'Amore, P.A. 2001. Cellular interactions in vascular growth and differentiation. Int. Rev. Cytol. 204, 1–48.

O'Reilly, M.S., Boehm, T., Shing, Y., Fukai, N., Vasios, G., Lane, W.S., Flynn, E., Birkhead, J.R., Olsen, B.R., Folkman, J. 1997. Endostatin: An endogenous inhibitor of angiogenesis and tumor growth. Cell 88, 277–285.

Oh, S.P., Seki, T., Goss, K.A., Imamura, T., Yi, Y., Donahoe, P.K., Li, L., Miyazono, K., ten Dijke, P., Kim, S., Li, E. 2000. Activin receptor-like kinase 1 modulates transforming growth factor-beta 1 signaling in the regulation of angiogenesis. Proc. Natl. Acad. Sci. USA 97, 2626–2631.

Olson, T.M., Michels, V.V., Urban, Z., Csiszar, K., Christiano, A.M., Driscoll, D.J., Feldt, R.H., Boyd, C.D., Thibodeau, S.N. 1995. A 30 kb deletion within the elastin gene results in familial supravalvular aortic stenosis. Hum. Molec. Gen. 4, 1677–1679.

Onda, M., Ishiwata, T., Kawahara, K., Wang, R., Naito, Z., Sugisaki, Y. 2002. Expression of lumican in thickened intima and smooth muscle cells in human coronary atherosclerosis. Exp. Molec. Pathol. 72, 142–149.

Ornitz, D.M., Xu, J., Colvin, J.S., McEwen, D.G., MacArthur, C.A., Coulier, F., Gao, G., Goldfarb, M. 1996. Receptor specificity of the fibroblast growth factor family. J. Biol. Chem. 271, 15292–15297.

Owens, G.K. 1995. Regulation of differentiation of vascular smooth muscle cells. Physiol. Rev. 75, 487–517.

Owens, G.K., Kumar, M.S., Wamhoff, B.R. 2004. Molecular regulation of vascular smooth muscle cell differentiation in development and disease. Physiol. Rev. 84, 767–801.

Ozaki, H., Watanabe, Y., Takahashi, K., Kitamura, K., Tanaka, A., Urase, K., Momoi, T., Sudo, K., Sakagami, J., Asano, M., Iwakura, Y., Kawakami, K. 2001. Six4, a putative myogenin gene regulator, is not essential for mouse embryonal development. Molec. Cell. Biol. 21, 3343–3350.

Paule, W.J. 1963. Electron microscopy of the newborn rat aorta. J. Ultrastruct. Res. 8, 219–235.

Pease, D.C., Paule, W.J. 1960. Electron microscopy of elastic arteries; the thoracic aorta of the rat. J. Ultrastruct. Res. 3, 469–483.

Peng, J., Zhang, L., Drysdale, L., Fong, G.H. 2000. The transcription factor EPAS-1/hypoxia-inducible factor 2alpha plays an important role in vascular remodeling. Proc. Natl. Acad. Sci. USA 97, 8386–8391.

Pepper, M.S. 2001. Extracellular proteolysis and angiogenesis. Thromb. Haemost. 86, 346–355.

Petajaniemi, N., Korhonen, M., Kortesmaa, J., Tryggvason, K., Sekiguchi, K., Fujiwara, H., Sorokin, L., Thornell, L.E., Wondimu, Z., Assefa, D., Patarroyo, M., Virtanen, I. 2002. Localization of laminin alpha4-chain in developing and adult human tissues. J. Histochem. Cytochem. 50, 1113–1130.

Ploplis, V.A., Carmeliet, P., Vazirzadeh, S., Van Vlaenderen, I., Moons, L., Plow, E.F., Collen, D. 1995. Effects of disruption of the plasminogen gene on thrombosis, growth, and health in mice. Circulation 92, 2585–2593.

Riquelme, C., Larrain, J., Schonherr, E., Henriquez, J.P., Kresse, H., Brandan, E. 2001. Antisense inhibition of decorin expression in myoblasts decreases cell responsiveness to transforming growth factor beta and accelerates skeletal muscle differentiation. J. Biol. Chem. 276, 3589–3596.

Risau, W., Flamme, I. 1995. Vasculogenesis. Annu. Rev. Cell. Dev. Biol. 11, 73–91.

Roose, J., Korver, W., Oving, E., Wilson, A., Wagenaar, G., Markman, M., Lamers, W., Clevers, H. 1998. High expression of the HMG box factor sox-13 in arterial walls during embryonic development. Nucleic Acids Res. 26, 469–476.

Rossant, J.a.H., Lorraine 2002. Signaling Pathways in Vascular Development. Annu. Rev. Cell. Dev. Biol. 18, 541–573.

Ruegg, C., Mariotti, A. 2003. Vascular integrins: Pleiotropic adhesion and signaling molecules in vascular homeostasis and angiogenesis. Cell Molec. Life Sci. 60, 1135–1157.

Sambhi, M.P., Swaminathan, N., Wang, H., Rong, H. 1992. Increased EGF binding and EGFR mRNA expression in rat aorta with chronic administration of pressor angiotensin II. Biochem. Med. Metab. Biol. 48, 8–18.

Santiago, F.S., Atkins, D.G., Khachigian, L.M. 1999. Vascular smooth muscle cell proliferation and regrowth after mechanical injury in vitro are Egr-1/NGFI-A-dependent. Am. J. Pathol. 155, 897–905.

Schwartz, S.M., Mecham, R.P. 1995. In: The Vascular Smooth Muscle Cell: Molecular and Biological Responses to the Extracellular Matrix San Diego: Academic Press, p. 410.

Shapiro, S.D., Griffin, G.L., Gilbert, D.J., Jenkins, N.A., Copeland, N.G., Welgus, H.G., Senior, R.M., Ley, T.J. 1992. Molecular cloning, chromsomal localization, and bacterial expression of a muring macrophage metalloelastase. J. Biol. Chem. 267, 4664–4671.

Shi, Y., Patel, S., Niculescu, R., Chung, W., Desrochers, P., Zalewski, A. 1999. Role of matrix metalloproteinases and their tissue inhibitors in the regulation of coronary cell migration. Arterioscler. Thromb. Vasc. Biol. 19, 1150–1155.

Shiojima, I., Walsh, K. 2002. Role of Akt signaling in vascular homeostasis and angiogenesis. Circ. Res. 90, 1243–1250.

Sibilia, M., Steinbach, J.P., Stingl, L., Aguzzi, A., Wagner, E.F. 1998. A strain-independent postnatal neurodegeneration in mice lacking the EGF receptor. EMBO J. 17, 719–731.

Sibilia, M., Wagner, E.F. 1995. Strain-dependent epithelial defects in mice lacking the EGF receptor. Science 269, 234–238.

Silverman, E.S., Khachigian, L.M., Santiago, F.S., Williams, A.J., Lindner, V., Collins, T. 1999. Vascular smooth muscle cells express the transcriptional corepressor NAB2 in response to injury. Am. J. Pathol. 155, 1311–1317.

Smyth, N., Vatansever, H.S., Murray, P., Meyer, M., Frie, C., Paulsson, M., Edgar, D. 1999. Absence of basement membranes after targeting the LAMC1 gene results in embryonic lethality due to failure of endoderm differentiation. J. Cell. Biol. 144, 151–160.

Soriano, P. 1994. Abnormal kidney development and hematological disorders in PDGF beta-receptor mutant mice. Genes Dev. 8, 1888–1896.

Sorokin, L.M., Pausch, F., Frieser, M., Kroger, S., Ohage, E., Deutzmann, R. 1997. Developmental regulation of the laminin alpha5 chain suggests a role in epithelial and endothelial cell maturation. Dev. Biol. 189, 285–300.

Spurbeck, W.W., Ng, C.Y., Strom, T.S., Vanin, E.F., Davidoff, A.M. 2002. Enforced expression of tissue inhibitor of matrix metalloproteinase-3 affects functional capillary morphogenesis and inhibits tumor growth in a murine tumor model. Blood 100, 3361–3368.

Strydom, D.J. 1998. The angiogenins. Cell. Molec. Life Sci. 54, 811–824.

Swaminathan, N., Vincent, M., Sassard, J., Sambhi, M.P. 1996. Elevated epidermal growth factor receptor levels in hypertensive Lyon rat kidney and aorta. Clin. Exp. Pharmacol. Physiol. 23, 793–796.

Tanaka, K., Hiraiwa, N., Hashimoto, H., Yamazaki, Y., Kusakabe, M. 2004. Tenascin-C regulates angiogenesis in tumor through the regulation of vascular endothelial growth factor expression. Int. J. Cancer 108, 31–40.

Tang, M., Wang, G., Lu, P., Karas, R.H., Aronovitz, M., Heximer, S.P., Kaltenbronn, K.M., Blumer, K.J., Siderovski, D.P., Zhu, Y., Mendelsohn, M.E. 2003. Regulator of G-protein signaling-2 mediates vascular smooth muscle relaxation and blood pressure. Nat. Med. 9, 1506–1512.

Theocharis, A.D., Karamanos, N.K. 2002. Decreased biglycan expression and differential decorin localization in human abdominal aortic aneurysms. Atherosclerosis 165, 221–230.

Threadgill, D.W., Dlugosz, A.A., Hansen, L.A., Tennenbaum, T., Lichti, U., Yee, D., LaMantia, C., Mourton, T., Herrup, K., Harris, R.C., et al. 1995. Targeted disruption of mouse EGF receptor: Effect of genetic background on mutant phenotype. Science 269, 230–234.

Thyberg, J., Hinek, A., Nilsson, J., Friberg, U. 1979. Electron microscopic and cytochemical studies of rat aorta. Intracellular vesicles containing elastin- and collagen-like material. Histochem. J. 11, 1–17.

Thyboll, J., Kortesmaa, J., Cao, R., Soininen, R., Wang, L., Iivanainen, A., Sorokin, L., Risling, M., Cao, Y., Tryggvason, K. 2002. Deletion of the laminin alpha4 chain leads to impaired microvessel maturation. Molec. Cell. Biol. 22, 1194–1202.

Timpl, R., Dziadek, M., Fujiwara, S., Nowack, H., Wick, G. 1983. Nidogen: A new, self-aggregating basement membrane protein. Eur. J. Biochem. 137, 455–465.

Trask, B.C., Trask, T.M., Broekelmann, T., Mecham, R.P. 2000. The microfibrillar proteins MAGP-1 and fibrillin-1 form a ternary complex with the chondroitin sulfate proteoglycan decorin. Molec. Biol. Cell. 11, 1499–1507.

Tremblay, K.D., Dunn, N.R., Robertson, E.J. 2001. Mouse embryos lacking Smad1 signals display defects in extra-embryonic tissues and germ cell formation. Development 128, 3609–3621.

Urness, L.D., Sorensen, L.K., Li, D.Y. 2000. Arteriovenous malformations in mice lacking activin receptor-like kinase-1. Nat. Genet. 26, 328–331.

van Gijn, M.E., Blankesteijn, W.M., Smits, J.F., Hierck, B., Gittenberger-de Groot, A.C. 2001. Frizzled 2 is transiently expressed in neural crest-containing areas during development of the heart and great arteries in the mouse. Anat. Embryol. 203, 185–192.

Vandromme, M., Rochat, A., Meier, R., Carnac, G., Besser, D., Hemmings, B.A., Fernandez, A., Lamb, N.J.C. 2001. Protein kinase B beta /Akt2 plays a specific role in muscle differentiation. J. Biol. Chem. 276, 8173–8179.

Wang, H.U., Chen, Z.F., Anderson, D.J. 1998. Molecular distinction and angiogenic interaction between embryonic arteries and veins revealed by ephrin-B2 and its receptor Eph-B4. Cell 93, 741–753.

Wang, J., Xiong, W., Yang, Z., Davis, T., Dewey, M.J., Chao, J., Chao, L. 1994. Human tissue kallikrein induces hypotension in transgenic mice. Hypertension 23, 236–243.

Wang, L.C., Kuo, F., Fujiwara, Y., Gilliland, D.G., Golub, T.R., Orkin, S.H. 1997. Yolk sac angiogenic defect and intra-embryonic apoptosis in mice lacking the Ets-related factor TEL. EMBO J. 16, 4374–4383.

Wight, T.N. 2002. Versican: A versatile extracellular matrix proteoglycan in cell biology. Curr. Opin. Cell. Biol. 14, 617–623.

Wolinsky, H., Glagov, S. 1967. A lamellar unit of aortic medial structure and function in mammals. Circ. Res. 20, 99–111.

Wu, L., Tanimoto, A., Murata, Y., Sasaguri, T., Fan, J., Sasaguri, Y., Watanabe, T. 2003. Matrix metalloproteinase-12 gene expression in human vascular smooth muscle cells. Genes Cells 8, 225–234.

Xin, X., Yang, S., Kowalski, J., Gerritsen, M.E. 1999. Peroxisome proliferator-activated receptor gamma ligands are potent inhibitors of angiogenesis *in vitro* and *in vivo*. J. Biol. Chem. 274, 9116–9121.

Xu, H., el-Gewely, M.R. 2001. P53-responsive genes and the potential for cancer diagnostics and therapeutics development. Biotechnology Annu. Rev. 7, 131–164.

Xu, W., Baribault, H., Adamson, E.D. 1998. Vinculin knockout results in heart and brain defects during embryonic development. Development 125, 327–337.

Yamaguchi, T.P., Harpal, K., Henkemeyer, M., Rossant, J. 1994. Fgfr-1 is required for embryonic growth and mesodermal patterning during mouse gastrulation. Genes Dev. 8, 3032–3044.

Yang, X., Castilla, L.H., Xu, X., Li, C., Gotay, J., Weinstein, M., Liu, P.P., Deng, C.X. 1999. Angiogenesis defects and mesenchymal apoptosis in mice lacking SMAD5. Development 126, 1571–1580.

Yao, L.Y., Moody, C., Schonherr, E., Wight, T.N., Sandell, L.J. 1994. Identification of the proteoglycan versican in aorta and smooth muscle cells by DNA sequence analysis, in situ hybridization and immunohistochemistry. Matrix Biol. 14, 213–225.

Basement membrane and extracellular matrix molecules in the skin

Julia Tzu,[1] Jie Li[2] and M. Peter Marinkovich[1]

[1]Program in Epithelial Biology, Department of Dermatology, Stanford University School of Medicine and Veterans' Association Palo Alto Health Care System, Palo Alto, California
[2]Department of Dermatology, University of Miami School of Medicine, Miami, Florida

Contents

Advances in Developmental Biology
Volume 15 ISSN 1574-3349
DOI: 10.1016/S1574-3349(05)15004-2

1. Introduction

The dermal-epidermal basement membrane zone (BMZ) lies at the interface between the epithelium and underlying mesenchymal tissue. It is a highly organized structure with both similarities and differences compared to BMZs present in other tissues. It is extremely dynamic, undergoing constant renewal, providing communication signals, and contributing to normal developmental processes. In the skin, the BMZ serves many functions. Its architectural role includes providing stable dermal-epidermal cohesion and mechanical integrity of the skin. It also regulates and serves as a source of ligands/signaling molecules that bind to cellular receptors such as integrins, to influence cellular proliferation, migration, and differentiation. In this manner, it also contributes to the characteristic polarized stratification found in the epidermis, whereby proliferation begins basally and differentiation progresses as the cells migrate away towards the surface.

The BMZ of the skin is composed mainly of collagen and laminin networks, connected by proteoglycans/glycosaminoglycans, all of which can undergo degradation/activation by metalloproteinases. Under electron microscopy, small, evenly distributed electron-dense hemidesmosomes, a complex of many proteins which links the keratinocyte cytoskeleton with the basal lamina, are found at the border of the keratinocyte adjacent to the BMZ. This is followed by a thin electron-poor lamina lucida region comprised mostly of the extracellular portion of transmembrane-anchoring filament proteins, such as collagen XVII/BP180 and $\alpha6\beta4$ integrin, extending from the hemidesmosomes. Underneath the lamina lucida, an electron dense lamina densa can be found, comprised of collagens, laminins, and proteoglycans. The sublamina densa consists of collagen-anchoring fibrils to the dermis (McMillan et al., 2003). Figure 1 illustrates the structural participants of the BMZ.

This chapter will provide an overview on the wide variety of molecules found in the extracellular matrix (ECM) of the skin, with particular focus on the BMZ. The role of each molecule in normal as well as disease and developmental processes in the skin will be explored.

2. Interactions of the basal keratinocyte and the BMZ

2.1. Integrins

Integrins are transmembrane heterodimeric cellular adhesion molecules involved in cell-ECM and cell-cell interactions which provide for physical anchorage as well as initiating signaling pathways involved in cell survival, proliferation, and migration. Integrins are composed of an α chain that provides for substrate specificity and a ß chain that mediates signal transduction.

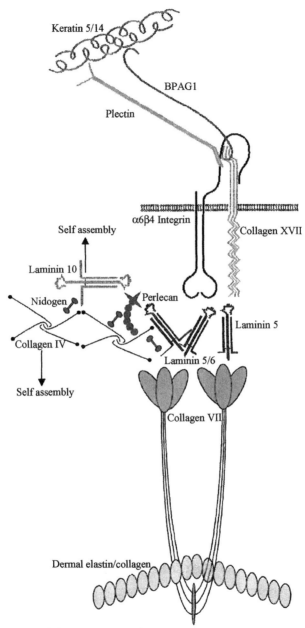

Fig. 1. General structure of the BMZ. Proceeding from the top of the figure, the BMZ is divided into the lamina lucida which consists of the extracellular extensions of the hemidesmosome (BP180 and α6β4integrins); the lamina densa which consists of an extensive network of collagen IV, laminins, perlecan, and nidogen; and the sublamina densa which consists of collagen VII fibrils.

Eighteen different α chains and eight different ß chains provide for a total of 24 known integrins, with different heterodimeric combinations yielding different substrate specificities (Hynes, 2002). Table 1 provides a list of various integrin receptors that associate with cell surface and ECM ligands.

Integrin activation involves the traditional outside-in signaling as well as inside-out signaling, whereby affinity of the integrin extracellular ligand-binding domains is increased through allosteric activation by the cytoplasmic domain via proteins such as talin and cytohesin-1 (Liddington and Ginsberg, 2002). Once activated, integrins effect signaling in a variety of ways. They can act synergistically with other growth factor receptors to phosphorylate downstream proteins such as raf, or MEK or to recruit protein kinases and their substrates together at the membrane in multi-matrix adhesion aggregates. These matrix adhesions then link the integrins with the actin cytoskeleton, with the force of any cytoskeletal structural changes transmitted towards the nucleus to effect changes in gene transcription. Integrin-mediated adhesion can also activate many receptor tyrosine kinases in the absence of growth factors. Moreover, integrins can activate latent growth factors, such as transforming growth factor (TGF)-β, to effect cellular changes (Danen et al., 2000).

The function of integrin signaling is ultimately to regulate cellular proliferation, migration, survival, and differentiation. Proliferation is mediated by driving cells into the S phase via rac and rho GTPase regulation of cyclin D1 and cki (Mettouchi et al., 2001; Welsh et al., 2001). Cell survival is sustained

Table 1

Major skin cells, extracellular matrix ligands, and their integrin receptors. Integrin receptors for the major cell types and extracellular matrix molecules. Different combinations of α and β chains create a variety of different integrin receptors, which are highly specific for only certain cell types and extracellular ligands

Cells/matrix	Integrin receptors
Cells	
Keratinocytes	$\alpha2\beta1$, $\alpha3\beta1$, $\alpha5\beta1$, $\alpha6\beta4$, $\alpha v\beta3$, $\alpha v\beta5$
Fibroblasts	$\alpha1\beta1$, $\alpha2\beta1$, $\alpha3\beta1$, $\alpha4\beta1$, $\alpha5\beta1$, $\alpha5\beta1$, $\alpha v\beta3$, $\alpha v\beta3$, $\alpha v\beta5$, $\alpha v\beta6$
Vascular endothelial cells	$\alpha1\beta1$, $\alpha2\beta1$, $\alpha3\beta1$, $\alpha5\beta1$, $\alpha v\beta3$, $\alpha v\beta5$
Extracellular matrix ligands	
Collagens	$\alpha1\beta1$, $\alpha2\beta1$
Fibronectin	$\alpha3\beta1$, $\alpha4\beta1$, $\alpha5\beta1$, $\alpha v\beta3$, $\alpha v\beta5$, $\alpha v\beta6$
Laminins	$\alpha1\beta1$, $\alpha2\beta1$, $\alpha3\beta1$, $\alpha6\beta1$, $\alpha6\beta4$, $\alpha v\beta3$, $\alpha v\beta5$
MMPs	$\alpha2\beta1$, $\alpha v\beta3$
Tenascins	$\alpha8\beta1$, $\alpha9\beta1$, $\alpha v\beta3$, $\alpha v\beta6$
Vitronectin	$\alpha v\beta1$, $\alpha v\beta3$, $\alpha v\beta5$, $\alpha IIb\beta3$

by integrin adhesion or ligand-binding, as demonstrated by cells which upregulate mediators of the apoptotic pathway upon loss of integrin activity (Boudreau et al., 1995; Stupack et al., 2001). Differentiation can also be regulated by integrin activation, as demonstrated by inhibition of keratinocyte terminal differentiation in suspension culture when α5β1 integrin ligand fibronectin is added (Watt et al., 1993).

The major integrins expressed in keratinocytes of human skin include laminin 5 binding 6β4 and 3β1. Minor epidermal integrins include collagen and laminin binding 2β1, fibronectin binding 5β1, tenascin-C binding 9β1, vitronectin binding vβ5, and wound-induced vβ6 (Palmer et al., 1993; Yokosaki et al., 1994; Marcinkiewicz et al., 2000). As such, integrins play a large role in normal and pathological skin processes.

The 6β4 integrin is a critical component of hemidesmosomal structure. The cytoplasmic portion of the 6β4 integrin is connected to the keratinocyte cytoskeleton via plakin proteins BP230 and plectins. The extracellular domain connects the hemidesmosome to laminins of the lamina densa (Zillikens, 1999). Structurally abnormal 6β4 leads to epidermolysis bullosa with pyloric atresia, an autosomal recessive blistering disease in humans. Because the fundamental defect occurs at the level of the hemidesmosome, composed of a number of different proteins one of which is α6β4 integrin, disruption in any of the constituents leads to detachment between the epidermis from the dermis with slight mechanical trauma (Wehrle-Haller and mhof, 2003). This is confirmed in 6 or β4 knockout (KO) mice. Furthermore, β1 KO mouse models also show blistering defects, while α3 KO show microblistering in limb regions. Although not a direct participant in hemidesmosome formation, these integrins may compromise hemidesmosome integrity by affecting basement organization (Raghavan et al., 2000).

The β1 integrins are normally expressed initially in basal keratinocytes but are downregulated as keratinocytes move suprabasally and differentiate (Watt and Jones, 1993). Mice that express abnormal levels of β1 in suprabasal regions, however, exhibit sporadic psoriatic lesions, possibly due to increased mitogen-activated protein kinase (MAPK) activity levels (Haase et al., 2001). Conditional KO mice that do not express β1 as previously mentioned, however, show additional impaired epidermal proliferation, as well as hair follicle formation defects (Raghavan et al., 2000).

The vitronectin-binding vβ5 integrin has been demonstrated to be associated with melanoma progression. The v subunit is expressed in all stages of melanoma, whereas the β5 unit is expressed mostly in melanoma in the vertical growth phase. Evidence for the role of β5 in the transition from radial to vertical growth phase is not in complete agreement. Melanoma cells transfected with v complementary deoxyribonucleic acid (cDNA) have been shown to promote tumor development in one study but inhibit tumor development in another study. The vβ5 may also promote malignant development by associating with matrix metalloproteinase (MMP)–2 and thereby help localize MMPs to the cell surface. Additionally, metastasis to other sites within the body is promoted by increased vß5 interaction with L1 adhesion receptors on endothelial cells, and inhibited by lack of vß5 expression or presence of antibodies directed to either one of these components. The vß5-L1 interaction results in extravasation of the melanoma cells through the blood vessel walls (McGary et al., 2002).

Abnormal integrin expression profiles are exhibited during wound healing. Integrins expression extends to suprabasal layers and 5ß1 and vß5; and vß6 (Clark, 1990; Larjava et al., 1993; Zambruno et al., 1995), integrins which bind to fibronectins, tenascins, and TGF-ß, are characteristically induced. *In vivo*, ß1-deficient keratinocytes display delayed wound healing due to impaired keratinocyte migration and re-epithelialization (Grose et al., 2002).

2.2. Collagen XVII

Similar to integrin α6β4, collagen XVII (BP180 or BPAG2) is another transmembrane protein that links the keratinocyte surface to the basal lamina. The intracellular portion of collagen XVII is connected to the cytoskeletal elements via BP230, and also interacts with the β4 chain of α6β4 integrins. The extracellular collagenous portion of the molecule interacts with the α chain of α6β4 integrins and perhaps laminin 5 (McMillan et al., 2003). It may function in propagating signals across the membrane and regulating keratinocyte differentiation. The extracellular domain may also be proteolytically clipped from cell surface, forming linear IgA disease antigen I (LAD-1) fragment, the function of which is still unclear. Not surprisingly, patients with collagen XVII null mutations lack the LAD–1 fragment. Mutations of collagen XVII give rise to the generalized atrophic benign epidermolysis bullosa (GABEB) variant of junctional epidermolysis bullosa (JEB), in which patients exhibit congenital generalized blistering without scarring, mild mucosal lesions, atrophic alopecia, pigment defects, and dental/nail changes. Collagen XVII is often undetectable in the skin of these patients. Because of the location of the defect, blistering arises at the level of the hemidesmosome, although hemidesmosomes are still present in these patients (Jonkman et al., 1995). Mutations in collagen XVII also give rise to a rare variant of epidermolysis bullosa simplex, in which blistering occurs at the level of the epidermis. In some autoimmune disorders, autoantibodies to collagen XVII give rise to bullous pemphigoid, cicatricial pemphigoid, linear IgA dermatosis, lichen planus pemphigoides, and pemphigoid gestations. The underlying autoantigen is the same for each of the different diseases, but the antibody and the epitope to which it is targeted are different. However, bullous pemphigoid and pemphigoid gestations are indistinguishable through molecular analysis and may represent the same disease entity (Van den Bergh and Giudice, 2003).

3. Lamina densa structural components

3.1. Glycoproteins

3.1.1. Laminins

Laminins are a large family of extracellular heterotrimeric glycoproteins that participate in diverse functions including cell adhesion, migration, differentiation, proliferation, and angiogenesis. They are synthesized by endothelial and epithelial cells, and are major constituents of the basement membrane lamina densa. The 15 known laminin isoforms, each with different tissue specificities, are formed by the unique combination of five known α, three known β, and three known γ chains. Laminin molecules exhibit a cross shape, with three short arms and one long arm. The C terminal binds to cell surface receptors and is involved in adhesion and signaling, while the N terminal interacts with ECM proteins to provide anchorage to the underlying dermis. Structural features include small globular domains involved in polymerization, epidermal growth factor (EGF) repeats that connect to nidogen and thus the collagen IV network, and an integrin-binding site at the α unit. Laminins also interact with non-integrins such as α-dystroglycan and syndecans (Ido et al., 2004).

Laminins found in skin include laminins 1 $(\alpha1\beta1\gamma1)$, 5 $(\alpha3\beta3\gamma2)$, 6 $(\alpha3\beta1\gamma1)$, 7 $(\alpha3\beta2\gamma1)$, 8$(\alpha4\beta1\gamma1)$, 10 $(\alpha5\beta1\gamma1)$, and 12 $(\alpha2\beta1\gamma3)$ (Schuler and Sorokin, 1995; Sorokin et al., 1997; Koch et al., 1999). The major ones found in the dermal-epidermal BMZ include laminin 1, 5, 6, and 10. Laminin 1 is involved in epithelial differentiation, adhesion, and possibly angiogenesis (Dixelius et al., 2004). The distribution of laminin 1 expression is actually more limited than what was once thought, as its detection was once carried out via an antibody (4C7) that detected the $\alpha5$ chain of laminin 10/11, not the $\alpha1$ chain of laminin 1 (Ferletta and Eckblom, 1999). Laminin 1 has a nidogen-binding $\gamma1$ chain that allows for self assembly, as is the case for most other laminins except for laminin 5. *In vivo* experiments using $\gamma1$ KO mice reveal disruption of BMZ development and embryonic lethality, while *in vitro* experiments using human skin organotypic co-cultures result in suppression of lamina densa and hemidesmosomal formation (Breitkreutz et al., 2004). However, it is difficult to ascribe this result to laminin 1 deficiency alone, as many other laminins are also dependent on the $\gamma1$ chain. The effect of $\alpha1$ chain deficiency has not yet been reported.

Laminin 5 (also known as epiligrin, kalinin, BM-600, and nicein) is involved in epithelial cell adhesion and migration, as well as platelet, leukocyte, and endothelial cell adhesion via interaction with α3β1, α6β4, and α2β1 integrins. Unlike other laminins that have a cross-like structure, laminin 5 has a characteristic dumbbell shape due to truncations at its γ2 and α3 chains that result from metalloproteinases and plasmin processing. In humans, mammalian Tolloid (mTLD) seems to be the main metalloproteinase that is responsible for cleaving the γ subchain (Veitch et al., 2003). The truncation also prevents nidogen-binding and hence association with other BMZ components. However, laminin 5 can complex with laminins 6 (α3β1γ1) and 7 (α3β2γ1) to bind nidogen (Champliaud et al., 1996) through the γ1 chain. JEB, in which blistering occurs at the level of the lamina lucida, is perhaps one of the most well-known skin pathologies associated with defective laminin 5 and 6 structure. A complete absence of laminin 5 is seen in Herlitz JEB, a fatal disease with a mortality of 40% by one year of age, characterized by widespread, slow-healing bullae. Reduced levels of laminin 5 are found in patients with non-Herlitz JEB. Most mutations occur in the LAMB3 gene encoding laminin 5 due to the spontaneous demethylation of 5-methylcytosine (McGowan and Marinkovich, 2000). Surprisingly, reduced levels of laminin 5 from mutation of the LAMC2 gene encoding the γ2 chain gives rise to intradermal blistering despite presence of mature hemidesmosomes (Spirito et al., 2001). Autoimmune diseases can also involve laminin 5. Autoantibodies against laminin 5 and 6 have been observed in patients with bullous systemic lupus erythematosis (Chan et al., 1999) as well as a variant of cicatricial pemphigoid termed "antilaminin cicatricial pemphigoid" (Chan et al., 1997). Recent evidence has also implicated laminin 5-α6β4 signaling with invasive squamous cell carcinoma development. Laminin 5-α6β4 expression is upregulated in squamous cell carcinoma (SCC) of many tissues including skin and increased with Ras-IκBα expression. Human keratinocytes deficient in either laminin 5 or α6β4 fail to initiate tumorigenesis upon oncogenic Ras/IkBa expression, but tumorigenesis is restored, however, with re-expression of those genes (Dajee et al., 2003).

The role of laminin 10 in skin development has been less well characterized compared to laminin 5. It is unique in that it is expressed in significant amounts in developing hair follicles when laminin 1 and 5 are downregulated. Laminin 10 is thought to play a role, perhaps via its interaction with α3β1 integrins, in hair follicle development. Although α5 KO mice are embryonic lethal (Miner et al., 1998), skin grafted from α5 KO mice failed to develop hair compared to wild-type (WT) control grafts, whereas addition of exogenous human laminin 10 restored hair follicle development in KO skin (Li et al., 2003). Laminin 10 is found in the dermal microvascular BMZ in addition to the dermal-epidermal BMZ. It is synthesized by human microvascular endothelial cells and is found in significant quantities in dermal granulation tissue of wounds; antibodies that block its integrin receptor α6β1 inhibit vessel branching. Taken together, this suggests its role in wound repair angiogenesis. Laminin 10 has also been demonstrated to provide survival signals via protein kinase B (PKB)/Akt activation in preventing apoptosis induced by serum depletion (Gu et al., 2002).

3.1.2. Tenascins

Tenascins are a family of large, multimeric glycoproteins first discovered approximately 20 years ago as part of the glioma mesenchymal ECM (Bourdon et al., 1983). They function in promoting weak cellular adhesion and are expressed in connective tissue cells, with different tenascins having distinct expression patterns. Family members include TN-C, TN-R, TN-W, TN-Y, and TN-X, sharing common motifs consisting of heptad and EGF-like repeats, a fibronectin type III binding domain, a C terminal globular domain, and an N terminal oligomerization domain. Alternative splicing also generates structural diversity, especially for TN-C and TN-R. Tenascins interact with a variety of ECM components including integrins, chondroitin, fibronectins, lecticans, and metallo/serine proteases (Bosman and Stamenkovic, 2003).

Tenascin C has a unique structure among the tenascins in that its assembly portion of the oligomerization domain allows it to form hexamers. It functions to inhibit cellular spreading by preventing the interaction of syndecan and integrin $\alpha5\beta1$ to fibronectin, which is usually required for activation of rho to form stress fibers. The inactivation of rho by tenascins is thought to be mediated by phosphoserine/threonine binding adaptor protein 12–3–3 tau (Martin et al., 2003).

TN-C expression appears mostly during organogenesis, but constitutive expression in the adult is found mostly in areas facing large amounts of mechanical stress, such as joints, ligaments, and arterial walls. It is thought that in addition to growth factors, mechanical stress also induces expression of TN-C and that this is important in the wound response. For example, transient TN-C expression is found in the dermis adjacent to skin wounds (Mackie et al., 1988). Numerous *in vivo* and *in vitro* experiments have also demonstrated the induction of TN-C expression upon stimulation with mechanical stress (Mikic et al., 2000; Chiquet et al., 2003). Deletion analysis suggests that the TN-C promoter consists of a serum-responsive and a static stress-responsive component, although responsiveness to other forms of stress may exist in other promoter regions depending on cell type (Chiquet-Ehrismann et al., 1994; Yamamoto et al., 1999). The significance of increased TN-C expression in face of mechanical stress may be to avoid excessive stretching during times of wound healing.

Inflammation also induces TN-C expression. Skin disorders such as psoriasis (Gerritsen et al., 1997), acne (Knaggs et al., 1994), blistering diseases (Schenk et al., 1995), and allergic reactions (Kusubata et al., 1999) all trigger transient local increases in TN-C expression. Gamma/ultraviolet light B (UVB) radiation induces while corticosteroids reduce TN-C expression (van der Vleuten et al., 1996).

TN-C also reappears during pathological states of tumor formation. It has been demonstrated that increasing expression of TN-C correlates with the higher grade of tumor and poorer prognosis. In melanoma for example, there exists a direct correlation of TN-C expression and stage (Tuominen et al., 1994). However, there has also been evidence that some of the most invasive tumors (prostate and colon adenocarcinomas) also have no TN-C expression (Sugawara et al., 1991; Xue et al., 1998). Despite the heterogenous data, these results suggest that there may be diagnostic value in TN-C detection as well as even therapeutic value. For example, radiolabeled TN-C antibodies can be provided for local radiotherapy against tumors. In fact, I131 labeled TN-C antibodies have been recently employed in phase II clinical trials for patients with malignant gliomas (Reardon et al., 2002).

TN-X, another tenascin with interesting associations to skin pathology, is located in the MHC III complex (Bristow et al.,1993) and also functions to inhibit cell spreading. It is expressed in the dermis of the skin as well as connective tissues in cardiac and skeletal muscles. TN-X KO mice exhibit a phenotype similar to Ehlers-Danlos syndrome (Mao et al., 2002), which in humans is known to result from type V collagen mutation and is characterized by hyperextensible joints and skin, vascular fragility, and poor wound healing. Recent evidence also suggests that Ehlers-Danlos in humans may also be caused by TN-X deficiency. TN-X interacts with decorin (Elefteriou et al., 2001); curiously, decorin deficient mice also exhibit similar phenotypes (Corsi et al., 2002). It is therefore suggested that TN-X may mediate its effects on collagen deposition via decorin.

3.2. Proteins

3.2.1. Collagens

Collagens are proteins which are major constituents of the ECM and whose main function is to provide structural integrity to tissue during mechanical stress. Twenty-six different collagens have been identified to date (Ihanamaki et al., 2004). All collagens consist of three polypeptide chains with at least one stretch of a triple helical collagenous domain composed of X-HydroxyPro-Gly triplet repeats and other shorter, non-collagenous N and C terminal domains or larger structural domains such as fibronectin type III repeats (Hulmes, 2002). Collagens can be further classified based on structural assembly. Collagens can spontaneously assemble into fibrils (I, II, III, V, XI), form networks (IV, VIII, X), associate with fibril surfaces (IX, XII, XIV), exist as transmembrane structures (XIII, XVII), or form periodic beaded filaments (VI).

Collagens found in the skin include collagens I, III, IV, V, VI, VII, XIV, and XVII. These are synthesized by fibroblasts, myofibroblasts, epithelial cells, and endothelial cells. Collagen I, heterotrimer of two one-chain and one two-chains, is a major component of the dermal ECM and is found in significant amounts in scar tissue. It is synthesized by fibroblasts, and like other fibrillar collagens, is secreted by the Golgi into the extracellular space, where it undergoes proteolysis by procollagen N and C proteinases. Hyperproduction and deposition of collagen I is found in keloids and also in scleroderma, a systemic sclerosis characterized in the skin by increased thickness and tightness (Ghosh, 2002). Collagen III is also a major component of dermal ECM and associates with collagen I and V to form heterotypic fibrils. Similar to collagen I, increased levels have been detected in hypertrophic scar tissue and is inducible by TGF-β (Wang et al., 1999). Mutant forms of collagen III have also been described in patients with specific variants of Ehlers-Danlos syndrome (Kroes et al., 2003). Collagen IV is the major collagen of the central BMZ and forms a complex structural scaffold critical for BMZ assembly. Six α chain isoforms have been identified, which combine to form $\alpha 1\alpha 1\alpha 2$ (broad distribution), $\alpha 3\alpha 4\alpha 5$, and $\alpha 5\alpha 5\alpha 6$ (found in skin) protomers (Poschl et al., 2004). In the skin, abnormal collagen IV distribution is found in basal cell carcinoma. Nodular forms are associated with downregulation and abnormal distribution of $\alpha 5$ chains, while more invasive forms show abnormal $\alpha 1$ chain deposition (Quatresooz et al., 2003). Increased levels in the serum have also been demonstrated in patients with stage IV melanoma and have been proposed as a marker for advanced melanoma (Burchardt et al, 2003). Collagen V is a minor component of the heterotypic I/III/V fibrils. Three α chain isoforms combine to form $\alpha 1\alpha 2\alpha 3$, $\alpha 1\alpha 1\alpha 1$ and $\alpha 1\alpha 1\alpha 2$. Collagen V abnormalities are also found in most patients with Ehlers-Danlos syndrome. Collagen V $\alpha 2$ KO mice exhibit increased levels and ectopic deposition of $\alpha 1$ homotrimers, increased α chain heterotrimer degradation, and abnormal heterotypic fibril formation leading to an abnormal ECM with skin thinning (Chanut-Delalande et al., 2004). Collagen VI is another major component of the dermal ECM, with three known α chain isoforms. Collagen VI interacts with collagens I, III, and IV and hence affects the structural properties of skin. Increased synthesis of collagen VI by fibroblasts has been found in cutis laxa, a condition of loose, sagging skin with impaired resilience (Watson et al., 2001). Collagen VII is a major component of BMZ anchoring fibrils and is composed of three identical α chains. Its NC1 domain interacts with multiple BMZ components such as fibronectin, laminin 5, and type I and IV collagen. Abnormalities with collagen VII result in impaired basement membrane organization and hence dermal-epidermal separation as seen in dystrophic epidermolysis bullosa (DEB). More than 100 mutations in collagen VII, occurring in collagenous or non-collagenous NC1/2 domains, have been described in patients with DEB (Chen et al., 2002). Collagen XVII was discussed in the previous section and will not be mentioned here.

3.3. Proteoglycans and glycosaminoglycans

Proteoglycans are complex molecules found in the ECM and/or cellular membranes composed of long, unbranched sugar chains known as glycosaminoglycans (GAGs), attached to a central protein core. GAGs, whose multitude of functions include serving as cofactors/coreceptors for cytokines/growth factors/chemokines, enzyme activity modulation, signaling pathological processes, and targets for pathogenic virulence factors, include heparin sulfate, keratin sulfate, heparin, and dermatan/chondroitin sulfate.

Proteoglycan interaction with other molecules may involve direct core protein interaction, GAG interaction, or a combination of GAG and core protein interaction. Proteoglycans are classified into families, each with specific numbers and types of GAGs attached to the core protein, although the GAG type may vary depending on external mileu (Trowbridge and Gallo, 2002). (1) Lecticans, whose function is predominantly to provide structural support, contain chondroitin sulfate side chains and interact with hyaluronan at the N terminus. Members include aggrecan, versican, neurocan, and brevican. (2) Small leucine-rich proteoglycans (SLRPs), which function predominantly in modulating collagen fibril formation, contain dermatan and keratin sulfate GAGs. Members include decorin, biglycan, fibromodulin, and keratocan. (3) Heparin sulfate proteoglycans are divided into those which are membrane-associated (syndecans, glypicans) and those that are found in the ECM (perlecan, agrin). They play important signaling functions and their interactions with growth factors have important growth modulating effects on cells and influence proliferation, differentiation, and migration (Bosman and Stamnkovic, 2003). Syndecan-4, for example, is a coreceptor for integrin previously shown to mediate rho-dependent stress fiber formation (Echtermeyer et al., 1999). Syndecan-4 is upregulated at injury sites and appears to influence wound healing response, as syndecan-4-deficient mice exhibit wound healing and impaired angiogenesis within granulation tissue (Echtermeyer et al., 2001).

The following descriptions will focus on dermatan sulfate, the major GAG found in skin, and its associated core protein decorin. These deserve further mention in relation to the previously discussed glycoprotein tenascin-X. Hyaluronan, another important GAG of the skin that has recently received much publicity as the new Food and Drug Administration-approved "anti-age skin filler" Restylane, will also be discussed.

3.4. Decorin

Decorins belong to the SLRP family of core proteins which associate with dermatan sulfate GAGs. It consists of an N terminal region with a dermatan side chain and a specific pattern of cystine residues; a central region of well conserved leucine-rich repeats, which serves as the main interface for binding other constituents; and a cystine-rich C terminal region (Reed and Iozzo, 2002). Intact decorins are found mostly in fetal skin whereas truncated forms arc found in adult skin. Decorins are found in all connective tissues and bind to collagens (I to VI), TGF-β, EGF, fibronectin, and thrombospondin. It maintains tissue integrity and modulates cell growth. In maintaining skin integrity, it mediates collagen fibrillogenesis by regulating lateral fusion of collagen fibrils. Under normal circumstances, the presence of decorin on the collagen fibril shafts inhibits uncontrolled lateral fusion into larger diameter fibers and promotes longitudinal fusion. In the thin and fragile skin of decorin homozygous KO mice, EM demonstrated collagen fibrils of irregular diameter and organization in the dermis. The reduced tensile strength of the skin may be caused by the abnormal collagen fibrils or the direct absence of proteoglycans. A patient with a variant of Ehler-Danlos syndrome has been described with decreased amounts of decorin in the skin (Fushimi et al., 1989). Decorin also causes suppression of cellular growth in many tumor lines in response to EGF (Santra et al., 1997). The surrounding stromata of tumors are often enriched with decorins as a means to counteract continued tumor growth (Moscatello et al., 1998). Its anti-tumorigenicity effect is also contributed by suppression of tumor neovascularization by reducing VEGF levels. Tumor cells which secrete decorin exhibited less angiogenic endothelial activity compared to tumor cells that did not secrete decorin (Grant et al., 2002). Decorins also exhibit antifibrotic properties by inhibiting TGF-β activity (Shimizukawa et al., 2003) which has been demonstrated in fetal tissue to facilitate the transition from scarless wound repair to scar formation (Shah et al., 1995). On the other hand, increased decorin levels have also been shown to be associated with scar formation in fetal wound models (Beanes et al., 2001).

3.5. Dermatan sulfate

Dermatan sulfate, also known as chondroitin B, is the primary GAGs found in the skin (dermis and ECM). It is a linear polysaccharide with disaccharides units containing N acetyl galactosamine or glucuronic acid, with additional complexity conferred by variably positioned iduronic acid and sulfation as well as variable total length. Core proteins with which it mostly associates in the skin include decorin, biglycan, versican, and thrombomodulin. The dermatan sulfate proteoglycans (DSPG) are able to interact with a large variety of matrix proteins, growth factors, cytokines, chemokines, and pathogen virulence factors (Trowbridge and Gallo, 2002). Known interactions with the dermatan sulfate GAG itself, however, are more limited and include tenascin-X (Elefteriou et al., 2001). As previously discussed, it is hypothesized that this interaction is what mediates the association with collagen fibrils, as decorin-deficient mice exhibit the same Ehlers Danlos-like phenotype as tenascin X-deficient mice (Mao et al., 2002). Tenascin-X deficiency mimics Ehlers-Danlos syndrome in mice through alteration of collagen deposition (Mao et al., 2002). Given the whole host of other molecules with which these DSPGs interact, it is also not surprising that they may also play a role in tumorigenesis and wound repair. During wound repair, dermatan sulfate is found to be released at high concentrations in quantities sufficient to activate FGF2, a mitogen that mediates mesenchymal cell migration, proliferation, and differentiation. In fact, wound fluid analysis suggests that dermatan sulfate may play more of a role in the FGF2 pathway than heparin sulfate, a more well studied binding partner of FGF2 (Penc et al., 1998). Dermatan sulfate release also appears to result in increased ICAM–1 expression and leukocyte adhesion to endothelial cells and increased circulating ICAM1 levels, indicating another role for these molecules in the inflammatory response to injury (Penc et al., 1999). Tumor metastasis is similarly supported. For example, melanoma and endothelial cell lines treated with chondroitinase b exhibited less proliferation and invasion (Denholm et al., 2001).

3.6. Hyaluronans

Hyaluronans are large molecular weight GAGs composed of linear disaccharide units of alternating glucuronic acid and N-acetylglucosamine. Unlike other GAGs, hyaluronans are not sulfated or bound to a protein core. They are expressed in all connective tissues, but especially in the skin where they are synthesized primarily by dermal fibroblasts but also by keratinocytes as well (Sayo et al., 2002). Although found in all layers of skin, they are localized mainly to the papillary dermis and BMZ. In the skin, they maintain structural integrity by regulation of water balance, functioning as a sieve and barrier, providing elastic and shearing properties, and providing organization via its interaction with other ECM components such as aggregan and versican. In addition to its structural role in the skin, hyaluronans also function to promote cellular migration, proliferation, and differentiation by its interaction with cellular receptors. For example, migration of Langerhans cells in the skin and hence the adaptive immunity response in the skin is promoted by hyaluronans (Mummert et al., 2003). High levels of hyaluronans and its receptor CD 44 are also thought to promote wound healing and tissue regeneration (Hu et al., 2003), as they are especially expressed at high levels during scarless wound healing in fetal skin.

Hyaluronans are unique in their rapid turnover time, with a half-life of <1 day (Tammi et al., 2002). They are synthesized by hyaluronan synthesizing enzymes (Has 1–3), each having different expression patterns during an organism's development and variable-synthesizing capabilities. Synthesis activity is often induced with growth factors such as PDGF-BB. Degradation is carried out by hyaluronidases (Hyal 1–3, PH-20), each isoform also exhibiting different expression patterns and hence functions, with some functions extending to gamete fertilization. The net effect of these anabolic and catabolic enzymes in tissue stroma regulates the amount and activity of hyaluronans.

Hyaluronan activity has been studied in rheumatoid arthritis and inflammatory and vascular diseases as well as in cancer. Increased levels in the serum have been observed in psoriatic arthritis, with skin involvement as measured by PASI correlating most with serum hyaluronan levels. Hence, hyaluronan levels have been proposed as a diagnostic tool for such diseases (Elkayam et al., 2000). Interestingly, hyaluronan accumulates abundantly in the basal cell layer of psoriasis skin samples, suggesting that hyaluronan may be involved in keratinocyte proliferation and turnover (Wells et al., 1991). It has also been found that hydrocortisone, which is used to decrease proliferation, also inhibits hyaluronan synthesis (Agren et al., 1995). Recently, it has been determined that high molecular weight hyaluronans are divided into smaller products to activate dendritic cells via toll receptors and hence facilitate the inflammatory response (Termeer et al., 2002).

Hyluronan levels are also increased in stroma of various epithelial and connective tissue tumors, with an increased fraction of hyaluronan compared to other GAGs. This could be due to increased production (by tumor or surrounding cells) or decreased clearance of the molecule. Transfected mesothelioma cells overexpressing hyaluronan exhibit a more malignant phenotype than non-transfected cells. Increased expression has also been found in melanoma (Heldin, 2003). The basis for the increased invasive property of tumor cells overexpressing hyaluronan could be either due to decreased immunogenicity of tumor cells from the hyaluronan coating or increased ECM substrate for tumor migration (Gately et al., 1984).

4. BMZ proteases and inhibitors

4.1. Metalloproteinases

ECM remodeling is an important aspect of normal physiological as well as pathological processes in the skin. Its degradative component is largely carried out by four groups of proteolytic enzymes classified based on amino acid or co-factors involved in catalysis: cysteine proteases, aspartic proteases, serine proteases, and metalloproteinases.

Metalloproteinases are prominent players of ECM proteolysis and are further divided into the serralysins, adamalysins, astracins, and matrixins (matrix metalloproteinase [MMP]). MMPs are perhaps most well characterized, with much attention recently focused on their contributions to tumor development. MMPs are further subclassified based on substrate specificity (to avoid the problem of substrate overlap, however, there has been recent efforts to change to a classification system based on structure): (1) collagenases, which include MMP-1, -8, and -13, degrade substrates including collagen (I, II, III V, IX), aggrecan, nidogen, tenascin, and fibrinogen; (2) gelatinases, which include MMP-2 and -9, are involved in degradation of collagens (IV, V, VII, X, XI, XIV), gelatins, elastins, proteoglycans, and fibronectin; (3) stromelysins, which include MMP-3, -10, and -11, degrade substrates including collagens (III, IV, V, VII, IX, X), elastin, fibronectin, nidogen; (4) matrilysins, which include MMP-7 and -26, and are the smallest MMPs, are known to degrade substrates including collagen IV, elastin, fibronectin, tenascin, aggrecan; and (5) membrane type MMPs (MT-MMPs), are unlike other metalloproteinases in that they contain a transmembrane domain with a cytoplasmic tail and are expressed on the surface of cell membranes; members include MMP-14, -15, -16, -17, -24, and -25 (Kerkela et al., 2003).

Metalloproteinases are produced as zymogens. The basic structure includes an N terminal secretory signal sequence followed by an active site binding prodomain which precludes premature activation, followed by a zinc-binding catalytic domain and "met-turn" motif. Some metalloproteinases, in addition, have hemopexin-binding domains, fibronectin-like repeats, or serine protease recognition motifs.

Under basal conditions, metalloproteinases are not produced in significant quantities. With growth factor (EGF, IL-1, TNF-α, TGF-β) stimulation, however, a variety of cells including fibroblasts, myofibroblasts, endothelial, epithelial, and malignant cells produce these molecules (Nagase and Woessner, 1999). The prodomain which maintains the inactive state can be cleaved either intracellularly or extracellularly. Once the metalloproteinase is secreted outside of the cell, localization to the cell surface and association with other surface molecules such as integrins, CD44, and proteoglycans can induce activation. Surface localization-dependent activation appears to be an advantageous property, given that this property will allow proteolysis to occur at the cell-ECM interface to facilitate migration. Outside of the cell, other MMPs can activate while $\alpha2$ macroglobulin and tissue inhibitor of metalloproteinase (TIMPs) can inhibit its activity. Four TIMPs have been discovered so far and are able to provide reversible inhibition of specific metalloproteinases by a 1:1 complex. Additionally, TIMPs, in some cases, may activate metalloproteinases, but can also function in promoting tumor angiogenesis/growth as well as apoptosis (Jiang et al., 2002). Most exist in secreted form, but TIMP3 is found attached to membranes (Gomez et al., 1997).

Because of its role in ECM remodeling, metalloproteinases contribute to physiological functions such as development and tissue repair. In wound-healing models, the addition of MMP inhibitors prevent keratinocyte migration and delay wound healing. *In vitro*, migration of keratinocytes depends on cleaving of procollagen I by MMP-1 and do not occur if MMP inhibitors are added (Stamenkovic, 2003). Metalloproteinases are also well known for their role in tumor development and metastasis and have been found in increased amounts in many tumors, either secreted by the tumor itself or by cells in the surrounding stroma. They are critical in promoting tumorigenesis not only in allowing cells to break through the ECM, but also in activating latent growth factors, releasing sequestered growth factors, and cleaving cell surface growth factors and adhesion receptors to allow for enhanced tumor cell development (McCawley and Matrisian, 2001). On the other hand, metalloproteinases may also release sequestered fragments that inhibit tumorigenesis.

In the skin, different cancers have different metalloproteinase and/or TIMP expression profiles. MMP-1, -2, -3, -7, -9, -10, -12, -13, and -14 production by epithelial cells and MMP-1, -2, -3, -9, -11, -12, -13, and -14; TIMP-1, -2, and -3 by stromal cells, for example, may be found in squamous cell carcinoma. MMP-3, -7, -10, -12, and -13, and TIMP-3 production by tumor cells and MMP-1, -2, -3, -9, -11, -12, and -14, and TIMP-1, -2, and -3 by stromal cells have been found in basal cell carcinoma (Kerkela and Saarialho-Kere, 2003). MMP-1, -2, -3, -9, -14, -15, and -16 have been found in invasive melanoma (Ntayi et al., 2004). It is thought that the different MMP profiles give rise to the different invasive properties of each cancer, and that such differences in expression pattern may have important diagnostic as well as therapeutic implications.

5. Concluding remarks

The strategic position of the BMZ makes it an excellent target to study processes that arise from interactions between the dermis and epidermis. In this chapter, we have discussed each of the main BMZ components in relation both to normal skin development as well as major pathologic categories such as wound healing, inflammatory disease, autoimmune and hereditary forms of blistering diseases, malignancy, and connective tissue disorders. As new molecules, ligand interactions, and signal pathways are discovered, we can expect to discover additional roles of the basement membrane in skin development and pathology. This will provide new diagnostic tools and additional points to intervene for therapeutic purposes.

References

Agren, U.M., Tammi, M., Tammi, R. 1995. Hydrocortisone regulation of hyaluronan metabolism in human skin organ culture. J. Cell. Physiol. 164, 240–248.

Beanes, S.R., Dang, C., Soo, C., Wang, Y., Urata, M., Ting, K., Fonkalsrud, E.W., Benhaim, P., Hedrick, M.H., Atkinson, J.B., Lorenz, H.P. 2001. Down-regulation of decorin, a transforming growth factor-beta modulator, is associated with scarless fetal wound healing. J. Pediatr. Surg. 36, 1666–1671.

Bosman, F.T., Stamenkovic, I. 2003. Functional structure and composition of the extracellular matrix. J. Pathol. 200, 423–428.

Boudreau, N., Sympson, C.J., Werb, Z., Bissell, M.J. 1995. Suppression of ICE and apoptosis in mammary epithelial cells by extracellular matrix. Science 267, 891–893.

Bourdon, M.A., Wikstrand, C.J., Furthmayr, H., Matthews, T.J., Bigner, D.D. 1983. Human glioma-mesenchymal extracellular matrix antigen defined by monoclonal antibody. Cancer Res. 43, 2796–2805.

Breitkreutz, D., Mirancea, N., Schmidt, C., Beck, R., Werner, U., Stark, H.J., Gerl, M., Fusenig, N.E. 2004. Inhibition of basement membrane formation by a nidogen-binding laminin gamma1-chain fragment in human skin-organotypic cocultures. J. Cell. Sci. 117, 2611–2622.

Bristow, J., Tee, M.K., Gitelman, S.E., Mellon, S.H., Miller, W.L. 1993. Tenascin-X: A novel extracellular matrix protein encoded by the human XB gene overlapping P450c21B. J. Cell. Biol. 122, 265–278.

Burchardt, E.R., Hein, R., Bosserhoff, A.K. 2003. Laminin, hyaluronan, tenascin-C and type VI collagen levels in sera from patients with malignant melanoma. Clin. Exp. Dermatol. 28, 515–520.

Champliaud, M.F., Lunstrum, G.P., Rousselle, P., Nishiyama, T., Keene, D.R., Burgeson, R.E. 1996. Human amnion contains a novel laminin variant, laminin 7, which like laminin 6, covalently associates with laminin 5 to promote stable epithelial-stromal attachment. J. Cell. Biol. 132, 1189–1198.

Chan, L.S., Lapiere, J.C., Chen, M., Traczyk, T., Mancini, A.J., Paller, A.S., Woodley, D.T., Marinkovich, M.P. 1999. Bullous systemic lupus erythematosus with autoantibodies recognizing multiple skin basement membrane components, bullous pemphigoid antigen 1, laminin-5, laminin-6, and type VII collagen. Arch. Dermatol. 135, 569–573.

Chan, L.S., Majmudar, A.A., Tran, H.H., Meier, F., Schaumburg-Lever, G., Chen, M., Anhalt, G., Woodley, D.T., Marinkovich, M.P. 1997. Laminin–6 and laminin–5 are recognized by autoantibodies in a subset of cicatricial pemphigoid. J. Invest. Dermatol. 108, 848–853.

Chanut-Delalande, H., Bonod-Bidaud, C., Cogne, S., Malbouyres, M., Ramirez, F., Fichard, A., Ruggiero, F. 2004. Development of a functional skin matrix requires deposition of collagen V heterotrimers. Mol. Cell. Biol. 24, 6049–6057.

Chen, M., Costa, F.K., Lindvay, C.R., Han, Y.P., Woodley, D.T. 2002. The recombinant expression of full-length type VII collagen and characterization of molecular mechanisms underlying dystrophic epidermolysis bullosa. J. Biol. Chem. 277, 2118–2124.

Chiquet, M., Renedo, A.S., Huber, F., Fluck, M. 2003. How do fibroblasts translate mechanical signals into changes in extracellular matrix production? Matrix Biol. 22, 73–80.

Chiquet-Ehrismann, R., Tannheimer, M., Koch, M., Brunner, A., Spring, J., Martin, D., Baumgartner, S., Chiquet, M. 1994. Tenascin-C expression by fibroblasts is elevated in stressed collagen gels. J. Cell. Biol. 127, 2093–2101.

Clark, R.A. 1990. Fibronectin matrix deposition and fibronectin receptor expression in healing and normal skin. J. Invest. Dermatol. 94, 128S–134S.

Corsi, A., Xu, T., Chen, X.D., Boyd, A., Liang, J., Mankani, M., Sommer, B., Iozzo, R.V., Eichstetter, I., Robey, P.G., Bianco, P., Young, M.F. 2002. Phenotypic effects of biglycan deficiency are linked to collagen fibril abnormalities, are synergized by decorin deficiency, and mimic Ehlers-Danlos-like changes in bone and other connective tissues. J. Bone Miner Res. 17, 1180–1189.

Dajee, M., Lazarov, M., Zhang, J.Y., Cai, T., Green, C.L., Russell, A.J., Marinkovich, M.P., Tao, S., Lin, Q., Kubo, Y., Khavari, P.A. 2003. NF-kappaB blockade and oncogenic Ras trigger invasive human epidermal neoplasia. Nature 421, 639–643.

Danen, E.H., et al. 2000. Dual stimulation of Ras/mitogen-activated protein kinase and RhoA by cell adhesion to fibronectin supports growth factor-stimulated cell cycle progression. J. Cell. Biol. 151, 1413–1422.

Denholm, E.M., Lin, Y.Q., Silver, P.J. 2001. Anti-tumor activities of chondroitinase AC and chondroitinase B: Inhibition of angiogenesis, proliferation and invasion. Eur. J. Pharmacol. 416, 213–221.

Dixelius, J., Jakobsson, I., Genersch, E., Bohman, S., Ekblom, P., Clacsson-Welsh, L. 2004. Laminin-1 promotes angiogenesis in synergy with fibroblast growth factor by distinct regulation of the gene and protein expression profile in endothelial cells. J. Biol. Chem. 279, 23766–23772.

Echtermeyer, F, Wilcox-Adelman, S., Saoncella, S., Denhez, F., Detmar, M., Goetinck, P. 2001. Delayed wound repair and impaired angiogenesis in mice lacking syndecan-4. J. Clin. Investig. 107, R9–R14.

Echtermeyer, F., Baciu, P.C., Saoncella, S., Ge, Y., Goetinck, P.F. 1999. Syndecan-4 core protein is sufficient for the assembly of focal adhesions and actin stress fibers. J. Cell. Sci. 112, 3433–3441.

Elefteriou, F., Exposito, J.Y., Garrone, R., Lethias, C. 2001. Binding of tenascin-X to decorin. FEBS Lett. 495, 44–47.

Elkayam, O., Yaron, I., Shirazi, I., Yaron, M., Caspi, D. 2000. Serum levels of hyaluronic acid in patients with psoriatic arthritis. Clin. Rheumatol. 19, 455–457.

Ferletta, M., Ekblom, P. 1999. Identification of laminin-10/11 as a strong cell adhesive complex for a normal and a malignant human epithelial cell line. J. Cell. Sci. 112, 1–10.

Fushimi, H., Kameyama, M., Shinkai, H. 1989. Deficiency of the core proteins of dermatan sulphate proteoglycans in a variant form of Ehlers-Danlos syndrome. J. Intern. Med. 226, 409–416.

Gately, C.L., Muul, L.M., Greenwood, M.A., Papazoglou, S., Dick, S.J., Kornblith, P.L., Smith, B.H., Gately, M.K. 1984. *In vitro* studies on the cell-mediated immune response to human brain tumors. II. Leukocyte-induced coats of glycosaminoglycan increase the resistance of glioma cells to cellular immune attack. J. Immunol. 133, 3387–3395.

Gerritsen, M.J., Elbers, M.E., de Jong, E.M., van de Kerkhof, P.C. 1997. Recruitment of cycling epidermal cells and expression of filaggrin, involucrin and tenascin in the margin of the active psoriatic plaque, in the uninvolved skin of psoriatic patients and in the normal healthy skin. J. Dermatol. Sci. 14, 179–188.

Ghosh, A.K. 2002. Factors involved in the regulation of type I collagen gene expression: Implication in fibrosis. Exp. Biol. Med. (Maywood) 227, 301–314.

Gomez, D.E., Alonso, D.F., Yoshiji, H., Thorgeirsson, U.P. 1997. Tissue inhibitors of metalloproteinases: Structure, regulation and biological functions. Eur. J. Cell. Biol. 74, 111–122.

Grant, D.S., Yenisey, C., Rose, R.W., Tootell, M., Santra, M., Iozzo, R.V. 2002. Decorin suppresses tumor cell-mediated angiogenesis. Oncogene 21, 4765–4777.

Grose, R., Hutter, C., Bloch, W., Thorey, I., Watt, F.M., Fassler, R., Brakebusch, C., Werner, S. 2002. A crucial role of beta 1 integrins for keratinocyte migration *in vitro* and during cutaneous wound repair. Development 129, 2303–2315.

Gu, J., Fujibayashi, A., Yamada, K.M., Sekiguchi, K. 2002. Laminin-10/11 and fibronectin differentially prevent apoptosis induced by serum removal via phosphatidylinositol 3-kinase/ Akt- and MEK1/ERK-dependent pathways. J. Biol. Chem. 277, 19922–19928.

Haase, I., Hobbs, R.M., Romero, M.R., Broad, S. 2001. A role for mitogen-activated protein kinase activation by integrins in the pathogenesis of psoriasis. J. Clin. Invest. 108, 527–536.

Heldin, P. 2003. Importance of hyaluronan biosynthesis and degradation in cell differentiation and tumor formation. Braz. J. Med. Biol. Res. 36, 967–973.

Hu, M., Sabelman, E.E., Cao, Y., Chang, J., Hentz, V.R. 2003. Three-dimensional hyaluronic acid grafts promote healing and reduce scar formation in skin incision wounds. J. Biomed. Mater. Res. 67B, 586–592.

Hulmes, D.J. 2002. Building collagen molecules, fibrils, and suprafibrillar structures. J. Struct. Biol. 137, 2–10.

Hynes, R.O. 2002. Integrins: Bidirectional, allosteric signaling machines. Cell 110, 673–687.

Ido, H., Harada, K., Futaki, S., Hayashi, Y., Nishiuchi, R., Natsuka, Y., Li, S., Wada, Y., Combs, A.C., Ervasti, J.M., Sekiguchi, K. 2004. Molecular dissection of the alpha-dystroglycan- and integrin-binding sites within the globular domain of human laminin-10. J. Biol. Chem. 279, 10946–10954.

Ihanamaki, T, Vuorio, E. 2004. Collagens and collagen-related matrix components in the human and mouse eye. Prog. Retin. Eye Res. 23, 403–434.

Jiang, Y., Goldberg, I.D., Shi, Y.E. 2002. Complex roles of tissue inhibitors of metalloproteinases in cancer. Oncogene 21, 2245–2252.

Jonkman, M.F., de Jong, M.C., Heeres, K., Pas, H.H., J.B., vanderMeer, Owaribe, K., A.M., MartinezdeVelasco, Niessen, C.M., Sonnenberg, A. 1995. 180-kD bullous pemphigoid antigen (BP180) is deficient in generalized atrophic benign epidermolysis bullosa. J. Clin. Invest. 95, 1345–1352.

Kerkela, E., Saarialho-Kere, U. 2003. Matrix metalloproteinases in tumor progression: Focus on basal and squamous cell skin cancer. Exp. Dermatol. 12, 109–125.

Knaggs, H.E., Layton, A.M., Morris, C., Wood, E.J., Holland, D.B., Cunliffe, W.J. 1994. Investigation of the expression of the extracellular matrix glycoproteins tenascin and fibronectin during acne vulgaris. Br. J. Dermatol. 130, 576–582.

Koch, M., Olson, P.F., Albus, A., Jin, W., Hunter, D.D., Brunken, W.J., Burgeson, R.E., Champliaud, M.F. 1999. Characterization and expression of the laminin gamma3 chain: A novel, non-basement membrane-associated, laminin chain. J. Cell. Biol. 145, 605–618.

Kroes, H.Y., Pals, G., van Essen, A.J. 2003. Ehlers-Danlos syndrome type IV: Unusual congenital anomalies in a mother and son with a COL3A1 mutation and a normal collagen III protein profile. Clin. Genet. 63, 224–227.

Kusubata, M., Hirota, A., Ebihara, T., Kuwaba, K., Matsubara, Y., Sasaki, Y., Kusakabe, M., Tsukada, T., Irie, S., Koyama, Y. 1999. Spatiotemporal changes of fibronectin, tenascin-C, fibulin-1, and fibulin-2 in the skin during the development of chronic contact dermatitis. J. Invest. Dermatol. 113, 906–912.

Larjava, H., Salo, T., Haapasalmi, K., Kramer, R.H., Heino, J. 1993. Expression of integrins and basement membrane components by wound keratinocytes. J. Clin. Invest. 92, 1425–1435.

Li, J., Tzu, J., Chen, Y., Zhang, Y.P., Nguyen, N.T., Gao, J., Bradley, M., Keene, D.R., Oro, A. E., Miner, J.H., Marinkovich, M.P. 2003. Laminin-10 is crucial for hair morphogenesis. EMBO J. 22, 2400–2410.

Liddington, R.C., Ginsberg, M.H. 2002. Integrin activation takes shape. J. Cell. Biol. 158, 833–839.

Mackie, E.J., Halfter, W., Liverani, D. 1988. Induction of tenascin in healing wounds. J. Cell. Biol. 107, 2757–2767.

Mao, J.R., Taylor, G., Dean, W.B., Wagner, D.R., Afzal, V., Lotz, J.C., Rubin, E.C., Bristow, J. 2002. Tenascin-X deficiency mimics Ehlers-Danlos syndrome in mice through alteration of collagen deposition. Nat. Genet. 30, 421–425.

Marcinkiewicz, C., Taooka, Y., Yokosaki, Y., Calvete, J.J., Marcinkiewicz, M.M., Lobb, R.R., Niewiarowski, S., Sheppard, D. 2000. Inhibitory effects of MLDG-containing heterodimeric disintegrins reveal distinct structural requirements for interaction of the integrin alpha 9beta 1 with VCAM-1, tenascin-C, and osteopontin. J. Biol. Chem. 275, 31930–31937.

Martin, D., Brown-Luedi, M., Chiquet-Ehrismann, R. 2003. Tenascin-C signaling through induction of 14-3-3 tau. J. Cell. Biol. 160, 171–175.

McCawley, L.J., Matrisian, L.M. 2001. Matrix metalloproteinases: They're not just for matrix anymore!. Curr. Opin. Cell. Biol. 13, 534–540.

McGary, E.C., Lev, D.C., Bar-Eli, M. 2002. Cellular adhesion pathways and metastatic potential of human melanoma. Cancer. Biol. Ther. 1, 459–465.

McGowan, K.A., Marinkovich, M.P. 2000. Laminins and human disease. Microsc. Res. Tech. 51, 262–279.

McMillan, J.R., Akiyama, M., Shimizu, H. 2003. Epidermal basement membrane zone components: Ultrastructural distribution and molecular interactions. J. Dermatol. Sci. 31, 169–177.

McMillan, J.R., Akiyama, M., Shimizu, H. 2003. Ultrastructural orientation of laminin 5 in the epidermal basement membrane: An updated model for basement membrane organization. J. Histochem. Cytochem. 51, 1299–1306.

Mettouchi, A., Klein, S., Guo, W., M., Lopez-Lago, Lemichez, E., Westwick, J.K., Giancotti, F.G. 2001. Integrin-specific activation of Rac controls progression through the G(1) phase of the cell cycle. Mol. Cell 8, 115–127.

Mikic, B., Wong, M., Chiquet, M., Hunziker, E.B. 2000. Mechanical modulation of tenascin-C and collagen-XII expression during avian synovial joint formation. J. Orthop. Res. 18, 406–415.

Miner, J.H., Cunningham, J., Sanes, J.R. 1998. Roles for laminin in embryogenesis: Exencephaly, syndactyly, and placentopathy in mice lacking the laminin alpha5 chain. J. Cell. Biol. 143, 1713–1723.

Moscatello, D.K., Santra, M., Mann, D.M., McQuillan, D.J., Wong, A.J., Iozzo, R.V. 1998. Decorin suppresses tumor cell growth by activating the epidermal growth factor receptor. J. Clin. Invest. 101, 406–412.

Mummert, D.I., Takashima, A., Ellinger, L., Mummert, M.E. 2003. Involvement of hyaluronan in epidermal Langerhans cell maturation and migration *in vivo*. J. Dermatol. Sci. 33, 91–97.

Nagase, H., Woessner, J.F. ,Jr. 1999. Matrix metalloproteinases. J. Biol. Chem. 274, 21491–21494.

Ntayi, C., Hornebeck, W., Bernard, P. 2004. [Involvement of matrix metalloproteinases (MMPs) in cutaneous melanoma progression]. Pathol. Biol. (Paris) 52, 154–159.

Palmer, E.L., Ruegg, C., Ferrando, R., Pytela, R., Sheppard, D. 1993. Sequence and tissue distribution of the integrin alpha 9 subunit, a novel partner of beta 1 that is widely distributed in epithelia and muscle. J. Cell. Biol. 123, 1289–1297.

Penc, S.F., Pomahac, B., Eriksson, E., Detmar, M., Gallo, R.L. 1999. Dermatan sulfate activates nuclear factor-kappab and induces endothelial and circulating intercellular adhesion molecule-1. J. Clin. Invest. 103, 1329–1335.

Penc, S.F., Pomahac, B., Winkler, T., Dorschner, R.A., Eriksson, E., Herndon, M., Gallo, R.L. 1998. Dermatan sulfate released after injury is a potent promoter of fibroblast growth factor-2 function. J. Biol. Chem. 273, 28116–28121.

Poschl, E., U., Schlotzer-Schrehardt, Brachvogel, B., Saito, K., Ninomiya, Y., Mayer, U. 2004. Collagen IV is essential for basement membrane stability but dispensable for initiation of its assembly during early development. Development 131, 1619–1628.

Quatresooz, P., Martalo, O., Pierard, G.E. 2003. Differential expression of alpha1 (IV) and alpha5 (IV) collagen chains in basal-cell carcinoma. J. Cutan. Pathol. 30, 548–552.

Raghavan, S., Bauer, C., Mundschau, G., Li, Q., Fuchs, E. 2000. Conditional ablation of beta1 integrin in skin. Severe defects in epidermal proliferation, basement membrane formation, and hair follicle invagination. J. Cell. Biol. 150, 1149–1160.

Reardon, D.A., Akabani, G., Coleman, R.E., Friedman, A.H., Friedman, H.S., Herndon, J.E., Cokgor, I., McLendon, R.E., Pegram, C.N., Provenzale, J.M., Quinn, J.A., Rich, J.N., Regalado, L.V., Sampson, J.H., Shafman, T.D., Wikstrand, C.J., Wong, T.Z., Zhao, X.G., Zalutsky, M.R., Bigner, D.D. 2002. Phase II trial of murine (131)I-labeled antitenascin monoclonal antibody 81C6 administered into surgically created resection cavities of patients with newly diagnosed malignant gliomas. J. Clin. Oncol. 20, 1389–1397.

Reed, C.C., Iozzo, R.V. 2002. The role of decorin in collagen fibrillogenesis and skin homeostasis. Glycoconj. J. 19, 249–255.

Santra, M., Mann, D.M., Mercer, E.W., Skorski, T., Calabretta, B., Iozzo, R.V. 1997. Ectopic expression of decorin protein core causes a generalized growth suppression in neoplastic cells of various histogenetic origin and requires endogenous p21, an inhibitor of cyclin-dependent kinases. J. Clin. Invest. 100, 149–157.

Sayo, T., Sugiyama, Y., Takahashi, Y., Ozawa, N., Sakai, S., Ishikawa, O., Tamura, M., Inoue, S. 2002. Hyaluronan synthase 3 regulates hyaluronan synthesis in cultured human keratinocytes. J. Invest. Dermatol. 118, 43–48.

Schenk, S., Bruckner-Tuderman, L., Chiquet-Ehrismann, R. 1995. Dermo-epidermal separation is associated with induced tenascin expression in human skin. Br. J. Dermatol. 133, 13–22.

Schuler, F., Sorokin, L.M. 1995. Expression of laminin isoforms in mouse myogenic cells *in vitro* and *in vivo*. J. Cell. Sci. 108, 3795–3805.

Shah, M., Foreman, D.M., Ferguson, M.W. 1995. Neutralisation of TGF-beta 1 and TGF-beta 2 or exogenous addition of TGF-beta 3 to cutaneous rat wounds reduces scarring. J. Cell. Sci. 108, 985–1002.

Shimizukawa, M., Ebina, M., Narumi, K., Kikuchi, T., Munakata, H., Nukiwa, T. 2003. Intratracheal gene transfer of decorin reduces subpleural fibroproliferation induced by bleomycin. Am. J. Physiol. Lung. Cell. Mol. Physiol. 284, L526–L532.

Sorokin, L.M., Pausch, F., Frieser, M., Kroger, S., Ohage, E., R., Deutzmann 1997. Developmental regulation of the laminin alpha5 chain suggests a role in epithelial and endothelial cell maturation. Dev. Biol. 189, 285–300.

Spirito, F., Chavanas, S., Chavanas, S., C., Prost-Squarcioni, Pulkkinen, L., Fraitag, S., Bodemer, C., Ortonne, J.P., Meneguzzi, G. 2001. Reduced expression of the epithelial adhesion ligand laminin 5 in the skin causes intradermal tissue separation. J. Biol. Chem. 276, 18828–18835.

Stamenkovic, I. 2003. Extracellular matrix remodelling: The role of matrix metalloproteinases. J. Pathol. 200, 448–464.

Stupack, D.G., Puente, X.S., Boutsaboualoy, S., Storgard, C.M., Cheresh, D.A. 2001. Apoptosis of adherent cells by recruitment of caspase-8 to unligated integrins. J. Cell. Biol. 155, 459–470.

Sugawara, I., Hirakoshi, J., Masunaga, A., Itoyama, S., Sakakura, T. 1991. Reduced tenascin expression in colonic carcinoma with lymphogenous metastasis. Invasion Metastasis 11, 325–331.

Tammi, M.I., Day, A.J., Turley, E.A. 2002. Hyaluronan and homeostasis: A balancing act. J. Biol. Chem. 277, 4581–4584.

Termeer, C., Benedix, F., Sleeman, J., Fieber, C., Voith, U., Ahrens, T., Miyake, K., Freudenberg, M., Galanos, C., Simon, J.C. 2002. Oligosaccharides of Hyaluronan activate dendritic cells via toll-like receptor 4. J. Exp. Med. 195, 99–111.

Trowbridge, J.M., Gallo, R.L. 2002. Dermatan sulfate: New functions from an old glycosaminoglycan. Glycobiology 12, 117R–125R.

Tuominen, H., Junttila, T., Karvonen, J., Kallioinen, M. 1994. Cell-type related and spatial variation in the expression of integrins in cutaneous tumors. J. Cutan. Pathol. 21, 500–506.

Van den Bergh, F., Giudice, G.J. 2003. BP180 (type XVII collagen) and its role in cutaneous biology and disease. Adv. Dermatol. 19, 37–71.

van der Vleuten, C.J., Snijders, C.G., E.M., deJong, van de Kerkhof, P.C. 1996. Effects of calcipotriol and clobetasol-17-propionate on UVB-irradiated human skin: An immunohistochemical study. Skin Pharmacol. 9, 355–365.

Veitch, D.P., Nokelainen, P., McGowan, K.A., Nguyen, T.T., Nguyen, N.E., Stephenson, R., Pappano, W.N., Keene, D.R., Spong, S.M., Greenspan, D.S., Findell, P.R., Marinkovich, M.P. 2003. Mammalian tolloid metalloproteinase, and not matrix metalloprotease 2 or membrane type 1 metalloprotease, processes laminin-5 in keratinocytes and skin. J. Biol. Chem. 278, 15661–15668.

Wang, X., Smith, P., Pu, L.L., Kim, Y.J., Ko, F., Robson, M.C. 1999. Exogenous transforming growth factor beta(2) modulates collagen I and collagen III synthesis in proliferative scar xenografts in nude rats. J. Surg. Res. 87, 194–200.

Watson, R.E., Ball, S.G., Craven, N.M., Boorsma, J., East, C.L., Shuttleworth, C.A., Kielty, C. M., Griffiths, C.E. 2001. Distribution and expression of type VI collagen in photoaged skin. Br. J. Dermatol. 144, 751–759.

Watt, F.M., Jones, P.H. 1993. Expression and function of the keratinocyte integrins. Dev. Suppl. 185–192.

Watt, F.M., Jones, P.H. 1993. Regulation of keratinocyte terminal differentiation by integrin-extracellular matrix interactions. J. Cell. Sci. 106, 175–182.

Wehrle-Haller, B., Imhof, B.A. 2003. Integrin-dependent pathologies. J. Pathol. 200, 481–487.

Wells, A.F., Lundin, A., Michaelsson, G. 1991. Histochemical localization of hyaluronan in psoriasis, allergic contact dermatitis and normal skin. Acta. Derm. Venereol. 71, 232–238.

Welsh, C.F., Villanueva, R.K., Liu, Y., Schwartz, M.A., Assoian, R.K. 2001. Timing of cyclin D1 expression within G1 phase is controlled by Rho. Nature. Cell Biol. 3, 950–957.

Xue, Y., Smedts, F., Latijnhouwers, M.A., Ruijter, E.T., Aalders, T.W., J.J., delaRosette, Debruyne, F.M., Schalken, J.A. 1998. Tenascin-C expression in prostatic intraepithelial neoplasia (PIN): A marker of progression? Anticancer Res. 18, 2679–2684.

Yamamoto, K., Smedts, F., Latijnhouwers, M.A., Ruijter, E.T., Aalders, T.W., J.J., delaRosette, Debruyne, F.M., Schalken, J.A. 1999. Induction of tenascin-C in cardiac myocytes by mechanical deformation. Role of reactive oxygen species. J. Biol. Chem. 274, 21840–21846.

Yokosaki, Y., Palmer, E.L., Prieto, A.L., Crossin, K.L., Bourdon, M.A., Pytela, R., Sheppard, D. 1994. The integrin alpha 9 beta 1 mediates cell attachment to a non-RGD site in the third fibronectin type III repeat of tenascin. J. Biol. Chem. 269, 26691–26696.

Zambruno, G., Marchisio, P.C., Marconi, A., Vaschieri, C., Melchiori, A., Giannetti, A., De Luca, M. 1995. Transforming growth factor-beta 1 modulates beta 1 and beta 5 integrin receptors and induces the de novo expression of the alpha v beta 6 heterodimer in normal human keratinocytes: Implications for wound healing. J. Cell. Biol. 129, 853–865.

Zillikens, D. 1999. Acquired skin disease of hemidesmosomes. J. Dermatol. Sci. 20, 134–154.

Integrin signaling and central nervous system development

Richard Belvindrah and Ulrich Müller

The Scripps Research Institute, Department of Cell Biology, La Jolla, California

Contents

1. Introduction

Integrins are a family of cell surface receptors, each consisting of an α and a β subunit. Genes for 8 different β and 18 different α subunits have been identified in the mammalian genome. They form at least 24 heterodimers. Each α and β subunit is a type I transmembrane glycoprotein consisting of a large extracellular domain, a single-span transmembrane domain, and a short cytoplasmic domain. The only exception is the integrin $\beta 4$ subunit that contains a cytoplasmic domain of approximately 1,000 amino acids.

Advances in Developmental Biology
Volume 15 ISSN 1574-3349
DOI: 10.1016/S1574-3349(05)15005-4

Alternative splicing of the extracellular and intracellular domains of several integrin subunits leads to further diversity. Each integrin heterodimer binds to specific sets of ligands, which include ECM glycoproteins, cell surface receptors, proteases, and pathogens (Bokel and Brown, 2002; Hynes, 2002; Müller, 2004).

Signals from within the cell can modulate the conformation of integrins, thereby affecting their ligand-binding activity (Hynes, 2002; Liddington and Ginsberg, 2002; Giancotti, 2003). This so-called "inside-out" signaling is incompletely understood, but recent experiments demonstrate that interactions of integrin cytoplasmic domains with proteins such as talin can induce high-affinity ligand binding (Cram and Schwarzbauer, 2004). In a process termed "outside-in" signaling, interactions of integrins with ligands trigger the activation of cellular second messenger systems (Hynes, 1992, 2002; Giancotti and Ruoslahti, 1999; Schwartz and Ginsberg, 2002; Brakebusch and Fassler, 2003). The integrin cytoplasmic domains do not have intrinsic catalytic activity and the activation of signaling pathways is likely achieved through the recruitment of proteins to the integrin cytoplasmic domains. Consistent with this model, a large variety of proteins have been described that can be recruited by the cytoplasmic domain of α and β chains, providing a likely explanation for the vast array of signaling pathways that can be activated by different integrins in distinct cell types (Liu et al. 2000; Brakebusch and Fassler, 2003).

Genetic studies have demonstrated that integrins, their ligands and downstream effectors are essential for the proper development and function of many if not all tissues. In the CNS, integrin-dependent signaling pathways regulate the behavior of proliferating precursor cells as well as of postmitotic neuronal and glial cells. In this chapter, we will highlight the recent advances in our understanding of the mechanisms by which integrins contribute to the regulation of cell behavior during the development of cortical structures in the CNS. We will also summarize findings that provide first evidence that integrins are important for the proper functioning of the adult CNS. The role of integrins in oligodendrocytes will not be addressed because this topic has been recently reviewed extensively (Decker et al. 2004; Baron et al. 2005).

2. Integrins and cell proliferation during CNS development

Neuronal and glial cells of the CNS are generated from neural stem cells that are located in specialized proliferation zones. The mechanisms that regulate the self-renewal of the stem cells and that determine which neuronal or glial subtypes are generated in distinct locations at particular developmental time points are only just beginning to be understood. Until recently, even the identity of the neural stem cells had remained elusive. Studies from several laboratories have now demonstrated that radial glial cells serve as neural stem cells in many if not all parts of the developing CNS. Radial glial cells were originally identified by their morphology and by shared properties with the glial cell, hence their name. In the developing cerebral cortex, the cell bodies of radial glial cells are located in the ventricular neuroepithelium; the cells extend a long process towards the brain surface and express glial marker proteins such as BLBP, RC2, and GFAP in developmentally regulated patterns. The processes of radial glial cells have long been known to serve as guideposts for cell migration. Neurons that are generated in the ventricular neuroepithelium attach to and migrate along the radial glial fibers towards the brain surface (Fig. 1). Elegant studies using *in vivo* time laps recordings, genetic marking of stem cells, and *in vitro* differentiation assays have demonstrated that radial glial cells also fulfill the hallmark of neural stem cells: they undergo self-renewal and generate differentiated neurons and glial cells. Therefore, radial glial cells serve a dual purpose for the generation of cellular subtypes in the CNS and for guiding their migration (Campbell and Gotz, 2002; Hatten, 2002; Nadarajah and Parnavelas, 2002; Fishell and Kriegstein, 2003; Rakic, 2003).

Consistent with a function for ECM molecules in regulating radial glial function, ECM glycoproteins such as fibronectin (FN), tenascin-C (TN-C), laminin 1 (LN–1), and the laminin α2 (LN-α2) subunit are expressed in the ventricular neuroepithelium where the cell bodies of radial glial cells are situated (Stewart and Pearlman, 1987; Chun and Shatz, 1988; Gates et al. 1995; Sheppard et al. 1995; Campos et al. 2004; Stettler and Galileo, 2004; Garcion et al. 2004). Using cells in culture, it has been shown that LN and FGF act synergistically to regulate the proliferation of neuroepithelial cells

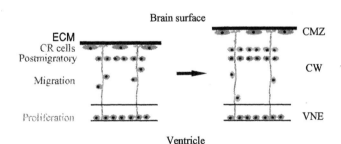

Fig. 1. Development of the cerebral cortex. Two stages of development are shown. Left picture: neurons (blue) are generated in the ventricular neuroepithelium (VNE) from proliferating precursor cells (dark green). The cells migrate along radial glial fibers (dark green) into the cortical wall (CW). The cells terminate migration close to the cortical marginal zone (CMZ) where the Cajal Retzius (CR) cells (orange) are situated below the meningeal basement membrane (ECM). The migrating neurons detach from the glial fibers and are assembled into cortical layers. Right picture: a second wave of migrating neurons (light green) bypasses the earlier born neurons and forms a layer close to the CR cell layer. Subsequent waves of neurons (not shown) will similarly bypass earlier born neurons leading to the formation of the multilayered cerebral cortex. (See Color Insert.)

(Drago et al. 1991). Recent studies with neurospheres derived from early postnatal animals have further demonstrated that FN can also affect proliferation (Jacques et al. 1998). While the function of FN and LN for stem cell behavior in the intact animal has not been investigated, genetic and *in vitro* studies have implicated TN-C in regulating cell proliferation in the CNS. In knock-out mice that lack the *Tnc* gene, cell proliferation in cortical areas of the CNS is reduced, suggesting that TN-C may have a stimulatory role during cell proliferation (Garcion et al. 2001). In support of this hypothesis, FGF-stimulated cell proliferation in neurospheres derived from *Tnc*-null animals is reduced. TN-C may directly act on radial glial cells because staining for RC2, a radial glial marker, is decreased in *Tnc*-null animals (Garcion et al. 2004). Surprisingly, TN-C does not affect the proliferation of neurospheres derived from wild type animals and inhibits the stimulatory effect of FN on cell proliferation (Jacques et al. 1998; Garcion et al. 2004). While the reasons for the different effects of TN-C in wild-type and *Tnc*-deficient neurospheres is not known, it seems possible that the precise concentration and ratio of different ECM molecules may be important. *In vivo*, ECM molecules may be present in varying amounts in distinct areas of the CNS, leading to local variations in the response of distinct precursor populations.

Several experiments have addressed the receptors that mediate effects of ECM glycoproteins on cell proliferation. Studies with LN fragments have demonstrated that synergistic activation of cell proliferation by LN and FGF can be observed with the proteolytic LN E8 fragment that contains the integrin-binding site (Drago et al. 1991). However, while proliferating cells in neurospheres express the LN receptor integrin $\alpha6\beta1$, antibodies against the $\alpha6$ subunit that inhibit interactions with LN do not affect neurosphere proliferation (Jacques et al. 1998). As one possibility, other LN receptors, such as additional integrins or dystroglycan (Henry and Campbell, 1999), may mediate effects on cell proliferation. In contrast to the findings with LN, effects of FN on cell proliferation appear to be dependent predominantly on integrins. Neurospheres express high levels of several integrins that are FN receptors, such as $\alpha5\beta1$ and $\alpha v\beta1$, and lower levels of $\alpha v\beta5$ and $\alpha v\beta8$. RGD peptides that block interactions of these integrins with FN inhibit the effect of FN on proliferation. A partial block is observed with $\beta1$ integrin antibodies, suggesting that the integrin $\alpha5\beta1$ and/or $\alpha v\beta1$ contribute. In further support of this model, neuroepithelial cells that are FACS-sorted according to high $\beta1$ expression levels have a higher propensity to generate neurospheres than cells with lower $\beta1$ expression levels (Campos et al. 2004). This is similar to findings in the skin, where stem cells express the highest $\beta1$ levels (Jones and Watt, 1993; Jones et al. 1995; Zhu et al. 1999). However, there are some conflicting results. While antibodies to $\beta1$ integrins can partially inhibit the effect of ECM components on proliferation, no such block is observed in neurospheres genetically modified to lack the integrin $\beta1$

subunit gene (*Itgb1*). The *Itgb1*-deficient cells produce similar amounts of neurospheres as wild-type cells and can be passaged with equal efficiency (Campos et al. 2004). The reason for this discrepancy is at present unclear. As one possibility, integrin functions may be more clearly revealed in acute perturbations (antibody addition); genetic perturbations may lead to the selection of cells that have activated pathways that compensate for the loss of $\beta 1$ integrins.

Experiments to define the *in vivo* function of $\beta 1$ integrins during cell proliferation in the CNS have been conducted using CRE/LOX-mediated gene ablation (Graus-Porta et al. 2001). Inactivation of the *Itgb1* gene in the precursors of neurons and glia in the CNS using *Itgb1-flox* mice and *nestin-CRE* mice leads not only to dramatic defects in the morphology of radial glial process (see the following text); the size of the cerebral and cerebellar cortices is also reduced in the mutants. The mechanism that leads to the reduction in the size of the cerebellar cortex has been studied further. The vast majority of cerebellar neurons are cerebellar granule cells, which are derived from cerebellar granule cell precursors that are situated in the external granule cell layer (EGL) (Fig. 2). Granule cells in the EGL self-renew and generate granule cells that subsequently migrate to the internal granule cell layer (IGL) where they terminally differentiate (Hatten, 1999). Granule cell precursors in the EGL proliferate in close apposition to the meningeal basement membrane that contains several LN isoforms. Proliferating granule cells, but not their differentiating descendants, express the LN receptor integrin $\alpha 6\beta 1$ and $\alpha 7\beta 1$ (Pons et al. 2001; Blaess et al. 2004). In mice that lack expression of $\beta 1$ integrins in the CNS, proliferation of granule cell precursors is drastically impaired (Blaess et al. 2004). Integrins appear to

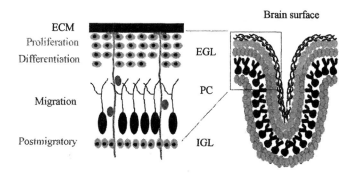

Fig. 2. Development of the cerebellar cortex. A cerebellar folium is shown to the right, and an enlargement of a small area of a folium (boxed) to the left. Granule cell precursors (green) proliferate in the external granule cell layer (EGL) that contacts the meningeal ECM. The cells give rise to premigratory neurons (orange) that subsequently differentiate further and migrate (blue cells) along Bergman glial fibers (orange) through the Purkinje cell layer (PC) (black) into the internal granule cell layer (IGL). (See Color Insert.)

regulate granule cell precursor proliferation by two mechanisms. First, integrins act autonomously in granule cell precursors to regulate their proliferation, suggesting that integrins activate signaling pathways that control the cell cycle machinery. In addition, the integrin ligand LN binds sonic hedgehog (SHH), a potent regulator of granule cell precursor proliferation, and SHH and LN cooperate to regulate granule cell precursor proliferation (Dahmane and Ruiz-i-Altaba, 1999; Wallace, 1999; Wechsler-Reya and Scott, 1999; Pons et al. 2001; Blaess et al. 2004). This suggests that integrins help to recruit SHH via LN to the SHH receptor, leading to its activation and stimulation of cell proliferation. Alternatively, LN and SHH may independently activate signaling pathways via their respective receptors that ultimately converge on the cell cycle machinery. Regardless of the mechanism, the findings clearly demonstrate a critical role for $\beta1$ integrins in regulating the proliferation of granule cell precursors.

Taken together, the data provide the first evidence that integrins and their ECM ligands regulate cell proliferation of some precursor populations in the developing CNS. However, the mechanisms by which integrins act remain to be defined. Do integrins assemble ECM constituents and signaling molecules into specialized "niches" that are required to maintain cell proliferation? Which cell cycle regulators are targets for integrin-activated signaling pathways? Are integrins required to control symmetric versus asymmetric cell division thereby affecting the self-renewal and differentiation of precursors? Are integrins required only in particular precursor populations? Refined genetic perturbations combined with cell culture experiments will undoubtedly clarify these issues.

3. Formation of cortical layers in the CNS: Radial glia and Cajal-Retzius cells

The mature cerebral cortex has a layered appearance because neuronal subtypes with similar functions and morphologies are assembled into distinct layers (Fig. 1). Developmental timing is essential for the development of these layers. At the onset of cortical development, the first cells that are generated in the ventricular neuroepithelium migrate towards the brain surface to form the preplate. Cells that are generated in subsequent waves of proliferation migrate along radial glial fibers, invade the preplate, and split it into a subplate and the marginal zone that contains the Cajal-Retzius cells. The earliest generated neuronal subtypes terminate their migration and detach from radial glial fibers as soon as they arrive close to the Cajal-Retzius cell layer. Later-born neurons bypass the earlier-born neuronal subtypes and form in more superficial layers. In this way, the distinct layers of the cerebral cortex are formed in an inside-out fashion (Fig. 1). Additional neurons that are born in the ganglionic eminence invade the cortical layers by tangential migratory routes and populate the emerging layers (Hatten, 2002; Nadarajah and Parnavelas, 2002; Marin and Rubenstein, 2003; Kriegstein and Noctor, 2004). The molecular mechanisms that regulate the formation of cell layers are only partially understood. Intense research efforts focus around the identification of the guidance cues for the migrating neurons, including signals that stimulate directional migration and stop-signals for migrating cells. Several lines of evidence suggest that integrins and their ECM ligands are involved.

Converging genetic evidence from several laboratories has provided convincing evidence that $\beta1$ integrins are instrumental in the formation of cortical layers. Mice carrying null mutations in the genes for the *Itga6* subunit (encoding the $\alpha6$ protein) or double mutants for the *Itga6/Itga3* subunit genes (encoding the $\alpha6$ and $\alpha3$ proteins, respectively) have been generated. The mice show defects in the formation of cortical layers, and in the assembly/maintenance of the meningeal basement membrane that surrounds the brain (Georges-Labouesse et al. 1998; De Arcangelis et al. 1999). To address the mechanism that causes the defect and to clarify whether integrins are required in the meningeal cells that surround the brain or in neural tissue, the *Itgb1* gene and, therefore, the expression of all $\beta1$ integrins including $\alpha3\beta1$ and $\alpha6\beta1$ has been inactivated in neural precursors using CRE/LOX-mediated gene inactivation (Graus-Porta et al. 2001). Expression of $\beta1$ integrins in the meningeal cell layer was not affected. Nevertheless, the mice show similar defects in cortical layers to *Itga3/Itga6*-deficient mice. While neurons assemble into layers, the layers portray a wavy appearance. In some areas, neurons invade the cortical marginal zone; in other areas, neurons accumulate within the cortical wall. When interpreted superficially, these findings suggest that $\beta1$ integrins regulate the migration of cortical neurons directly. However, further studies have demonstrated that neuronal migration per se and interactions between neurons and glial guideposts are not detectably affected in the mutant mice. The defects are instead caused by perturbations in the cortical marginal zone. In the absence of $\beta1$ integrins, the meningeal basement membrane shows interruptions, radial glial fibers frequently do not project to the meningeal basement membrane, and the formation of the Cajal-Retzius cell layer is perturbed. As a consequence, Cajal-Retzius cells form ectopia within the cortical wall. The positioning of neurons in cortical layers precisely follows the defects in the cortical marginal zone. In areas where the Cajal-Retzius layer and meningeal basement membrane are interrupted, neurons invade the cortical marginal zone. In areas where Cajal-Retzius cell ectopia form within the cortical wall, neurons accumulate below the ectopia (Graus-Porta et al. 2001). Taken together, the findings suggest that integrin-mediated interactions with the meningeal basement membrane are essential to maintain glial processes, the Cajal-Retzius cell layer, and the basement membrane. Positioning defects of cortical neurons are likely a secondary consequence of these perturbations. This model predicts that mutations that affect ECM components that are constituents of the meningeal basement membrane may cause similar defects in cortical layers, a prediction that has been confirmed by studies in several laboratories. Mice with mutations in genes encoding perlecan, nidogen/entactin, and LN have similar cortical defects to mice with a CNS-restricted *Itgb1* knock-out (Arikawa-Hirasawa et al. 1999; Costell et al. 1999; Dong et al. 2002; Halfter et al. 2002). Intriguingly, inactivation of the gene encoding the integrin downstream effector focal adhesion kinase (FAK) in neural precursors also leads to similar cortical defects (Beggs et al. 2003). Taken together, the findings suggest that an ECM-integrin-FAK signaling pathway is essential for the anchorage of the radial glial process at the meningeal basement membrane and to maintain the Cajal-Retzius cell layer. Defects in the anchorage of glial fibers are also observed in the retina, hippocampus, and cerebellum of *Itgb1*-deficient mice (Foerster et al. 2002 our unpublished data [2004]), providing strong evidence that $\beta1$ integrins are essential in maintaining radially oriented glial fibers throughout the CNS.

4. Formation of cortical layers in the CNS: Integrins, reelin, and cell migration

Several studies suggest that integrins may modulate the migratory behavior of neurons during the formation of cortical layers. These studies have also established a potential link between integrins and the ECM glycoprotein reelin, a key regulator of cortical development. In reelin-deficient mice, the preplate does not split and the formation of cortical layers is perturbed. Subsequent studies have shown that reelin binds to the extracellular domain of the VLDLR and APOER2 receptors. The cytoplasmic domains of these receptors in turn bind to the adapter protein dab-1 and reelin can induce phosphorylation of dab-1. Mice that lack the dab-1 or VLDLR/APOER2 proteins show comparable cortical defects to reelin-deficient mice, providing strong evidence that the proteins act in a common pathway. Despite the substantial biochemical and genetic evidence that link reelin, VLDLR/ APOER2, and dab-1 into a pathway, the molecular and cellular mechanisms by which reelin affects cortical development have remained unclear. Reelin is expressed in Cajal-Retzius cells and it has been proposed to serve as a stop signal for migrating neurons (although other mechanisms have been proposed). According to the stop-signal model, contact of migrating neurons with reelin changes their adhesive behavior, leading to the termination of migration and the detachment of migrating cells from glial guideposts (Magdaleno and Curran, 2001; Rice and Curran, 2001; Herz and Bock, 2002; Tissir and Goffinet, 2003). Some results have recently been presented that raise the possibility that integrins regulate adhesive interactions between neurons and glia and may mediate effects of reelin on migrating neurons.

Consistent with a role for integrins in cortical neurons, the positioning of tectal neurons is perturbed in chickens upon infection with retroviruses expressing antisens messenger ribonucleic acids (mRNAs) of the integrin $\beta 1$ and $\alpha 6$ subunits (Galileo et al. 1992; Zhang and Galileo, 1998). In cells in culture, antibodies to the integrin $\alpha 3$ subunit reduce the rate of migration of neurons along radial glial fibers without obvious defects in attachment. In contrast, antibodies to the αv subunit lead to detachment of neurons from glial fibers. To further substantiate these findings, Anton and colleagues (1999) have analyzed the CNS of *Itga3*-deficient mice. The authors have reported that cortical layers are abnormal. In subsequent studies, the authors have analyzed the effect of $\alpha 3$ further, in particular the relation between reelin and the integrin $\alpha 3 \beta 1$ (Dulabon et al. 2000). Recombinant reelin, when added to cells in culture, can slow the migration of cortical neurons along radial glial fibers. Reelin-soaked beads, when implanted into the cortex, can inhibit the migration of neurons past the beads, an effect that is not observed in the cortex of *Itga3*-deficient mice. Integrin $\alpha 3 \beta 1$ and reelin can also be co-immunoprecipitated from brain extracts (Dulabon et al. 2000). Furthermore, in mice that lack the dab-1 protein, neurons appear to be situated in closer apposition to radial glial fibers than normal, indicative of increased adhesion. Antibodies to the integrin $\alpha 3$ subunit reverse the phenotype, suggesting that dab-1 regulates adhesion via the integrin $\alpha 3 \beta 1$ (Sanada et al. 2004). Taken together, the studies have been interpreted as a demonstration of a function for the integrin $\alpha 3 \beta 1$ in cortical migration and reelin signaling.

However, there are several contradictory findings that need further clarification. First, while the phenotype of mice that lack reelin, dab-1, or VLDLR/APOER2 is very similar if not identical, the phenotype of *Itga3*-deficient mice differs substantially. Mutations in the genes encoding reelin, dab-1, and VLDLR/APOER2 lead to defects in preplate splitting, a process that is not affected in *Itga3*-deficient mice (Schmid et al. 2004). Therefore, early reelin functions do not require the $\alpha 3 \beta 1$ protein. Second, the extent to which layering defects in *reelin* and *Itga3*-deficient mice are similar has not been clearly addressed. In particular, *Itga3*-deficient mice die shortly after birth due to kidney defects (Kreidberg et al. 1996), preventing a detailed analysis of the organization of mature cortical layers. It would be interesting to analyze the development of cortical structures in mice with a CNS-restricted knock-out of the *Itga3* gene that are expected to survive into adulthood. Third, the $\alpha 3$ protein is strongly expressed in cells close to the ventricular neuroepithelium, but downregulated if not absent in cells close to the marginal zone where the reelin protein is expressed (D'Arcangelo et al. 1995; Ogawa et al. 1995; Sanada et al. 2004). It seems surprising that precisely those cells that are confronted with reelin express no or strongly reduced levels of $\alpha 3 \beta 1$. Fourth, while $\beta 1$ integrins and reelin can be co-immunoprecipitated from brain extracts (Dulabon et al. 2000), the interaction is not necessarily direct. Reelin is a very large glycoprotein that may bind (specifically or non-specifically) to other proteins in extracts that can interact with integrins. Further biochemical experiments are necessary to directly demonstrate that reelin is a bona-fide integrin α3β1 ligand. Fifth, the extent to which reelin acts as a stop signal for migrating neurons is unclear. Curran and colleagues have used transgenic mice to express reelin ectopically in the ventricular neuroepithelium in the same expression domain where the integrin α3 protein is highly expressed. Ectopic reelin expression does not affect the migration of cortical neurons, suggesting that reelin does not act as a stop-signal for their migration (Magdaleno et al. 2002). Sixth, as outlined in the previous section, in mice that lack all β1 integrins in the developing cerebral cortex, neuron-glia interactions are not affected *in vivo* or upon culturing of the cells *in vitro*. The cells are still able to migrate, to terminate migration, and to assemble into layers (albeit with a wavy appearance caused by defects in the cortical marginal zone as previously outlined) (Graus-Porta et al. 2001). Finally, while antibodies to the integrin αv lead to *in vitro* detachment of neurons from glia (Anton et al. 1999), neurons still migrate in the cortex of mice with a CNS-restricted knock-out of the *Itgav* subunit gene (McCarty et al. 2005).

How can these different findings be reconciled? As one possibility, β1 integrins (and αv integrins) may not be essential for migration but modulate the rate of migration. A slight decrease in the rate of migration may not have been detected in the *Itgb1*-deficient mice (and has not been evaluated in *Itgav*-deficient mice). Alternatively, different β1 integrins may have antagonistic roles in cell migration and reelin signaling. Inactivation of two integrins that affect migration negatively and positively, respectively, may lead to no net effects on cell migration. Alternatively, neurons that lack all β1 integrins (but not those that only lack the integrin α3β1) may use different receptors to compensate for the loss of β1 integrins. Clearly, further studies will be necessary to distinguish between these possibilities and to provide further genetic, cell biological, and biochemical evidence that clarify the extent to which integrins regulate glial-guided migration and are an important component of the reelin signaling pathway.

5. The adult CNS: Integrin functions during adult neurogenesis

Integrins and their ECM ligands are also expressed in the adult CNS, and recent studies have implicated them as regulators of adult neurogenesis. During adulthood, neurogenesis mainly occurs in two zones of the CNS: the subventricular zone (SVZ) of the lateral ventricle and the dentate gyrus of the hippocampus (Alvarez-Buylla and Garcia-Verdugo, 2002; Gage, 2002). Integrin functions have been studied in the SVZ (Fig. 3). Neural stem cells in the SVZ undergo slow asymmetric division that lead to their self-renewal and to the generation of transient amplifying cells that differentiate

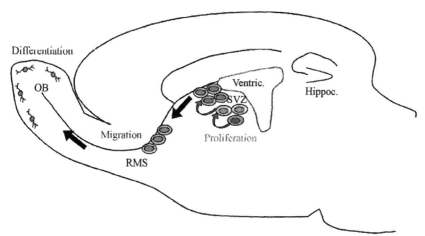

Fig. 3. Adult neurogenesis. A diagram of a saggital brain section is shown. Neuroblasts (yellow) that are generated from stem cells (blue) in the subventricular zone (SVZ) migrate along the rostral migratory stream (RMS) toward the olfactory bulb (OB) where they differentiate. (See Color Insert.)

into neuronal precursors. The precursor cells migrate tangentially along the rostral migratory stream (RMS) from the SVZ to the olfactory bulb (Lois and Alvarez-Buylla, 1993). Once the cells reach the center of the olfactory bulb, they change their direction and migrate radially to invade more superficial layers of the olfactory bulb where they differentiate into interneurons of the granular and periglomerular layers (Luskin, 1993). During migration in the RMS, the cells are linked into chains by direct cell-cell interactions. Recent experiments suggest that integrins and their ECM ligands affect migration in the RMS. TN-C and the LN α5 and γ1 subunits are expressed in the RMS (Gates et al. 1995; Jankovski and Sotelo, 1996; Murase and Horwitz, 2002). Expression of different integrin subunits has also been reported. The α1, β8, and β1 integrin subunits are particularly prominently expressed in the RMS of embryonic animals, co-expressed with the αv subunit at early postnatal ages, but subsequently downregulated. αv persists into adulthood, where it is co-expressed with the β3 and β6 subunits. Consistent with the expression patterns, antibodies to the α1, β1, and αv subunits reduce migration in slices derived from early postnatal animals, while antibodies to αv and β3 are affective in slices derived from older animals (Murase and Horwitz, 2002). Studies by Emsley and Hagg provide evidence that β1 integrins, and particularly the α6β1 heterodimer, are expressed not only prenatally, but also in the adult. Antibodies that inhibit α6β1 function block migration when injected into the RMS of adult mice (Emsley and Hagg, 2003). While both studies support a function for integrins in the RMS, further studies will be essential to define the function of different integrins, and the mechanisms by which they affect RMS formation and maintenance.

6. The adult CNS: Integrins and synaptic plasticity

Recent findings provide evidence that integrins also regulate the formation and function of synapses in the CNS and are important components of the molecular network that controls synaptic plasticity. Expression of integrin subunits has been detected in the CNS by *in situ* hybridization, and several subunits, including the $\alpha3$, $\alpha8$, and $\beta8$, have been localized by immunohistochemistry to synaptic sites. Studies with antibodies and small molecular weight inhibitors of integrin function have provided evidence that $\beta1$ and αv integrins are required for the stabilization of long-term potentiation (LTP) in the hippocampus, but not for basic synaptic transmission (Clegg, 2000). Further *in vitro* studies have implicated integrins, in particular the $\beta3$ integrin subunit and possibly the integrin $\alpha v\beta3$, in the maturation of hippocampal synaptic connections (Chavis and Westbrook, 2001). Recent genetic findings support the *in vitro* findings. Mice with single, double, and triple mutations in the integrin $\alpha3$, $\alpha5$, and $\alpha8$ subunit genes have been analyzed (Chan et al. 2003). $\alpha3+/-$ mice fail to maintain LTP that is generated in hippocampal CA1 neurons. No such defects have been observed in $\alpha5+/-$ or $\alpha8+/-$ mice. However, LTP is nearly completely abolished in triple heterozygous mice. No structural defects have been observed in the hippocampus. Taken together, these data suggest important roles for these three integrins in the generation and maintenance of LTP. Intriguingly, mice with a mutation in the gene for the integrin-associated protein (IAP) show significant reduction in the magnitude in LTP (Chang et al. 1999, 2001). IAP binds to integrins containing the $\beta3$ subunit, but also to $\alpha2\beta1$ (Brown and Frazier, 2001), raising the possibility that these integrins also regulate synaptic functions in the CNS. However, further genetic studies are necessary to address this issue.

7. Outlook

Our knowledge of the function of integrins and their ligands in the CNS has dramatically increased over the last few years. Importantly, we are now in a position to address the function of different integrins not only with cells in culture, but also *in vivo* using genetic perturbations. Cell biological and genetic studies support the view that integrins regulate many aspects of CNS development, ranging from effects of the proliferation of neural stem cells and neuronal precursors to the formation of the radial glial scaffold. The ongoing challenge is to further define the function of different integrin heterodimers in the CNS, to understand the integrin-dependent signaling pathways that are activated, and analyze how they interact with other signaling pathways to regulate the development and function of both neurons and glial cells.

References

Alvarez-Buylla, A., Garcia-Verdugo, J.M. 2002. Neurogenesis in adult subventricular zone. J. Neurosci. 22, 629–634.

Anton, E.S., Kreidberg, J.A., Rakic, P. 1999. Distinct functions of alpha3 and alpha(v) integrin receptors in neuronal migration and laminar organization of the cerebral cortex. Neuron 22, 277–289.

Arikawa-Hirasawa, E., Watanabe, H., Takami, H., Hassell, J.R., Yamada, Y. 1999. Perlecan is essential for cartilage and cephalic development. Nat. Genet. 23, 354–358.

Baron, W., Colognato, H., ffrench-Constant,, C. 2005. Integrin-growth factor interactions as regulators of oligodendroglial development and function. Glia 49, 467–479.

Beggs, H.E., Schahin-Reed, D., Zang, K., Goebbels, S., Nave, K.A., Gorski, J., Jones, K.R., Sretavan, D., Reichardt, L.F. 2003. FAK deficiency in cells contributing to the basal lamina results in cortical abnormalities resembling congenital muscular dystrophies. Neuron 40, 501–514.

Blaess, S., Graus-Porta, D., Belvindrah, R., Radakovits, R., Pons, S., Littlewood-Evans, A., Senften, M., Guo, H., Li, Y., Miner, J.H., Reichardt, L.F., Mueller, U. 2004. Beta1 integrins are critical for cerebellar granule cell precursor proliferation. J. Neurosci. 24, 3402–3412.

Bokel, C., Brown, N.H. 2002. Integrins in development: Moving on, responding to, and sticking to the extracellular matrix. Dev. Cell 3, 311–321.

Brakebusch, C., Fassler, R. 2003. The integrin-actin connection, an eternal love affair. EMBO J. 22, 2324–2333.

Brown, E.J., Frazier, W.A. 2001. Integrin-associated protein (CD47) and its ligands. Trends Cell. Biol. 11, 130–135.

Campbell, K., Gotz, M. 2002. Radial glia: Multi-purpose cells for vertebrate brain development. Trends Neurosci. 25, 235–238.

Campos, L.S., Leone, D.P., Relvas, J.B., Brakebusch, C., Fassler, R., Suter, U., ffrench-Constant, C. 2004. Beta1 integrins activate a MAPK signaling pathway in neural stem cells that contributes to their maintenance. Development 131, 3433–3444.

Chan, C.S., Weeber, E.J., Kurup, S., Sweatt, J.D., Davis, R.L. 2003. Integrin requirement for hippocampal synaptic plasticity and spatial memory. J. Neurosci. 23, 7107–7116.

Chang, H.P., Lindberg, F.P., Wang, H.L., Huang, A.M., Lee, E.H. 1999. Impaired memory retention and decreased long-term potentiation in integrin-associated protein-deficient mice. Learn. Mem. 6, 448–457.

Chang, H.P., Ma, Y.L., Wan, F.J., Tsai, L.Y., Lindberg, F.P., Lee, E.H. 2001. Functional blocking of integrin-associated protein impairs memory retention and decreases glutamate release from the hippocampus. Neuroscience 102, 289–296.

Chavis, P., Westbrook, G. 2001. Integrins mediate functional pre- and postsynaptic maturation at a hippocampal synapse. Nature 411, 317–321.

Chun, J.J., Shatz, C.J. 1988. A fibronectin-like molecule is present in the developing cat cerebral cortex and is correlated with subplate neurons. J. Cell. Biol. 106, 857–872.

Clegg, D.O. 2000. Novel roles for integrins in the nervous system. Mol. Cell. Biol. Res. Commun. 3, 1–7.

Costell, M., Gustafsson, E., Aszodi, A., Morgelin, M., Bloch, W., Hunziker, E., Addicks, K., Timpl, R., Fassler, R. 1999. Perlecan maintains the integrity of cartilage and some basement membranes. J. Cell. Biol. 147, 1109–1122.

Cram, E.J., Schwarzbauer, J.E. 2004. The talin wags the dog: New insights into integrin activation. Trends. Cell. Biol. 14, 55–57.

D'Arcangelo, G., Miao, G.G., Chen, S.C., Soares, H.D., Morgan, J.I., Curran, T. 1995. A protein related to extracellular matrix proteins deleted in the mouse mutant reeler. Nature 374, 719–723.

Dahmane, N., Ruiz-i-Altaba, A. 1999. Sonic hedgehog regulates the growth and patterning of the cerebellum. Development 126, 3089–3100.

De Arcangelis, A., Mark, M., Kreidberg, J., Sorokin, L., Georges-Labouesse, E. 1999. Synergistic activities of alpha3 and alpha6 integrins are required during apical ectodermal ridge formation and organogenesis in the mouse. Development 126, 3957–3968.

Decker, L., Baron, W., ffrench-Constant, C. 2004. Lipid rafts: Microenvironments for integrin-growth factor interactions in neural development. Biochem. Soc. Trans. 32, 426–430.

Dong, L., Chen, Y., Lewis, M., Hsieh, J.C., Reing, J., Chaillet, J.R., Howell, C.Y., Melhem, M., Inoue, S., Kuszak, J.R., De Geest, K., Chung, A.E. 2002. Neurologic defects and selective disruption of basement membranes in mice lacking entactin–1/nidogen–1. Lab. Invest. 82, 1617–1630.

Drago, J., Nurcombe, V., Bartlett, P.F. 1991. Laminin through its long arm E8 fragment promotes the proliferation and differentiation of murine neuroepithelial cells in vitro. Exp. Cell. Res. 192, 256–265.

Dulabon, L., Olson, E.C., Taglienti, M.G., Eisenhuth, S., McGrath, B., Walsh, C.A., Kreidberg, J. A., Anton, E.S. 2000. Reelin binds alpha3 beta1 integrin and inhibits neuronal migration. Neuron 27, 33–44.

Emsley, J.G., Hagg, T. 2003. alpha6 beta1 integrin directs migration of neuronal precursors in adult mouse forebrain. Exp. Neurol. 183, 273–285.

Fishell, G., Kriegstein, A.R. 2003. Neurons from radial glia: The consequences of asymmetric inheritance. Curr. Opin. Neurobiol. 13, 34–41.

Foerster, E., Tielsch, A., Saum, B., Weiss, K.H., Johanssen, C., Graus Porta, D., Müller, U., Frotscher, M. 2002. Reelin, disabled 1, and beta1-class integrins are required for the formation of the radial glial scaffold in the hippocampus.. Proc. Natl. Acad. Sci. USA 99, 13178–13183.

Gage, F.H. 2002. Neurogenesis in the adult brain. J. Neurosci. 22, 612–613.

Galileo, D.S., Majors, J., Horwitz, A.F., Sanes, J.R. 1992. Retrovirally introduced antisense integrin RNA inhibits neuroblast migration in vivo. Neuron 9, 1117–1131.

Garcion, E., Faissner, A., ffrench-Constant, C. 2001. Knockout mice reveal a contribution of the extracellular matrix molecule tenascin-C to neural precursor proliferation and migration. Development 128, 2485–2496.

Garcion, E., Halilagic, A., Faissner, A., ffrench-Constant, C. 2004. Generation of an environmental niche for neural stem cell development by the extracellular matrix molecule tenascin C. Development 131, 3423–3432.

Gates, M.A., Thomas, L.B., Howard, E.M., Laywell, E.D., Sajin, B., Faissner, A., Gotz, B., Silver, J., Steindler, D.A. 1995. Cell and molecular analysis of the developing and adult mouse subventricular zone of the cerebral hemispheres. J. Comp. Neurol. 361, 249–266.

Georges-Labouesse, E., Mark, M., Messaddeq, N., Gansmuller, A. 1998. Essential role of alpha6 integrins in cortical and retinal lamination. Curr. Biol. 8, 983–986.

Giancotti, F.G. 2003. A structural view of integrin activation and signaling. Dev. Cell 4, 149–151.

Giancotti, F.G., Ruoslahti, E. 1999. Integrin signaling. Science 285, 1028–1032.

Graus-Porta, D., Blaess, S., Senften, M., Littlewood-Evans, A., Damsky, C., Huang, Z., Orban, P., Klein, R., Schittny, J.C., Muller, U. 2001. Beta1-class integrins regulate the development of laminae and folia in the cerebral and cerebellar cortex. Neuron 31, 367–379.

Halfter, W., Dong, S., Yip, Y.P., Willem, M., Mayer, U. 2002. . A critical function of the pial basement membrane in cortical histogenesis. J. Neurosci. 22, 6029–6040.

Hatten, M.E. 1999. Central nervous system neuronal migration. Annu. Rev. Neurosci. 22, 511–539.

Hatten, M.E. 2002. New directions in neuronal migration. Science 297, 1660–1663.

Henry, M.D., Campbell, K.P. 1999. Dystroglycan inside and out. Curr. Opin. Cell. Biol. 11, 602–607.

Herz, J., Bock, H.H. 2002. Lipoprotein receptors in the nervous system. Annu. Rev. Biochem. 71, 405–434.

Hynes, R.O. 1992. Integrins: Versatility, modulation, and signaling in cell adhesion. Cell 69, 11–25.

Hynes, R.O. 2002. Integrins: Bidirectional, allosteric signaling machines. Cell 110, 673–687.

Jacques, T.S., Relvas, J.B., Nishimura, S., Pytela, R., Edwards, G.M., Streuli, C.H., ffrench-Constant, C. 1998. Neural precursor cell chain migration and division are regulated through different beta1 integrins. Development 125, 3167–3177.

Jankovski, A., Sotelo, C. 1996. Subventricular zone-olfactory bulb migratory pathway in the adult mouse: Cellular composition and specificity as determined by heterochronic and heterotopic transplantation. J. Comp. Neurol. 371, 376–396.

Jones, P.H., Harper, S., Watt, F.M. 1995. Stem cell patterning and fate in human epidermis. Cell 80, 83–93.

Jones, P.H., Watt, F.M. 1993. Separation of human epidermal stem cells from transit amplifying cells on the basis of differences in integrin function and expression. Cell 73, 713–724.

Kreidberg, J.A., Donovan, M.J., Goldstein, S.L., Rennke, H., Shepherd, K., Jones, R.C., Jaenisch, R. 1996. Alpha3 beta1 integrin has a crucial role in kidney and lung organogenesis. Development 122, 3537–3547.

Kriegstein, A.R., Noctor, S.C. 2004. Patterns of neuronal migration in the embryonic cortex. Trends Neurosci. 27, 392–399.

Liddington, R.C., Ginsberg, M.H. 2002. Integrin activation takes shape. J. Cell. Biol. 158, 833–839.

Liu, S., Calderwood, D.A., Ginsberg, M.H. 2000. Integrin cytoplasmic domain-binding proteins. J. Cell. Sci. 113(Pt. 20), 3563–3571.

Lois, C., Alvarez-Buylla, A. 1993. Proliferating subventricular zone cells in the adult mammalian forebrain can differentiate into neurons and glia. Proc. Natl. Acad. Sci. USA 90, 2074–2077.

Luskin, M.B. 1993. Restricted proliferation and migration of postnatally generated neurons derived from the forebrain subventricular zone. Neuron 11, 173–189.

Magdaleno, S., Keshvara, L., Curran, T. 2002. Rescue of ataxia and preplate splitting by ectopic expression of Reelin in reeler mice. Neuron 33, 573–586.

Magdaleno, S.M., Curran, T. 2001. Brain development: Integrins and the reelin pathway. Curr. Biol. 11, R1032–R1035.

Marin, O., Rubenstein, J.L. 2003. Cell migration in the forebrain. Annu. Rev. Neurosci. 26, 441–483.

McCarty, J.H., Lacy-Hulbert, A., Charest, A., Bronson, R.T., Crowley, D., Housman, D., Savill, J., Roes, J., Hynes, R.O. 2005. Selective ablation of alpha(v) integrins in the central nervous system leads to cerebral hemorrhage, seizures, axonal degeneration and premature death. Development 132, 165–176.

Müller, U. 2004. Integrins and Extracellular Matrix in Animal Models. In: *Handbook Exp. Pharm.* (J. Behrens, W.J. Nelson, Eds.), New York: Springer, 165, pp. 217–241.

Murase, S., Horwitz, A.F. 2002. Deleted in colorectal carcinoma and differentially expressed integrins mediate the directional migration of neural precursors in the rostral migratory stream. J. Neurosci. 22, 3568–3579.

Nadarajah, B., Parnavelas, J.G. 2002. Modes of neuronal migration in the developing cerebral cortex. Nat. Rev. Neurosci. 3, 423–432.

Ogawa, M., Miyata, T., Nakajima, K., Yagyu, K., Seike, M., Ikenaka, K., Yamamoto, H., Mikoshiba, K. 1995. The reeler gene-associated antigen on Cajal-Retzius neurons is a crucial molecule for laminar organization of cortical neurons. Neuron 14, 899–912.

Pons, S., Trejo, J.L., Martinez-Morales, J.R., Marti, E. 2001. Vitronectin regulates Sonic hedgehog activity during cerebellum development through CREB phosphorylation. Development 128, 1481–1492.

Rakic, P. 2003. Developmental and evolutionary adaptations of cortical radial glia. Cereb. Cortex 13, 541–549.

Rice, D.S., Curran, T. 2001. Role of the reelin signaling pathway in central nervous system development. Annu. Rev. Neurosci. 24, 1005–1039.

Sanada, K., Gupta, A., Tsai, L.H. 2004. Disabled-1-regulated adhesion of migrating neurons to radial glial fiber contributes to neuronal positioning during early corticogenesis. Neuron 42, 197–211.

Schmid, R.S., Shelton, S., Stanco, A., Yokota, Y., Kreidberg, J.A., Anton, E.S. 2004. alpha3 beta1 integrin modulates neuronal migration and placement during early stages of cerebral cortical development. Development 131, 6023–6031.

Schwartz, M.A., Ginsberg, M.H. 2002. Networks and crosstalk: Integrin signaling spreads. Nat. Cell. Biol. 4, E65–E68.

Sheppard, A.M., Brunstrom, J.E., Thornton, T.N., Gerfen, R.W., Broekelmann, T.J., McDonald, J.A., Pearlman, A.L. 1995. Neuronal production of fibronectin in the cerebral cortex during migration and layer formation is unique to specific cortical domains. Dev. Biol. 172, 504–518.

Stettler, E.M., Galileo, D.S. 2004. Radial glia produce and align the ligand fibronectin during neuronal migration in the developing chick brain. J. Comp. Neurol. 468, 441–451.

Stewart, G.R., Pearlman, A.L. 1987. Fibronectin-like immunoreactivity in the developing cerebral cortex. J. Neurosci. 7, 3325–3333.

Tissir, F., Goffinet, A.M. 2003. Reelin and brain development. Nat. Rev. Neurosci. 4, 496–505.

Wallace, V.A. 1999. Purkinje-cell-derived Sonic hedgehog regulates granule neuron precursor cell proliferation in the developing mouse cerebellum. Curr. Biol. 9, 445–448.

Wechsler-Reya, R.J., Scott, M.P. 1999. Control of neuronal precursor proliferation in the cerebellum by Sonic Hedgehog. Neuron 22, 103–114.

Zhang, Z., Galileo, D.S. 1998. Retroviral transfer of antisense integrin alpha6 or alpha8 sequences results in laminar redistribution or clonal cell death in developing brain. J. Neurosci. 18, 6928–6938.

Zhu, A.J., Haase, I., Watt, F.M. 1999. Signaling via beta1 integrins and mitogen-activated protein kinase determines human epidermal stem cell fate in vitro. Proc. Natl. Acad. Sci. USA 96, 6728–6733.

Extracellular matrix and inner ear development and function

Dominic Cosgrove[1] and Michael Anne Gratton[2]

[1]*National Usher Syndrome Center, Boys Town National Research Hospital, Omaha, Nebraska*
[2]*Department of Otorhinolaryngology (Head and Neck Surgery), University of Pennsylvania, Philadelphia, Pennsylvania*

Contents

Advances in Developmental Biology
Volume 15 ISSN 1574-3349
DOI: 10.1016/S1574-3349(05)15006-6

1. Introduction

The functional role of extracellular matrix (ECM) molecules in the inner ear has historically received little attention. In contrast, the distribution of ECM proteins in the cochlea, particularly those of the basement membrane and tectorial membrane, is well documented. In order to familiarize the reader with the inner ear, a brief description of cochlear anatomy follows. The inner ear is comprised of two systems: the cochlea, which provides information relating to hearing, and the vestibular system, which provides sensory information critical to balance. This chapter focuses upon the cochlear portion of the inner ear.

Sound originates when a body vibrates, sending a particle wave coursing through the air. However, the perception of sound requires the transformation of the particle wave to a neural impulse. The purpose of the outer, middle, and inner ears is to gather sound waves and convert them initially into mechanical pressures and finally into electrical, neural impulses. The interpretation of the neural impulse is a task of the auditory cortex. The organ of Corti of the inner ear is a highly developed, intricate component of the auditory system, and serves two functions: to convert mechanical energy into electrical energy and to dispatch a coded version of the original sound to the brain that includes information not only about frequency but intensity and timing as well.

Cochlear signaling is initiated by translation of sound pressure waves by the middle ear into the movement of fluid in the cochlear duct. This wavelike fluid movement causes vibratory motion of the basilar membrane. The cochlear hair cells, which are attached to the basilar membrane, in turn, ride with the wave generated along the basilar membrane. These specialized neuroepithelial cells have a highly organized array of interconnected stereocilia on their apical surface. The tips of the stereocilia are embedded into the tectorial membrane. The tectorial membrane, radiating from the spiral limbus, is fixed to the bony modiolus. Thus, the motile basilar membrane moves relative to the more immotile tectorial membrane, causing the stereocilia of the hair cells to be deflected. These deflections open or close the mechanosensitive channels at the apical surface of the hair cells. Opening of these channels allows divalent ions from the scala media to enter the hair cells, resulting in the depolarization event

required to initiate signaling across the synapse. These signals are trans-
duced across the spiral ganglion, through the cochlear (8th) nerve, and
translated by the brain to what we perceive as sound.

The inner ear is a compartmentalized structure comprised of numerous
unique and highly specialized cells and tissues that coordinately function
in translating sound pressure waves into electrical impulses. The inner
ear should be envisioned as a snail's shell with 2.5 turns (Fig. 1). The
basal portion of the inner ear, nearest to the stapes of the middle ear,
translates high frequency sound. The middle portion of the inner ear
spiral codes mid frequencies while the apical turn translates low frequency
sound. In the central core of the inner ear spiral is a bony structure
called the modiolus, which houses the 8th nerve fibers and their spiral
ganglion cell bodies. The middle compartment, the membranous labyrinth,
houses the end-organ of the peripheral auditory system. The basic structures

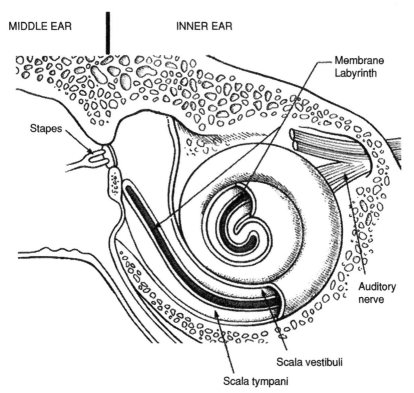

Fig. 1. Schematic of the cochlea showing orientation relative to the stapes and auditory nerve.
A portion of the bony labyrinth has been removed to reveal the scala vestibuli, scala tympani,
and the membranous labyrinth. The membranous labyrinth, also termed the cochlear duct or
scala media, contains the organ of Corti, the end organ of hearing.

of the cochlea, which are denoted on the mid-modiolar cross-section from an adult mouse, are shown in Fig. 2B. The spiral limbus radiates from the modiolus and, thus, is rigidly fixed to the bony structure. The apical surface of the limbus provides the attachment site for Reissner's membrane and the gelatinous tectorial membrane. The basal and lateral surface forms the inner sulcus (part of the "organ of Corti") and the attachment site for the basilar membrane. The basilar membrane spans across the organ of Corti to the spiral ligament which forms the "external sulcus" up to the spiral prominence. Just above the spiral prominence is the stria vascularis, a specialized tissue that serves, in part, as a divalent ion pump. The activity of the ion pump serves to maintain a high electrochemical potential in the scala media, the middle compartment of the three compartments of the inner ear formed by the basilar membrane and Reissner's membrane. This electrical potential, termed the endocochlear potential, is critical for the depolarization of neuroepithelium in the initiation of CNS signals from the organ of Corti. The organ of Corti sits atop the basilar membrane. This tissue supports the neuroepithelium, or cochlear hair cells, and may also function in the recirculation of divalent ions through a series of gap junctions.

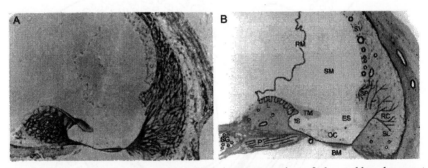

Fig. 2. Matrix in the inner ear with schematic representation of the cochlear basement membranes in the mouse. The middle turn of a mid-modiolar cross-section from an adult mouse was stained with Jones silver methenamine (A) or with eosin and hematoxylin (B). The histochemical staining of panel A highlights the ECM of the spiral limbus, basilar membrane, and spiral ligament. In panel B, the anatomical locations of the cochlear basement membranes are delineated by a wide black line. Inner ear structures relevant to this review are indicated. SG = spiral ganglion; P = perineurium; L = spiral limbus; IS = inner sulcus; BM = basilar membrane; ES = external sulcus; SP = spiral prominence; RC = root cell processes; SV = stria vascularis; SL = spiral ligament; RM = Reissner's membrane; TM = tectorial membrane. Figure 2B is adapted from Rodgers, K.D., Barritt, L.C., Miner, J.H., Cosgrove, D.E., 2001. The laminins in the murine inner ear: developmental transitions and expression in cochlear basement membranes. Hear Res. 158, 41, with permission from Elsevier.

2. Extracellular matrix of the inner ear

The cochlear duct has been referred to as the membranous labyrinth (Fig. 1), and indeed as revealed by Jones silver methenamine (Fig. 2A), the structures of the inner ear are rich in ECM. In the mature inner ear, the portion of the spiral limbus facing the inner sulcus is largely acellular, and comprised of ECM. Atop the limbus and radiating across the organ of Corti is a highly specialized mass of gelatinous ECM, the tectorial membrane. The stereocilia of the hair cells are embedded in the tectorial membrane. This serves to anchor the stereocilia to the immobile limbus, allowing differential deflection of the tip-links through the movement of the basilar membrane.

The lateral wall of the cochlea comprises the spiral ligament. This tissue contains five sub-types of fibrocytes (types I through V), classified according to their specialized ion transport properties (Spicer et al., 1991). This network of spiral ligament fibrocytes interspersed with ECM proteins is easily visualized by use of Jones silver methenamine histochemistry (Fig. 2A). The fibrocytes are connected by gap junctions and are thus implicated to function in the recirculation of divalent ions from scala vestibuli or scala tympani (Fig. 1) to the stria vascularis in the cochlea (Fig. 2A). Fibrocytes located in the region of the attachment to the basilar membrane, called tension fibrocytes, are thought to regulate the stiffness and hence to affect the frequency tuning of the basilar membrane.

3. Distribution of extracellular matrix proteins

The connective tissue regions of the cochlea share many of the same ECM proteins as demonstrated by numerous studies employing primarily immunohistochemical and ultrastructural techniques. The predominant matrix protein in the spiral limbus, spiral ligament, and basilar membrane as well as the tectorial membrane is type II collagen (Yoo and Tomoda, 1988; Ishibe et al., 1989; Ishii et al., 1992; Slepecky et al., 1992; Thalmann, 1993; Kaname et al., 1994; Tsuprun and Santi, 1999). Although some controversy existed regarding the presence of type II collagen in the basilar membrane (Tomoda et al., 1984; Ishibe et al., 1989), an elegant study by Tsuprun and Santi (1999) used Fourier analysis of scanning and transmission electron microscopy images to show aggregates of type II collagen bundles in the basilar membrane (Fig. 3). In fact, they found that a continuous track of cross-striated collagen fibrils extended from the spiral limbus, across the basilar membrane to terminate in the spiral ligament. Fourier transformations showed that the striations apparent in the collagen bundles shown in inserts of Figures 3B and 3E have a D-periodicity with a 67-nm axial repeat.

Proteoglycans, mainly members of the chondroitin-sulfate family, were noted to co-localize with the collagen fibrils in the spiral limbus, spiral

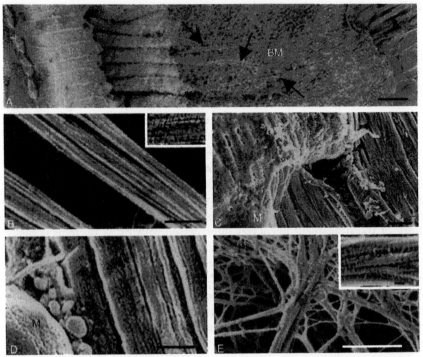

Fig. 3. The SEM views of the fractured tissue of the basilar membrane and spiral ligament. (A) Fibrous bundles (arrows) in the pars pectinata zone of the basilar membrane (BM) viewed from the upper scala media side (bar = 10 μm). (B) Each bundle consists of collagen fibrils (bar = 1 μm). The insert shows the D-periodicity of the cross-striations (bar = 100 μm). (C, D) Fibrous bundles in the basilar membrane viewed from the lower scala tympani side near the pars arcuata zone. (bar = 1 μm). M = mesothelial cell. (E) In the spiral ligament a network of randomly oriented collagen bundles of varying size (bar = 1 μm). The insert shows the orientation and striation of the individual collagen fibrils within a bundle (bar = 100 μm). Reprinted from Tsuprun, V.L., Santi, P., 1999. Ultrastructural and immunohistochemical identification of the extracellular matrix of the extracellular matrix of the chinchilla cochlea. Hear Res. 129, 41, with permission from Elsevier.

ligament, and basilar membrane (Thalmann, 1993; Munyer and Schulte, 1994; Tsuprun and Santi, 1999). The density and size of the proteoglycans particles differed among the connective tissues of the inner ear. These randomly oriented particles of chondroitin sulfate are thought to contribute to the structural integrity of the ECM.

Two glycoproteins, fibronectin and tenascin, have been immunolocalized in the spiral limbus, basilar membrane, and spiral ligament of rodent cochlea (Santi et al., 1989; Woolf et al., 1992; 1996; Keithley et al., 1993; Swartz and Santi, 1997). In the basilar membrane, the glycoproteins compose the ground substance surrounding the collagen fibrils and separating the two layers of radially oriented collagen bundles (Tsuprun and Santi, 1999). Reactivity in the mesothelial cells underlying the basilar membrane exceeded that in the ground substance suggesting that the mesothelial cells are the site of the fibronectin and tenascin production (Swartz and Santi, 1997). The glycoproteins in the spiral limbus and ligament appeared to be confined to the region of attachment to the basilar membrane and were not observed in the body of these connective tissues.

Despite the common content of extracellular proteins previously described, the connective tissues of the inner ear display regional differences in distribution and organization of their collagen fibrils. Moreover, additional matrix molecules have been found in some, but not all, of the inner ear connective tissues. The unique composition of the ECM in the spiral limbus, basilar membrane, spiral ligament, and tectorial membrane is most likely related to the role for each of these tissues in processing the incoming sound pressure wave and its conversion to a neural impulse.

The least complex ECMs both in terms of organization and protein content are those of the spiral limbus and spiral ligament. These tissues consist primarily of fibrocytes that are specialized for ion transport and type II collagen. Type II collagen fibrils form the base of the ECM (Iishi et al., 1992; Slepecky et al., 1992; Thalmann, 1993). However, with the exception of the regions of attachment to the basilar membrane, the fibrils do not show any preferences in orientation (Tsuprun and Santi, 1999). As shown in Fig. 3E, the individual fibrils form an amorphous mesh that envelopes the individual fibrocytes. The fibrils remain as individual fibrils although occasional small bundles are observed. There is a suggestion that in aged ears, the formation of the collagen bundles may occur as the cellular content of the spiral ligament decreases (Spicer and Schulte, 2002). Ultrastructurally, small heparan sulfate proteoglycan proteins are found in close association with the type II collagen fibrils. Recently, mRNA expression of two matricellular proteins, SPARC and SC1, was localized to the spiral limbus and subregions of the spiral ligament in embryonic and neonatal rat inner ears (Mothe and Brown, 2001). The presence these calcium-binding glycoproteins in tissues specialized for ion transport supports the notion that the spiral ligament and spiral limbus are actively involved in cochlear homeostasis.

The basilar membrane which underlies and supports the organ of Corti is primarily composed of radially oriented collagen bundles surrounded by a homogeneous ground substance (Figs. 3A and 3C). The stiffness and mass of the basilar membrane, critical parameters in cochlear micromechanics, are determined in part by the composition and structural organization of the ECM molecules. As with the spiral limbus and ligament, type II collagen has been detected in the basilar membrane using both biochemical and immunohistochemical techniques (Yoo and Tomada, 1988; Ishibe et al., 1992, Slepecky et al., 1992, Thalmann, 1993). It is the predominant form of collagen, forming approximately 40% of the collagen fibrils. Type V and

IX collagens are present in lesser amounts (Slepecky et al., 1992; Thalmann, 1993). In comparison to the adjoining spiral limbus and ligament tissues, the basilar membrane shows a complex organizational structure as displayed in Fig. 3B and D. Collagen fibrils, 10 to 12 nm in diameter, oriented in the radial direction, form one layer in the pars arcuata zone which adjoins the spiral limbus and underlies the inner hair cell region of the organ of Corti. However, in the pars pectinata region, which extends laterally from the pillar cell region of the organ of Corti to the spiral ligament, the collagen fibrils are arranged in two layers (Iurato, 1962; Katori et al., 1993, Mikuni et al., 1994, 1995; Tsuprun and Santi, 1999). Several types of various sized and randomly oriented proteoglycan particles co-localize with the distribution of the type II collagen in the basilar membrane. The relationship between the type II collagen fibrils and the proteoglycan particles is shown in Fig. 4. Immuno-histochemically, these particles have been identified as the chondroitin–4 and –6 sulfate proteoglycans and the small chondroitin sulfate particle, decorin, as well as keratan sulfate (Ishii et al., 1992; Munyer and Schulte, 1994; Swartz and Santi, 1997; Tsuprun and Santi, 1999). The collagen fibrils of the collagen layers are surrounded by an amorphous ground substance. This same substance also separates the two collagen layers of the pars pectinata. The glycoproteins, tenascin and especially fibronectin, immunolocalize to the ground substance region. Recently, emilin–2, a glycoprotein ascribed with elastin-like properties, was identified in the ground substance of the basilar membrane (Amma et al., 2003). The expression of emilin–2 in the mouse was found to be highest during the period of onset and maturation of hearing suggesting a role for emilin in matrix assembly and cochlea biome-chanics. Based upon mRNA expression and the relative degree of immuno-reactivity, the mesothelial cells underlying the basilar membrane are implicated in the biosynthesis of the glycoprotein content of the basilar membrane. The amount of the basilar membrane occupied by the ground substance varies along the length of the organ of Corti. Hence, the ground substance is thought to influence the mass and stiffness of the basilar mem-brane thereby contributing to the localization of sound frequency in differing regions of the cochlea.

Another connective tissue of the inner ear is the tectorial membrane. The tectorial membrane is a highly structured ECM extending from the spiral limbus to cover the hair cells of the organ of Corti (Fig. 2B). The tips of the taller stereocilia on the outer hair cells are embedded into the tectorial membrane. As the basilar membrane and tectorial membrane move in response to the sound pressure wave, the stereocilia bend. This deflection opens ion channels and initiates the transduction of sound pressure waves into neural impulses. Structurally, the tectorial membrane can be delineated into a number of distinct zones. The upper surface of the central core or body tectorial membrane is covered by a network of fibrils, the "covernet" zone. Hensen's stripe, a ridge on the lower surface of the tectorial membrane

Fig. 4. The TEM view of a radial section of the basilar membrane after treatment with cuprolinic blue to contrast the negatively charged glycosaminoglycan residues of the proteoglycans. Randomly oriented proteoglycans (arrows) are associated with the collagen fibrils (CF). Within the ground substance (GS), fibrous, nodular structures (arrowheads) thought to be fibronectin are oriented parallel to the collagen fibrils. Bar = 200 nm. The inset demonstrates a Fourier transformation from an image of ground substance containing the fibrous structure. The diffraction peak (arrow) corresponds to an average space of 35 nm between adjacent fibrous structures. Reprinted from Tsuprun, V.L., Santi, P., 1999. Ultrastructural and immunohistochemical identification of the extracellular matrix of the extracellular matrix of the chinchilla cochlea. Hear Res 129, 45, with permission from Elsevier.

just above the inner hair cell region, is thought, based on its location, to be involved in deflection of the inner hair cell stereocilia. The sulcul zone located in the lower portion of the tectorial membrane above the inner sulcus epithelium of the organ of Corti contains an organized complex of fibrils. Finally, Hardesty's membrane and the marginal net extend from the tip of the tectorial membrane to the lateral edge of the outer hair cell region.

Historically the molecules of the tectorial membrane were thought to be secreted by the interdental cells of the spiral limbus. However, recent work investigating the temporal and spatial mRNA expression of tectorial membrane collagens and glycoproteins during embryological development of the inner ear identified the inner sulcus region and emerging sensory cells as the primary expression sites (Khetapal et al., 1994; Rau et al., 1999; Legan et al., 2000).

In contrast to the previously described connective tissues of the inner ear, the tectorial membrane displays a greater diversity as well as abundance in collagen content (Hasko and Richardson, 1988; Slepecky et al., 1992; Thalmann, 1993; Tsuprun and Santi, 1997). As in the basilar membrane, the type II collagen predominates, accounting for approximately 40% of the tectorial membrane collagen with lesser amounts of types V, IX, and XI (Richardson et al., 1987; Slepecky et al., 1992; Thalmann, 1993). The ultrastructural view of Fig. 5A shows that two filament structures, designated types A and B, are visible within the tectorial membrane (Hasko and Richardson, 1988, Tsuprun and Santi, 1997). The thick type A fibrils are composed of straight unbranched 20-nm diameter collagen fibrils. Type A fibers radiate through the central core of the tectorial membrane arranged either in parallel rows or in aggregates having tetragonal pattern when viewed in the radial plane (Tsuprun and Santi, 1997). Numerous proteoglycans attached to the bands of collagen fibrils with a D-repeat of 62 to 63 nm, are present in the tectorial membrane (Hasko and Richardson, 1988; Slepecky et al., 1992; Tsuprun and Santi, 1997). The majority of these proteoglycans are oriented orthogonally to the collagen fibril axis although they can aggregate into linear arrays forming crystalline structures in the lower zone of the tectorial membrane. The rod-shape and size (50 to 65 nm long, 10 nm wide) of the individual proteoglycan molecules and their orthogonal association with the type A collagen fibrils is consistent with small globular proteins with one or two glycosaminoglycan chains (Tsuprun and Santi, 1997). Immunohistochemically, both keratan and chondroitin sulfate proteoglycans have been localized to the tectorial membrane (Munyer and Schulte, 1994; Swartz and Santi, 1997). Type B fibrils, the second of the tectorial membrane fibril structures, are described as a "striated-sheet" (Fig. 5B). This matrix is composed of fine, 7 to 9 nm diameter, light and dark staining filaments that are periodically linked to one another by cross bridges (Hasko and Richardson, 1988; Tsuprun and Santi, 1997; Goodyear and Richardson, 2002). The central core of the tectorial membrane is composed of type A collagen fibrils embedded in the striated-sheet type B matrix. In addition, the border zones of the tectorial membrane including the covernet, the marginal band, Hardesty's membrane, and the marginal band contain abundant type B fibrils. The type B fibrils which may comprise as much as 50% of total tectorial membrane protein are thought to consist of non-collagenous glycoproteins (Richardson et al., 1987). Three non-collagenous

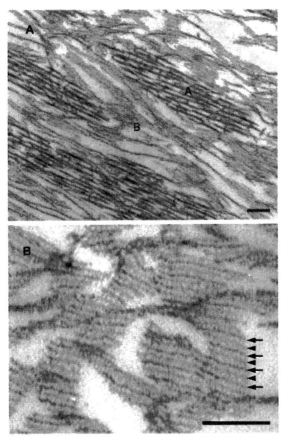

Fig. 5. Ultrastructure of the fibrillar matrix systems in the central zone of the tectorial membrane of the mouse. (A) Type A fibers composed of aggregates of collagen fibrils, roughly oriented in parallel (A = collagen fibrils), are embedded in a striated sheet of fine filaments (B = striated sheet matrix). (B) The striated sheet matrix is formed by crosslinked 7 to 9 nm diameter fibrils (arrows) thought to be tectorin. Bar = 200 nm. Adapted from Goodyear, R.J., Richardson, G.P. 2002. Extracellular matrices associated with the apical surfaces of sensory epithelia in the inner ear: molecular and structural diversity. J Neurobiol. 52, 222, with permission from Wiley-Liss, Inc., a subsidiary of John Wiley & Sons, Inc.

glycoproteins: α-tectorin, β-tectorin, and otogelin (Fig. 6) have been identified as components of the mouse tectorial membrane (Cohen-Salmon et al., 1997; Legan et al., 1997). The distribution of these matrix proteins is shown using immunofluorescence histochemistry in Fig. 6. In the mouse, the mRNA expression of *TECTA*, the gene encoding tectorin, is highest during the first few days postnatally suggesting a role for tectorin in tectorial membrane morphogenesis (Maeda et al., 2001). A genetic mutation or dysregulation of tectorin expression results in sensorineural hearing loss

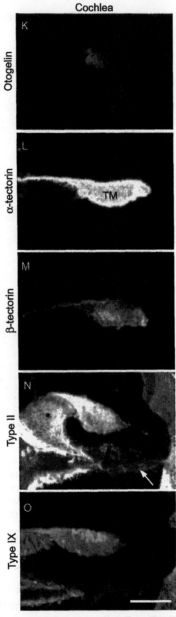

Fig. 6. Distribution of matrix molecules in the tectorial membrane of the mouse. (K) Otogelin
is present in the central zone of the tectorial membrane at low levels. (L) a-tectorin can be
detected throughout all zones of the tectorial membrane (TM). (M) In contrast, β-tectorin is
localized to the margins of the tectorial membrane which demonstrate type B but not type A,
fibrils. (N) Type II collagen fibrils are present not only in the central zone of the tectorial
membrane but also in the underlying spiral limbus (asterisk) as well as the basilar membrane
(arrow) and the bone of the otic capsule. (O) The distribution of type IX collagen fibrils is

consistent with loss of the cochlear amplifier function of the outer hair cells (Legan et al., 2000; Knipper et al., 2001).

4. Extracellular matrix otopathology

Mutations in several genes encoding ECM proteins are associated with both dominant and recessive modes of inherited hearing loss. These deafness-related genes include *COCH, COL2A1, COL11A1* or *COL11A2*, and *TECTA* which underlie autosomal dominant disorders such as Stickler and Marshall syndromes, DFNA9 and DFNA13, as well as autosomal recessive disorders such as OSMED and DFNB21.

4.1. COCH/*DFNA9*

COCH is a novel cochlear gene which encodes the ECM protein, cochlin, also called Coch–5B2. Cochlin has extensive homology to the collagen-binding domains of von Willebrand factor (Robertson et al., 1997). The expression of *COCH* is high during inner ear development in both the putative organ of Corti and vestibular regions suggesting a structural role for cochlin (de Kok et al., 1999). Immunohistochemistry indicates that cochlin is assembled in the endoplasmic reticulum and Golgi complex and then secreted to accumulate in the basilar membrane mirroring the distribution of fibronectin (Grabski et al., 2003). Mutations in the *COCH* gene have been linked to DFNA9, an autosomal dominant progressive sensorineural hearing loss with late onset in the high frequencies and associated progressive vestibular dysfunction (de Kok et al., 1999; Khetarpal, 2000). A detailed study by Khetarpal (2000) of human temporal bones from DFNA9 patients confirmed the eosinophilic deposition in membranous labyrinth noted by Fransen and Van Camp (1999) as well as degeneration of the 8th nerve. However, Khetarpal also found that the structure of type II collagen in the spiral ligament had been altered and that the spiral ligament displayed numerous glycosaminoglycan particles associated with a network of microfibrils. Grabski et al. (2003) and Robertson et al. (2003) showed that mutated cochlin was still assembled and secreted indicating that the pathology results from the inability of the mutated cochlin to correctly assemble with other ECM proteins. The late onset and progressive nature of the hearing loss associated with *COCH* mutations in DFNA9 patients suggests that accumulation of the aberrant cochlin may occur slowly over time.

---

limited to the upper zones of the tectorial membrane. Bar = 100 μm. Adapted from Goodyear, R.J., Richardson, G.P. 2002. Extracellular matrices associated with the apical surfaces of sensory epithelia in the inner ear: molecular and structural diversity. J. Neurobiol. 52, 216, with permission from Wiley-Liss, Inc., a subsidiary of John Wiley & Sons, Inc.

4.2. Stickler and Marshall syndrome

Stickler syndrome is an autosomal dominant disorder caused by mutations in *COL2A1*, *COL11A1*, or *COL11A2* genes (Snead and Yates, 1999). The syndrome is characterized by a wide range of variably penetrating phenotypes that include retinal detachment, orofacial defects, arthritis, and high frequency sensorineural hearing loss. Marshall syndrome, which is a phenotypically similar disorder, is caused by a dominant mutation in the *COL11A1* gene (Meisler et al., 1998; Annunen et al., 1999). In the cochlea, *COL11A1* and *COL11A2* mRNA while initially diffusely expressed in the cochlear duct, transiently localize to the greater epithelial ridge and cochlear lateral wall during postnatal days 1 to 5. However, by postnatal day 13, expression of the *COL11A1* and *COL11A2* genes is found only in the inner sulcus, Claudius' cells, and cells of Boettcher (Shpargel et al., 2004). In the murine cochlea, type II collagen is found throughout all regions with ECM while type XI collagen is expressed in the tectorial membrane. Mutations in the collagen genes underlying the Stickler and Marshall syndrome are thought to produce a dominant negative effect upon collagen assembly; however, the specific cochlear mechanism(s) responsible for the deafness phenotype is not yet known (Griffith et al., 1998; Symko-Bennett et al., 2003).

4.3. Otospondylo-megaepiphyseal dysplasia

Otospondylo-megaepiphyseal dysplasia (OSMED) is an autosomal recessive skeletal dysplasia that can be accompanied by severe hearing loss. The hearing loss is sensorineural and non-progressive in nature. The phenotype overlaps that of the Stickler and Marshall syndromes but is distinguished by disproportionately short limbs in the affected individuals. The defect is due to a missense mutation of the *COL11A2* gene that causes aggregation of collagen fibrils as well as an increase in the fibril diameter (Van Steensel et al., 1997).

After screening seven affected families, Melkoniemi et al. (2000) found different mutations of *COL11A2* in each family, all of which were predicted to cause complete absence of the α2 chain of type XI collagen. However, Li et al. (2001) implicated altered synthesis of type II collagen as the basis of the cartilaginous abnormalities found in mice having a targeted disruption of *COL11A2*.

4.4. DFNA13

The non-syndromic deafness, DFNA13, presents as a moderate-degree hearing loss that affects primarily the mid-frequencies resulting in a U-shaped audiogram. The DFNA13 locus has been linked to a mutation in the *COL11A2* gene (McGuirt et al., 1999; Deleenhear et al., 2001). The mutation found on chromosome 6p was predicted to alter the triple-helix domain of type XI collagen. McGuirt et al. (1999) reported that mice with a targeted disruption of *COL11A2* show disorganized collagen fibrils in the tectorial membrane.

4.5. TECTA/*DFNB21*/*DFNA8A12*

Mutations in the gene *TECTA* that encode tectorin are associated with the autosomal recessive deafness, DFNB21, as well as the autosomal dominant deafness, DFNA8A12. The mutation has been linked to chromosome 11q23 (Naz et al., 2003). The resultant hearing loss can be progressive or stable in nature and mild or severe in degree dependent upon whether the mutation is a splice variant or a missense as well as the domain of the missense (Verhoeven et al., 1998; Allosio et al., 1999; Mustapha et al., 1999; Moreno-Pelayo et al., 2001). Since mice that have a null allele of *TECTA* display a tectorial membrane that is detached from the organ of Corti and a moderate to severe hearing loss, it has been predicted that null alleles of tectorin, particularly α-tectorin, might have a similar effect upon cochlear morphology and function (Legan et al., 1997; Naz et al., 2003).

5. Basement membranes of the inner ear

The inner ear is very rich in basement membranes (Takahashi and Hokunan, 1992; Cosgrove et al., 1996a; Cosgrove and Rodgers, 1997). The location of these basement membranes is depicted in Fig. 2B. Extensive networks of basement membranes are associated with the spiral ganglion cell bodies and the perineurium surrounding their associated axons. A continuous track of basement membrane forms a loop-like structure extending along the spiral limbus, ensheathing each of the interdental cells lining the apical surface of the limbus. This same basement membrane then tracks down the inner sulcus, across the basilar membrane, across the external sulcus to the spiral prominence, where it radiates into the spiral ligament, surrounding the root cell processes. The epithelial basement membrane continues beneath the spiral prominence epithelium, but terminates at the stria vascularis. The epithelial cells of the stria vascularis, called marginal cells, are unique in that they do not have a basement membrane underlying their basal aspects. In the stria, basement membranes with a complex composition surround each capillary. Finally, Reissner's membrane also contains a basement membrane thought to be involved in fluid transport between the endolymph and the perilymph of the scala media and scala vestibuli, respectively (Satoh et al., 1998).

6. Distribution of basement membrane proteins

Studies aimed at defining the distribution of specific molecular components in the membranous labyrinth have relied largely on immunofluorescent analysis using antibody preparations specific for the protein of interest. Such analysis has been used to detect the presence of the classical and novel collagen type IV isoforms, laminin isoforms, fibronectin, entactin, and proteoglycans in the cochlea (Santi et al., 1989a,b; Takahashi and Hokunan, 1992; Woolf et al., 1992; Cosgrove et al., 1996a,c; Satoh et al., 1998; Tsuprun and Santi, 2001).

Specific basement membrane proteins show a preferential distribution in the cochlear basement membranes. Fibronectin is a major component of the basilar membrane, but does not appear in any other cochlear basement membranes (Santi et al., 1989b; Woolf et al., 1992). Research conducted at Boys Town National Research Hospital examined the differences in the cochlear distribution of the classical and novel type IV collagen chains (Cosgrove et al., 1996a). *COL4A1* and *COL4A2* chains are abundantly expressed in the perineurium of the osseous spiral lamina and basement membranes surrounding the spiral ganglion cell bodies, while expression in other cochlear basement membranes is minimal. In contrast, *COL4A3*, *COL4A4*, and *COL4A5* chains are highly expressed in the continuous basement membrane track depicted in Fig. 2B as well as surrounding all capillaries of the stria, spiral ligament, and limbus. Finally, heparan sulfate proteoglycan (HSPG), while found in all cochlear basement membranes (Satoh et al., 1998), shows local differences in its amount and molecular form, which results in heterogeneous ultrastructural profiles (Tsuprun and Santi, 2001). For instance, in the spiral ligament capillary basement membrane, particles of HSPG are randomly scattered throughout the lamina densa. However, in the strial capillary basement membrane, an abundance of thin HSPG particles is present in the thick homogenous single layer of lamina densa. The differences noted in HSPG content between these two capillary beds were speculated by Tsuprun and Santi (2001) to reflect functional differences in their filtration properties.

The second most abundant of the basement membrane proteins are the laminins. This family of glycoproteins is comprised of five known α chains, three known β chains, and three known γ chains. A member of each of these chains (α, β, and γ) combines to form the 15 known laminin heterotrimers designated laminin-1 to laminin-15 (Miner et al., 1997; Koch et al., 1999, Libby et al., 2000). Expression of the different laminins can vary greatly from one basement membrane to another even within the same tissue (Sanes et al., 1990). Rodgers et al. (2001) examined the laminin chain composition of the basement membranes in the inner ear of the mouse during both development and in adulthood using antibodies developed against the known laminin chains. The results (Fig. 7, lower panel) showed a very complex expression in different segments of adult cochlear basement membranes. By co-localization inference, these results provide a relatively comprehensive interpretation of the laminin heterotrimers present in cochlear basement membranes, and suggest that 9 of the 15 known laminin heterotrimers are present in the cochlea. A few of these heterotrimers, such as laminin-2 and -4, have limited distribution, while others such as laminin-11 are more ubiquitous. The locations in which the different laminins are expressed may reflect their role as key signals not only in the continuous synthesis and assembly of basement membranes, but also in the maintenance of cochlear homeostasis.

Ultrastructural examination of the cochlear basement membranes has revealed heterogeneity in the assembly of proteoglycans (Tsuprun and Santi, 2001) of which HSPG is the predominant molecule. Quasi-regular, linear arrays were found in the basement membrane of the inner and outer sulcus, the spiral prominence, and root cells. In contrast, proteoglycans were scattered throughout the basement membranes surrounding cochlear nerve fibers and capillaries. The differing composition and arrangement of the proteoglycans may relate to the demands of the various basement membrane tracks in the cochlea for filtration or structural support.

7. Basement membrane components: Developing versus mature cochlea

Rodents are altricial mammals, unable to hear at birth. The rodent inner ear undergoes extensive terminal cytodifferentiation postnatally. The endocochlear potential necessary to drive auditory transduction in the cochlear hair cells does not develop until 11 to 12 days of age (Ehret, 1976). Mice do not have adult-like auditory sensitivity until approximately 21 days of age. The series of mid-modiolar sections from mice cochlea shown in Fig. 8, illustrate the cochlea at various stages of postnatal development.

During this period of postnatal development in the rodent, basement membrane composition is in a dynamic state of synthesis and turnover. The result is a broad transition in basement membrane composition throughout the cochlea. During cochlear maturation, expression of basement membrane proteins is coordinately regulated in a development-specific manner that is spatially and developmentally specific for the various basement membrane proteins and their individual chains. At birth, in rodents, most of the cochlear basement membranes are undergoing active assembly. An abundant expression of fibronectin surrounding most cochlear tissues is noted in newborn rodents, with a reduced expression in all regions except the basilar membrane noted by 14 days after birth (Woolf et al., 1992). Cosgrove and Rodgers (1997) examined the expression patterns of major basement membrane-associated proteins (laminin-1, entactin, and HSPG) as a function of postnatal cochlear development. While fibronectin follows the previously described developmental regulation (Woolf et al., 1992), other proteins are coordinately regulated in a developmental-specific manner that is spatially and developmentally distinct from that for fibronectin. A prominent, but transient, expression of HSPG, laminin-1, and entactin was noted in the membrane spanning the interdental cells of the limbus, extending down the

D. Cosgrove and M. A. Gratton

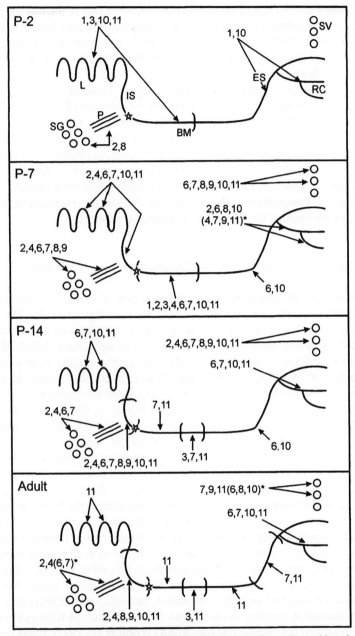

Fig. 7. Summary of developmental transitions of laminin isoforms in cochlear basement membranes. Based on co-localization of the laminin chains in the murine inner ear, the laminin heterotrimers that may be present in cochlear basement membranes at the indicated developmental time points are shown. Brackets are used to define the "zones" of variable laminin composition along the continuous track of basement membrane extending from the

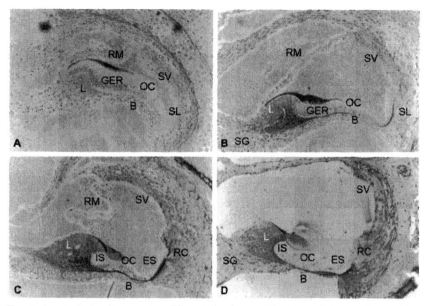

Fig. 8. Anatomical references for different developmental stages of the murine cochlea. Cochleae from 1-day (A), 3-day (B), 8-day (C), and 28-day old (D) mice were embedded in plastic resin and cut in the mid-modiolar plane at 2 μm. The sections were stained using the Jones silver methenamine technique to localize ECM proteins, then counterstained with hematoxylin. The anatomical structures of interest to this review are denoted. L = spiral limbus; GER = greater epithelial ridge; RM = Reissner's membrane; B = basilar membrane; OC = organ of Corti; SV = stria vascularis; SL = spiral ligament; SG = spiral ganglion; IS = inner sulcus; ES = external sulcus; RC = root cell processes. Reprinted from Cosgrove, D.E., Kornak, J.M., Samuelson, G., 1996a. Expression of basement membrane type IV collagen chains during postnatal development in the murine cochlea. Hear Res 100, 22, with permission from Elsevier.

inner sulcus and across the basilar membrane. The expression of laminin-1 and entactin, extended from the basilar membrane track into that of the root processes of the spiral ligament. This transient basement membrane, which assembles in 4 days and persists only through 14 days after birth, coincides with the peak period of cytodifferentiation for cells comprising the organ of Corti (Ito et al., 1995). These cells are located above this region of transient

spiral limbus to the external sulcus. The star indicates the position of the habenula perforata. SG = spiral ganglion; L = limbus; P = perineurium; IS = inner sulcus; BM = basilar membrane; ES = external sulcus; SP = spiral prominence; RC = root cell processes; SV = stria vascularis; RM = Reissner's membrane. Reprinted from Rodgers, K.D., Barritt, L.C., Miner, J.H., Cosgrove, D.E., 2001. The laminins in the murine inner ear: developmental transitions and expression in cochlear basement membranes. Hear Res 158, 47, with permission from Elsevier.

expression of HSPG, laminin-1, and entactin. In the adult, the novel collagen chains (COL4A3, COL4A4, and COL4A5) occupy the same basement membrane location.

Moreover, during the same interval of time (0 to 14 days), expression of type IV collagen and laminin is also dynamic. At birth, only the classical type IV collagen chains are expressed. The novel collagen genes do not activate until 3 days after birth. The alteration in type IV collagen isoforms is largely complete by 14 days after birth. At 14 days, protein levels of COL4A3, COL4A4, and COL4A5 exceed that of COL4A1 and COL4A2 in all cochlear basement membranes except the basement membranes of the spiral ganglion cells and processes (Cosgrove et al., 1996a). The laminin chains show dynamic changes in expression during development in every system thus far examined (Schugar et al., 1997; Miner et al., 1998; Powell et al., 1998; Simon-Assman et al., 1998; Yoshiba et al., 1998; Miner and Li, 2000). This dynamic expression is thought to drive programs of differentiation by way of signaling through cell surface receptors, such as the integrin family of heterodimeric glycoprotein receptors (Buttery and ffrench-Constant, 1999; Muschler et al., 1999; Tomastis et al., 1999). Recently, complex postnatal transitions in laminin expression have been described (Rodgers et al., 2001) using antibodies developed against the known laminin chains. The results, which are summarized in cartoon form in Fig. 7, illustrate dynamic changes in expression of the laminin chains during postnatal development. By col-localization inference, it appears that four laminin heterotrimers are present in the cochlear basement membrane at birth. By 7 days after birth, 10 of the 15 known laminin trimers are expressed. However, from 7 to 14 days after birth, downregulation occurs. The laminin expression for 9 heterotrimers in the adult cochlea basement membrane is site-specific. For instance, the basement membrane spanning the interdental cells and basilar membrane contains laminin-11; however, laminin-2, -4, -8, -9, -10, and -11 are expressed in the basement membrane in the inner sulcus basement membrane and the strial capillary basement membranes. The number and complexity of transitions in the laminin heterotrimers present in the cochlear basement membranes suggest that laminins have a prominent role in modulating postnatal cochlear maturation.

A change in the basement membrane location in the cochlear lateral wall is also noted during cochlear development in the rodent. The basement membrane which separated the developing stria vascularis from the spiral ligament disappears by 5 days after birth (Sagara et al., 1995). In the mature lateral wall, basement membranes are found only around capillaries of the stria and spiral ligament as well as under the epithelial and root processes of the external sulcus. Interestingly, the strial capillary basement membrane consists of single homogeneous lamina densa while all other cochlear basement membranes are comprised of a lamina lucida and densa.

8. Basement membrane otopathology

Alterations in the structure and function of basement membranes have recently been recognized as mediators of pathology in a variety of organ systems. Diseases associated with basement membrane pathology include: Alport syndrome, muscular dystrophy, epidermolysis bullosa, and forms of membranous glomerulonephropathy (Barker et al., 1990; Aberdam et al., 1994; Mochizuki et al., 1994; Xu et al., 1994). To date the only known human syndromic hearing losses attributable to mutations in basement membrane proteins are Alport syndrome and Usher syndrome type 2a. Despite advances during the past decade in the genetic basis for these syndromes and subsequent aberrant expression of basement membrane proteins, key questions remain concerning the mechanism of the hearing loss.

8.1. Alport syndrome

Familial nephritis or Alport syndrome is a genetic disorder of the basement membrane. Alport syndrome has a relatively common (approximately 1 in 5,000) occurrence. The basement membrane pathology results in a juvenile onset and progressive glomerulonephritis culminating in renal failure (Atkin et al., 1988; Wester et al., 1995) that requires dialysis or renal transplant. Many affected individuals exhibit a progressive, high frequency sensorineural hearing loss (Myers and Tyler, 1972). The incidence of hearing loss has increased as dialysis and kidney transplant have increased the lifespan of Alport patients (M. Madaio, personal communication, 2002). Ocular defects are less commonly reported (Streeten et al., 1987; Thompson et al., 1987; Sargona et al., 1999).

Otopathology in Alport syndrome is characterized as a moderate, progressive, high frequency, sensori-neural hearing loss. The hearing loss is not due to a specific mutation in the type IV collagen gene (Kim et al., 1995). Human temporal bone studies have reported a wide range of pathological features, including collapse of Reissner's membrane, absence of the organ of Corti, loss of cochlear neurons, and strial defects (Myers and Tyler, 1972; Arnold, 1984; Smith et al., 1992). Cosgrove et al. (1998) characterized the inner ear of a mouse model for Alport syndrome produced by targeted mutagenesis of the COL4A3 gene (Cosgrove et al., 1996b). Ultrastructurally, all inner ear basement membranes seemed thinned, with the exception of strial capillary basement membranes, which were grossly thickened relative to normal wild-type littermates (Fig. 9). Cosgrove et al. (1998) concluded that the clearly identifiable thickening of the strial capillary basement membrane formed a likely explanation for the progressive hearing loss accompanying

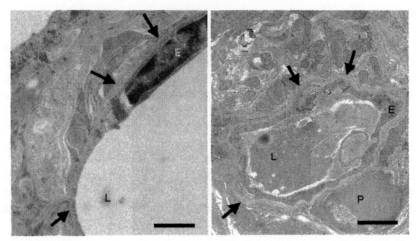

Fig. 9. Capillaries in the stria of a wild type and COL4A3, Alport mouse. The strial capillary basement membrane (arrows) of the Alport mice is significantly thicker than that of normal littermates. The capillary lumen is often constricted by a misshapen endothelial cell. Bar = 1 μm. L = lumen; E = endothelial cell; P = pericyte.

Alport syndrome. The thickening of the strial capillary basement membrane is consistent with an earlier report, in which a human cochlea was removed and fixed immediately postmortem (Weidauer and Arnold, 1976).

A recent study found that the COL4A3 mutant mice were more susceptible to noise-induced hearing loss than their normal, wild-type littermates. The accompanying drop in the voltage of the endocochlear potential strongly suggests lateral wall dysfunction is responsible for the noise susceptibility. Since the markedly thickened strial capillary basement membrane is the most clearly identifiable pathology in the Alport cochlea, the authors proposed that strial capillary basement membrane defects are responsible for the predisposition to noise-induced damage in these animals (Gratton et al., 2002).

8.2. Usher syndrome type 2a

Usher syndrome type 2a, an autosomal recessive disorder, was recently linked to mutations in a novel gene encoding a newly described protein, usherin, that has ECM motifs (Eudy et al., 1998). It has since been determined that usherin is in fact a basement membrane protein. This novel protein is widely expressed in both cochlear and retinal basement membranes as well as in a variety of other tissues (Bhattacharya et al., 2002; Pearsall et al., 2002). People with Usher syndrome type 2a have a high frequency sensori-neural hearing loss that may be somewhat progressive in nature (Sadehgi et al., 2004). The mechanism by which basement membrane dysfunction leads to hearing loss in Usher syndrome type 2a remains unknown.

8.3. Diabetes mellitus

Diabetes mellitus is a metabolic disorder whose basic defect is an absolute or relative lack of insulin. In its complete form, it is manifested by hyperglycemia, accelerated atherosclerosis, and microvascular disease of the retina and kidney as well as neuropathy. It affects approximately 5% of the U.S. population and is comprised of two subtypes, according to the classification system proposed by the American Diabetes Association (1997).

A relationship between diabetes mellitus and sensori-neural hearing loss was suggested during the mid–1800s. Although a large body of evidence has accumulated to support this view, the relationship between the pathologies remains controversial. However, the symmetric, progressive, high frequency hearing loss reported in some patients with diabetes mellitus is viewed by most to be the sequel of inner ear microangiopathy. Human temporal bone studies have revealed a wide variety of pathologies affecting every compartment of the cochlea. (Jorgensen, 1961; Makishima and Tanaka, 1971). The sole study to compare audiological status with histological observations (Wackym and Linthicum, 1985) noted greater hearing loss in the diabetic group whose microangiopathy involved inner ear vessels, particularly those of the endolymphatic sac and basilar membrane.

The application of animal models for diabetes mellitus, both type 1 and type 2, to the study of hearing loss and diabetes has confirmed the association between the two clinical entities. Several studies have employed rat strains having a genetic disposition for diabetes (Triana et al., 1991; Rust et al., 1992; Ishikawa et al., 1995; McQueen et al., 1999). An alternative has been to induce diabetes mellitus via alloxan (Saito et al., 1984) or streptozotocin (Smith et al., 1995). While human studies pertaining to the role of diabetes mellitus have provided conflicting data, animal models of diabetes mellitus have consistently demonstrated cochlear pathology. Strial capillary basement membranes were found to be thickened in the streptozoticin-injected rat, a model for type 1 diabetes mellitus (Smith et al., 1995), as well as in the spontaneously hypertensive-corpulent NID (SHR/N-cp) rat, a genetically based model of type 2 diabetes mellitus (McQueen et al., 1999). Interestingly, the capillary basement membrane thickening in the latter study was exacerbated by exposure to long-term noise. The basement membrane thickening is due to accumulation of type IV collagen and laminin accompanied by an as yet unexplained decrease in proteoglycans (Sumida et al., 1997). Although the rat has been the preferred animal model for the study of diabetes, several mouse models of diabetes mellitus have been described. Tachibana and Nakai (1986) examined the cochlea of the spontaneously diabetic KK mouse. At three months of age, these mice demonstrated glycosuria and hyperglycemia as well as early cochlear pathology. These changes were confined to the marginal cells of the stria vascularis, but increased by eight months of age to include the strial capillaries and their basement membranes as well intermediate cells.

8.4. Presbycusis

Decreased function of the stria vascularis has been implicated in the pathogenesis of presbycusis, also known as age-related hearing loss (Johnsson and Hawkins, 1972; Nadol, 1979; Keithley et al., 1992; Saitoh, 1995). In the gerbil, the age-related changes in stria vascularis and spiral ligament overwhelm the mild-moderate loss of hair cells and spiral ganglion cells of the eighth nerve. Indeed, even in the young gerbil, the elaborate capillary network of the stria vascularis shows focal lesions (Gratton and Schulte, 1995). Capillary atrophy increases with age and is thought to underlie the overall degeneration and dysfunction of the stria vascularis that is linked with advancing age (Gratton and Schulte, 1995; Gratton et al., 1996, 1997).

Ultrastructural examination of the stria vascularis revealed that 65% to 85% of strial capillaries in aged gerbils had thickened strial capillary basement membranes (Thomopoulos et al., 1997). The thickened capillary basement membrane was found to contain increased amounts of laminin with a concomitant decrease in expression of dystroglycan (Sakaguchi et al., 1997; Heaney and Schulte, 2002). The increase in laminin led investigators to postulate that oversynthesis rather than decreased degradation of basement membrane proteins formed the mechanism for the accumulation of basement membranes in the aged strial capillaries. The thickening of the basement membranes was a gradual process. In middle-aged gerbils, the association of basement membrane thickening with thin, degenerating processes of the pericytes suggested that pericytes were involved in the genesis of the capillary lesion. In later stages, the strial capillary basement membrane had become laminated and was several-fold thicker than that of young gerbils as shown in Fig. 10. Eventually the capillary lumen was found to narrow and fill with debris. By this stage, in most instances, the pericytes and endothelial cells were not discernible. As a secondary consequence of the capillary atrophy, alterations leading to necrosis occur in the marginal cell of the stria vascularis. Thomopoulos et al. (1997) attributed the cytopathologic changes in the strial marginal cell, which has a very high metabolic rate, to the effect of decreased permeability imposed by the thickened strial capillary basement membranes.

8.5. Systemic lupus erythematosus (SLE)

SLE is a chronic inflammatory disease caused by an overproduction of antibodies from pathological B cells. Evidence suggests multifactorial causes for SLE including genetic, hormonal, immunologic, and environmental. The prevalence of SLE is 1:2000, an incidence that has nearly tripled in the past four decades (Lawrence et al., 1998). SLE affects multiple body organs, with

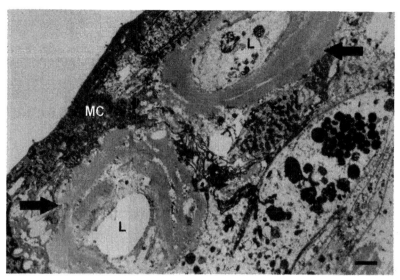

Fig. 10. Two capillaries in the stria of an aged gerbil display grossly thickened basement membranes (arrows). The basement membrane of the lower capillary shows lucent spaces and degenerating pericyte processes between the basement membrane lamina. The lumen of the upper capillary is occluded by debris while the endothelial cell is thinned. Bar = 1 μm. L = lumen; MC = marginal cell.

most clinical manifestations of SLE mediated either directly or indirectly by the formation of antinuclear antibodies, generation of circulating immune complexes, and activation of the immune complement system. Arthritic symptoms are reported by 90% of SLE patients, while 50% have nephritis.

SLE is one of the diseases whose reversible auditory manifestations fall under the clinical entity "autoimmune sensor-neural hearing loss" as proposed by McCabe (1979). The diagnosis of autoimmune inner ear disease is based primarily on the responsiveness to steroid treatments. The prevalence of hearing loss in SLE patients is not well documented and often conflicting (Bowman et al., 1986; Andonopoulos et al., 1995). A recent examination of temporal bones from seven SLE patients (Sone et al., 1999) reported varying degrees of cochlear pathology ranging from fibrotic tissue with new bone formation, to concretions in the stria vascularis, to loss of hair cells and spiral ganglion cells.

In an attempt to determine the underlying mechanism of autoimmune hearing loss, some investigators have analyzed animal models of SLE. Three murine models of SLE, with diverse genetic backgrounds, have displayed elevated auditory brainstem response thresholds and degeneration of the stria vascularis. These strains include the MRL-Fas[lpr], the C3H-Fas[lpr], and the (NZB x W)F1 strains. Cochlear pathology, specifically degeneration of the stria vascularis, has been reported in each of these strains (McMenomey et al.,1992, Ruckenstein et al.,1993, 1999b,c). Of particular interest for this review are the NZB strains. The NZB/kl strain demonstrated marked thickening of the strial capillary basement membranes (Nariuchi et al.,1994; Sone et al.,1995). Similar findings were described in the more common (NZB xW)F1 strain (Ruckenstein et al., 1999a). However, neither the composition of the thickened basement membrane nor the responsible mechanisms have been investigated.

8.6. Laminin α2 mutation (muscular dystrophy)

The *dy* mouse exhibits a mutation in the gene encoding laminin α2. This mouse is used as a model of congenital muscular dystrophy since the underlying defect in this disease is localized to the laminin-2 heterotrimer. A 25-dB elevation in ABR thresholds across frequencies was noted in comparison to age-matched C57 controls by Pillers et al. (2002). Light microscopy revealed a variety of morphological alterations in the *dy* mouse that ranged in severity from complete degeneration of the membranous labyrinth to structural integrity in the supporting cells in the organ of Corti. The authors concluded that the hearing loss was due to decreased mechanical efficiency in the cochlea.

9. Summary

The structures of the inner ear are replete with both structural ECM as well as intricate networks of basement membranes. Since the cochlea functions to translate physical sound waves into sensori-neural input, the structure of the ECM is centrally important to inner ear architecture. This matrix evolves as the cochlea matures with the appropriate physical characteristics to assist in the amplification of neural energy. Defects in this delicate structural balance are evident in mutant mice for tectorin α and β subunits in which structural changes in the tectorial membrane result in reduced efficiency or ablation of the cochlear amplifier function of the outer hair cells. The role of basement membranes in cochlear function is less clear. The dynamic transitions in basement membrane composition during development suggest cochlear basement membranes may function to direct differentiation and/or terminal cytodifferentiation of the complex cellular compartments of the cochlear epithelium. Direct evidence for such a functional role has not yet been provided. Thickening of the strial capillary basement membranes has been described in a number of pathologies that are accompanied by high-frequency sensori-neural hearing loss. Whether thickened strial capillary basement membranes result in loss of strial function awaits definitive proof.

References

Aberdam, D., Galliano, M-F., Vailly, J., Pulkkinene, L., Bonifas, J., Christiano, A.M., Tryggvason, K., Uitto, J., Epstein, E.H. ,Jr, Ortonne, J-P., Meneguzzi, G. 1994. Herlitz's junctional epidermolysis bulbosa is linked to mutations in the gene (LAMC2) for the a2 subunit of nicein/kalini laminin–5. Nat. Genet. 6, 299–403.

Amma, L.L., Goodyear, R., Faris, J.S., Jones, I., Ng, L., Richardson, G., Forrest, D. 2003. An emilin family extracellular matrix protein identified in the cochlear basilar membrane. Mol. Cell. Neurosci. 23, 460–472.

Andonopoulos, A.P., Naxakis, S., Goumas, P., Lygatsikas, C. 1995. Sensorineural hearing disorders in systemic lupus erythematosus. A controlled study. Clin. Exp. Rheumatol. 13, 137–141.

Annunen, S., Korkko, J., Czarny, M., Warman, M.L., Brunner, H.G., Kaariainen, H., Mulliken, J.B., Tranebjaerg, L., Brooks, D.G., Cox, G.F., Cruysberg, J.R., Curtis, M.A., Davenport, S.L., Friedrich, C.A., Kaitila, I., Krawczynski, M.R., Latos-Bielenska, A., Mukai, S., Olsen, B.R., Shinno, N., Somer, M., Vikkula, M., Zlotogora, J., Prockop, D.J., Ala-Kokko, L. 1999. Splicing mutations of 54-bp exons in the COL11A1 gene cause Marshall syndrome, but other mutations cause overlapping Marshall/Stickler phenotypes. Am. J. Hum. Genet. 65, 974–983.

Arnold, W. 1984. Inner ear and renal diseases. Ann. Otol. Rhinol. Laryngol. 112, 119–124.

Atkin, C.L., Gregory, M.C., Border, W.A. 1988. Alport Syndrome. In: *Diseases of the Kidney* (R.W. Schrier, C.W. Gottscha, Eds.), Boston: Little Brown, 4th Ed. pp. 617–641.

Barker, D., Hostikka, S.L., Zhou, J., Chow, L.T., Oliphant, A.R., Gerkin, S.C., Gregory, M.C., Skolnick, M.H., Atkin, C.L.T, ryggvasson, K. 1990. Identification of mutations in the COL4A5 collagen gene in Alport syndrome. Science 348, 1224–1227.

Bhattacharya, G., Miller, C., Kimberling, W.J., Jablonski, M.M., Cosgrove, D. 2002. Localization and expression of usherin: A novel basement membrane protein defective in people with Usher's syndrome type IIa. Hear Res. 163, 1–11.

Bowman, C.A., Linthicum, F.H., Jr., Nelson, R.A., Mikami, K., Quismorio, F. 1986. Sensorineural hearing loss associated with systemic lupus erythematosus. Otolaryngol. Head Neck Surg. 94, 197–204.

Buttery, P.C., ffrench-Constant, C. 1999. Laminin-2/integrin interactions enhance myelin membrane formation. Mol. Cell Neurosci. 14, 199–212.

Cosgrove, D.E., Kornak, J.M., Samuelson, G. 1996a. Expression of basement membrane type IV collagen chains during postnatal development in the murine cochlea. Hear Res. 100, 21–32.

Cosgrove, D.E., Meehan, D.T., Grunkemeyer, J.A., Kornak, J.M., Sayers, R., Hunter, W.J., Samuelson, G.C. 1996b. Collagen COL4A3 knockout: A mouse model for autosomal Alport syndrome. Genes Devel. 10, 2981–2992.

Cosgrove, D.E., Samuelson, G., Pinnt, J. 1996c. Immunohistochemical localization of basement membrane collagens and associated proteins in the murine cochlea. Hear Res. 97, 54–65.

Cosgrove, D.E., Rodgers, K.D. 1997. Expression of the major basement membrane-associated proteins during postnatal development in the murine cochlea. Hear Res. 105, 159–170.

Cosgrove, D.E., Samuelson, G., Meehan, D.T., Miller, C., McGee, J., Walsh, E.J., Siegel, M. 1998. Ultrastructural, physiological, and molecular defects in the inner ear of a gene-knockout mouse model for autosomal Alport syndrome. Hear Res. 121, 84–98.

de Kok, Y.J., Bom, T.M., Kemperman, M.H., van Beusekom, E., van der Velde-Visser, S.D., Robertson, N.G., Morton, C.C., Huygen, P.L., Verhagen, W.iI., Brunner, H.G., Cremers, C.W., Cremers, F.P. 1999. A Pro51Ser mutation in the COCH gene is associated with late onset autosomal dominant sensorineural hearing loss with vestibular defects. Hum. Mol. Genet. 8, 361–366.

De Leenhear, E.M., Kunst, H.H., McGuirt, W.T., Prasad, S.D., Brown, M.R., Huygen, P.L., Smith, R.J., Cremers, C.W. 2001. Autosomal dominant inherited hearing impairment caused by a missense mutations in COL11A2 (DFNA13). Arch. Otolaryngol. Head Neck Surg. 127, 13–17.

Ehret, G. 1976. Development of absolute auditory thresholds in the house mouse Mus musculus. J. Am. Audiol. Soc. 1, 179–184.

Eudy, J.D., Weston, M.D., Yao, S.-F., Hoover, D.M., Rehm, H.L., Ma-Edmonds, M., Yan, D., Ahmad, I., Cheng, J.J., Ayuso, C., Cremers, C.W.R.J., Davenport, S.L.H., Moller, C. G., Talmadge, C.B., Beisel, K.W., Tamayo, M.L., Morton, C.C., Swaroop, A., Kimberling, W.J., Sumegi, J. 1998. Mutation of a gene encoding a protein with extracellular matrix motifs in Usher syndrome type IIa. Science 280, 1753–1757.

Fransen, E., Van Camp, G. 1999. The COCH gene: A frequent cause of hearing impairment and vestibular dysfunction? Br. J. Audiol. 33, 297–302.

Goodyear, R.J., Richardson, G.P. 2002. Extracellular matrices associated with the apical surfaces of sensory epithelia in the inner ear: Molecular and structural diversity. J. Neurobiol. 52, 212–227.

Grabski, R., Szul, T., Sasaki, T., Timple, R., Mayne, R.., Nicks, B., Sztul, E. 2003. Mutations in COCH that result in non-syndromic autosomal dominant deafness (DFNA9) affect matrix deposition of cochlin. Hum. Genet. 113, 406–416.

Gratton, M.A., Schulte, B.A. 1995. Alterations in microvasculature are associated with atrophy of the stria vascularis in quiet-aged gerbils. Hear. Res. 82, 44–52.

Gratton, M.A., Schmiedt, R.A., Schulte, B.A. 1996. Age-related decreases in endocochlear potential are associated with vascular abnormalities in the stria vascularis. Hear. Res. 102, 181–190.

Gratton, M.A., Schulte, B.A., Smythe, N.M. 1997. Quantification of strial area in quiet-reared young and aged gerbils. Hear. Res. 114, 1–9.

Gratton, M.A., Cosgrove, D.E., Smyth, B.J., Ruckenstein, M.J. 2002. "Alport Syndrome" mice exhibit sensitivity to acoustic overstimulation Abstracts Assoc. Res. Otolaryngol. 25, 251.

Hasko, J.A., Richardson, G.P. 1988. The ultrastructural organization and properties of the mouse tectorial membrane matrix. Hear. Res. 35, 21–38.

Ishibe, T., Cremer, M.A., Yoo, T.J. 1989. Type II collagen distribution in the ear of the guinea pig fetus. Ann. Otol. Rhinol. Laryngol. 98, 648–654.

Ishii, K., Schroter-Kermani, C., Xu, D., Merker, H.J., Jahnke, V. 1992. Extracellular matrix in the rat spiral limbus. Eur. Arch. Otorhinolaryngol. 249, 224–230.

Ishikawa, T., Naito, Y., Taniguchi, K. 1995. Hearing impairment in WBN/Kob rats with spontaneous diabetes mellitus. Diabetol 38, 649–655.

Ito, M., Spicer, S.S., Schulte, B.A. 1995. Cytological changes related to maturation of the organ of Corti and the opening of Corti's tunnel. Hear. Res. 88, 107–123.

Iurato, S. 1962. Submicroscopic structure of the membranous labyrinth. III. The supporting structure of Corti's organ. Z. Zellforsch. 56, 40–96.

Johnsson, L.G., Hawkins, J.E., Jr. 1972. Sensory and neural degeneration with aging, as seen in microdissections of human inner ear. Ann. Otol. Rhinol. Laryngol. 81, 179–193.

Jorgensen, M.B. 1961. The ear in diabetes mellitus: Histological studies. Arch. Otolaryngol. 74, 373–381.

Kaname, H., Yoshihara, T., Ishii, T., Tatsuoka, H., Chiba, T. 1994. Ultrastructural and immunocytochemical study of the subepithelial fiber component of the guinea pig inner ear. J. Electron. Microsc. 43, 394–397.

Katori, Y., Hozawa, K., Kikuch, T., Tonosaki, A., Takasaka, T. 1993. Fine structure of the lamina basilaris of guinea pig cochlea. Acta. Otolaryngol. (Stockh) 113, 715–719.

Keithley, E.M., Ryan, A.F., Feldman, M.L. 1992. Cochlea degeneration in aged rats of four strains. Hear Res. 59, 171–178.

Keithley, E.M., Ryan, A.F., Woolf, N.K. 1993. Fibronectin-like immunoreactivity of the basilar membrane. J. Comp. Neurol. 327, 612–617.

Khetarpal, U. 2000. DFNA9 is a progressive audiovestibular dysfunction with a microfibrillar deposit in the inner ear. Laryngoscope 110, 1379–1384.

Khetapal, U., Robertson, N.G., Yoo, T.J., Morton, C.C. 1994. Expression and localization of COL2A1 mRNA and type II collagen in human fetal cochlea. Hear. Res. 79, 59–73.

Kim, K.H., Kim, Y., Gubler, M.C., Steffes, M.W., Lane, P.H., Kashtan, C.E., Crosson, J.T., Mauer, S.M. 1995. Structural-functional relationships in Alport syndrome. J. Am. Soc. Nephrol. 5, 1659–1672.

Knipper, M., Richardson, G., Mack, A., Muller, M., Goodyear, R., Limberger, A., Rohbock, K., Kopschall, I., Zenner, H.P., Zimmermann, U. 2001. Thyroid hormone-deficient period prior to the onset of hearing is associated with reduced levels of beta-tectorin protein in the tectorial membrane: Implication for hearing loss. J. Biol. Chem. 276, 39046–39052.

Koch, M., Olson, P.F., Albus, A., Jin, W., Hunter, D.D., Brunken, W.J., Burgeson, R.E., Champliaud, M.-F. 1999. Characterization and expression of laminin 3 chain: A novel non-basement membrane-associated, laminin chain. J. Cell. Biol. 145, 605–617.

Lawrence, R.C., Helmick, C.G., Arnett, F.C., Deyo, R.A., Felson, D.T., Giannini, E.H., Heyse, S.P., Hirsch, R., Hochberg, M.C., Hunder, G.G., Liang, M.H., Pillemer, S.R., Steen, V.D., Wolfe, F. 1998. Estimates of the prevalence of arthritis and selected musculoskeletal disorders in the United States. Arthritis. Rheum. 41, 778–799.

Legan, P.K., Lukashkina, V.A., Goodyear, R.J., Kossi, M., Russell, I.J., Richardson, G.P. 2000. A targeted deletion in alpha-tectorin reveals that the tectorial membrane is required for the gain and timing of cochlear feedback. Neuron 28, 273–285.

Li, S.W., Takanosu, M., Arita, M., Bao, Y., Ren, Z.X., Maier, A., Prockop, D.J., Mayne, R. 2001. Targeted disruption of Col11a2 produces a mild cartilage phenotype in transgenic mice: Comparison with the human disorder otospondylomegaepiphyseal dysplasia (OSMED). Dev. Dyn. 222, 141–152.

Libby, R.T., Champliaud, M-F, Claudepierre, T., Xu, Y., Gibbons, E.P., Koch, M., Burgeson, R.E., Hunter, D.D., Brunken, W.J. 2000. Laminin expression in adult and developing retina: Evidence of two novel CNS laminins. J. Neurosci. 20, 6517–6528.

McCabe, B.F. 1979. Autoimmune sensorineural hearing loss. Ann. Otol. Rhinol. Laryngol. 88, 585–589.

McGuirt, W.T., Prasad, S.D., Griffith, A.J., Kunst, H.H., Green, G.E., Shpargel, K.B., Runge, C., Huybrechts, C., Mueller, R.F., Lynch, I., King, M.C., Brunner, H.G., Cremers, C.W., Takanosu, M., Li, S.W., Arita, M., Mayne, R., Prockop, D.J., Van Camp, G., Smith, R.J. 1999. Mutations in COL11A2 cause non-syndromic hearing loss (DFNA13). Nat. Genet. 23, 413–419.

McMenomey, S.O., Russell, N.J., Morton, J.I., Trune, D.R. 1992. Stria vascularis ultrastructural pathology in the C3H/lpr autoimmune strain mouse: A potential mechanism for immune-related hearing loss. Otoarlyngol. Head Neck Surg. 106, 288–295.

McQueen, C.T., Baxter, A., Smith, T.L., Raynor, E., Yoon, S.M., Prazma, J., Pillsbury, H.C. 1999. Non-insulin-dependent diabetic microangiopathy in the inner ear. J. Laryngol. Otol. 113, 13–18.

Maeda, Y., Fukushima, K., Kasai, N., Maeta, M., Nishizaki, K. 2001. Quantification of TECTA and DFNA5 expression in the developing mouse cochlea. Neuroreport 12, 223–226.

Meisler, M.H., Griffith, A.J., Warman, M., Tiller, G., Sprunger, L.K. 1998. Gene symbol: COL11A1. Disease: Marshall syndrome. Hum. Genet. 102, 498.

Melkoniemi, M., Brunner, H.G., Manouvrier, S., Hennekam, R., Superi-Furga, A., Kääriäi-nen, H., Pauli, R.M., van Essen, T., Warman, M.L., Bonaventure, J., Miny, P, Ala-Kokko, L. 2000. Autosomal recessive disorder otospondylmagaepiphyseal dysplasia is associated with loss-of-function mutations in the COL11A2 gene. Am. J. Hum. Genet. 66, 368–377.

Mikuni, H., Fukuda, S., Kucuk, B., Inuyama, Y. 1995. The three-dimensional fibrillar arrange-
    ment of the basilar membrane in the mouse cochlea. Eur. Arch. Otorhinolaryngol. 252,
    495–498.
Miner, J.H., Patton, B.L., Lentz, S.I., Gilbert, D.J., Snider, W.D., Jenkens, N.A., Copeland, N.
    G., Sanes, J.R. 1997. The laminin a chains: Expression, developmental transitions, and
    chromosomal locations of a1–5, identification of heterotrimeric laminins 8–11, and cloning
    of a novel a3 isoform. J. Cell. Biol. 137, 685–701.
Miner, J.H., Li, C. 2000. Defective glomerulogenesis in the absence of laminin alpha5 demon-
    strates a developmental role for the kidney glomerular basement membrane. Dev. Biol. 217,
    278–289.
Mochizuki, T., Lemmink, H.H., Mariyama, M., Antignac, C., Gubler, M-C., Pirson, Y.,
    Verellen-Dumoulin, C., Chan, B., Schroder, C.H., Smeets, H.J., Reeders, S.T. 1994. Identi-
    fication of mutations in the α3 IV and α4 IV collagen genes in autosomal recessive Alport
    syndrome. Nat. Genet. 8, 77–81.
Moreno-Pelayo, M.A., del Castillo, I., Villamar, K.M., Romero, L., Hernandez-Calvin, F.J.,
    Herraiz, C., Barbera, R., Navas, C., Moreno, F. 2001. A cysteine substitution in the zona
    pellucida domain of alpha-tectorin results in autosomal dominant, postlingual, progressive,
    mid-frequency hearing loss in a Spanish family. J. Med. Genet. 38, 13.
Mothe, A.J., Brown, I.R. 2001. Expression of mRNA encoding extracellular matrix glycopro-
    teins SPARC and SC1 is temporally and spatially regulated in the developing cochlea of the
    rat inner ear. Hear. Res. 155, 161–174.
Munyer, P.D., Schulte, B.A. 1994. Immunohistochemical localization of keratan sulfate and
    chondroitin 4-and 6-sulfate proteoglycans in the subregions of the tectorial and basilar
    membranes. Hear. Res. 79, 83–93.
Muschler, J., Lochter, A., Roskelley, C.D., Yurchenco, P., Bissell, M.J. 1999. Divisions of
    labor among the 64 integrin, 1 integrins, and an E3 laminin receptor to signal morphogenesis
    and -casein expression in mammary epithelial cells. Mol. Biol. Cell. 10, 2817–2828.
Mustapha, M., Weil, D., Chardenoux, S., Elias, S., El-Zir, E., Beckmann, J.S., Loiselet, J., Petit,
    C. 1999. An a-tectorin gene defect causes a newly identified autosomal recessive form
    of sensorineural pre-lingual non-syndromic deafness, DFNB21. Hum. Mol. Genet. 8,
    409–412.
Myers, G.J., Tyler, H.R. 1972. The etiology of deafness in Alport's syndrome. Arch. Otolar-
    yngol. 96, 333–340.
Nadol, J.B. ,Jr 1979. Electron microscopic findings in presbycusic degeneration of the basal turn
    of the human cochlea. Otolaryngol. Head Neck Surg. 87, 818–836.
Nariuchi, H., Sone, M., Tago, C., Kurata, T., Saito, K. 1994. Mechanisms of hearing distur-
    bance in an autoimmune model mouse NZB/k1. Acta. Otolaryngol. S514, 125–131.
Naz, S., Alasti, F., Mowjoodi, A., Riazuddin, S., Sanati, M.H., Friedman, T.B., Griffith, A.J.,
    Wilcox, E.R., Riazuddin, S. 2003. Distinctive audiometric profile associated with DFNB21
    alleles of TECTA. J. Med. Genet. 40, 360–363.
Pearsall, N., Bhattacharya, G., Wisecarver, J., Adams, J., Cosgrove, D., Kimberling, W. 2002.
    Usherin expression is highly conserved in mouse and human tissues. Hear. Res. 174, 55–63.
Pillers, D.M., Kempton, J.B., Duncan, N.M., Pang, J., Dwinnell, S.J, Trune, D.J. 2002. Hearing
    loss in the laminin-deficient dy mouse model of congenital muscular dystrophy. Mol. Genet.
    Metab. 76, 217–224.
Powell, S.K., Williams, C.C., Nomizu, M., Yamada, Y., Kleinman, H.K. 1998. Laminin-like
    proteins are differentially regular during cerebellar development and stimulate granule cell
    neurite outgrowth in vitro. J. Neurosci. Res. 54, 233–247.
Rau, A., Legan, P.K., Richardson, G.P. 1999. Tectorin mRNA expression is spatially and
    temporally restricted during mouse inner ear development. J. Comp. Neurol. 405, 271–280.
Roberston, N.G., Skvorak, A.B., Yin, Y., Weremowicz, S., Johnson, K.R., Kovatch, K.A.,
    Battey, J.F., Bieber, F.R., Morton, C.C. 1997. Mapping and characterization of a novel

cochlear gene in human and mouse: A positional candidate gene for a deafness disorder, DFNA9. Genomics 15, 345–354.

Rodgers, K.D., Barritt, L.C., Miner, J.H., Cosgrove, D.E. 2001. The laminins in the murine inner ear: Developmental transitions and expression in cochlear basement membranes. Hear Res. 158, 39–50.

Ruckenstein, M.J., Gaber, L., Marion, T. 1999a. Cochlear and Renal Pathology in the NZB x W Mouse. Abstracts Assoc. Res. Otolaryngol. 22, 253.

Ruckenstein, M.J., Keithley, E.M., Bennett, T., Powell, H.C., Baird, S., Harris, J.P. 1999b. Ultrastructural pathology in the stria vascularis of the MRL-Fas[lpr] mouse. Hear Res. 131, 22–28.

Ruckenstein, M.J., Milburn, M.Hu,L. 1999c. Strial dysfunction in the MRL- Fas[lpr] mouse. Otolaryngol. Head Neck Surg. 121, 452–456.

Ruckenstein, M.J., Mount, R.J., Harrison, R.V. 1993. The MRL-lpr/lpr mouse: A potential model of autoimmune inner ear disease. Acta. Otolaryngol. 113, 160–165.

Rust, K.R., Prazma, J., Triana, R.J., Michaelis, O.E., IV, Pillsbury, H.C. 1992. Inner ear damage secondary to diabetes mellitus. II. Changes in aging SHR/N-cp rats. Arch. Otolaryngol. Head Neck Surg. 118, 397–400.

Sagara, T., Furukawa, H., Makishima, K., Fujimoto, S. 1995. Differentiation of the rat stria vascularis. Hear Res. 83, 121–132.

Saito, H., Iida, T., Uede, K. 1984. Changes of ground substance in the inner ear in alloxan diabetic mice. Arch. Otolaryngol. 239, 81–85.

Saitoh, Y., Hosokawa, M., Shimada, A., Wantanabe, Y, Yasuda, N., Murakami, Y., Takeda, T. 1995. Age-related cochlear degeneration in senescence-accelerated mouse. Neurobiol. Aging 16, 129–136.

Sakaguchi, N., Spicer, S.S., Thomopoulos, G.N., Schulte, B.A. 1997. Increased laminin deposition in capillaries of the stria vascularis of quiet-aged gerbils. Hear Res. 105, 44–56.

Sanes, J.R., Engvall, E., Butkowski, R.Hunter,D.D. 1990. Molecular heterogeneity of basal laminae: Isoforms of laminin and collagen IV at the neuromuscular junction and elsewhere. J. Cell. Biol. 111, 1685–1699.

Santi, P., Larson, J., Furcht, L., Economou, T.S. 1989a. Immunohistochemical localization of fibronectin in the chinchilla cochlea. Hear Res. 39, 91–102.

Santi, P.A., Larson, J.T., Furcht, L.T., Economou, T.S. 1989b. Immunohistochemical localization of fibronectin in the chinchilla cochlea. Hear Res. 39, 91–102.

Sargona, M.F., Çelika, H.H., Sener, C. 1999. Ultrastructure of the lens epithelium in Alport's syndrome. Ophthalmologica 213, 30–33.

Satoh, H., Kawasaki, K., Kihara, I., Nakano, Y. 1998. Importance of type IV collagen, laminin, and heparan sulfate proteoglycan in the regulation of labyrinthine fluid in the rat cochlear duct. Eur. Arch. Otorhinolaryngol. 255, 285–288.

Schuger, L., Skubitz, A.P., Zhang, J., Sorokin, L., He, L. 1997. Laminin 1 chain synthesis in the mouse developing lung: Requirement for epithelial-mesenchymal contact and possible role in bronchial smooth muscle development. J. Cell. Biol. 139, 553–562.

Shpargel, KB, Makishima, T, Griffith, A.J. 2004. Col11a1 and Col11a2 mRNA expression in the developing mouse cochlea: Implications for the correlation of hearing loss phenotype with mutant type XI collagen genotype. Acta. Otolaryngol. 124, 242–248.

Slepecky, N.B., Savage, J.E., Cefaratti, L.K., Yoo, T.J. 1992. Electron-microscopic localization of type II, IX and V collagen in the organ of Corti of the gerbil. Cell Tissue Res. 267, 413–418.

Smith, R.J.H., Steel, K.P., Barkway, C., Soucek, S., Michaels, L. 1992. A histologic study of nonmorphogenetic forms of hereditary hearing impairment. Arch. Otolaryngol. Head Neck Surg. 118, 1085–1094.

Smith, T.L., Raynor, E., Prazma, J., Buenting, J.E., Pillsbury, H.C. 1995. Insulin-dependent diabetic microangiopathy in the inner ear. Laryngoscope 105, 236–240.

Snead, M.P., Yates, J.R. 1999. Clinical and molecular genetics of Stickler syndrome. J. Med. Genet. 36, 353–359.

Sone, M., Naruichi, H., Saito, K., Yanagita, N. 1995. A substrain of the NZB mouse as an animal model of autoimmune inner ear disease. Hear Res. 83, 26–36.

Sone, M., Schachern, P.A., Paparella, M.M., Morizono, N. 1999. Study of systemic lupus erythematosus in temporal bones. Ann. Otol. Rhinol. Laryngol. 108, 338–344.

Spicer, S.S., Schulte, B.A. 1991. Differentiation of inner ear fibrocytes according to their ion transport related activity. Hear Res. 56, 53–64.

Spicer, S.S., Schulte, B.A. 2002. Spiral ligament pathology in quiet-aged gerbils. Hear Res. 172, 172–185.

Streeten, B.W., Robinson, M.R., Wallace, R., Jones, D.B. 1987. Lens capsule abnormalities in Alport's syndrome. Arch. Ophthalmol. 105, 1693–1697.

Sumida, Y., Ura, H., Yano, Y., Misaki, M., Shima, T. 1997. Abnormal metabolism of type-IV collagen in normotensive non-insulin-dependent diabetes mellitus patients. Horm. Res. 48, 23–28.

Swartz, D.J., Santi, P.A. 1997. Immunohistochemical localization of keratan sulfate in the chinchilla inner ear. Hear Res. 109, 92–101.

Tachibana, M., Nakae, S. 1986. The cochlea of the spontaneously diabetic mouse. Arch. Otorhinolaryngol. 243, 238–241.

Takahashi, M., Hokunan, K. 1992. Localization of type IV collagen and laminin in the guinea pig inner ear. Ann. Otol. Rhinol. Laryngol. 101, 58–62.

Thalmann, I. 1993. Collagen of accessory structures of organ of Corti. Connect. Tissue Res. 29, 191–201.

Thompson, S.M., Deady, J.P., Willshaw, H.E., White, R.H.R. 1987. Ocular signs in Alport's syndrome. Eye 1, 146–153.

Triana, R.J., Suits, G.W., Garrison, S., Prazma, J., Brechtelsbauer, B., Michaelis, O.E. ,IV, Pillsbury, H.C. 1991. Inner ear damage secondary to diabetes mellitus. I. Changes in adolescent SHR/N-cp rats. Arch. Otolaryngol. Head-Neck Surg. 117, 635–640.

Tsuprun, V.L., Santi, P. 1997. Ultrastructural organization of proteoglycans and fibrillar matrix of the tectorial membrane. Hear Res. 110, 107–118.

Tsuprun, V.L., Santi, P. 1999. Ultrastructural and immunohistochemical identification of the extracellular matrix of the extracellular matrix of the chinchilla cochlea. Hear Res. 129, 35–49.

Tsuprun, V., Santi, P. 2001. Proteoglycans arrays in the cochlear basement membrane. Hear Res. 157, 65–76.

Van Steensel, M.A., Buma, P., de Waal Malefijt, M.C., van den Hoogen, F.H., Brunner, H.G. 1997. Oto-spondylo-megaepiphyseal dysplasia (OSMED): Clinical description three patients homozygous for a missense mutation in the COL11A2 gene. Am. J. Med. Genet. 13, 315–323.

Verhoeven, K., Van Laer, L., Kirschofer, K., Legan, P.K., Hughes, D.C., Schatterman, I., Verstreken, M, Van Hauwe, P., Coucke, P., Chen, A., Smith, R.J., Somers, T., Offeciers, F. E., Van de Heyning, P., Richardson, G.P., Wachtler, F., Kimberling, W.J., Willems, P.J., Govaerts, P.J, Van Camp, G. 1998. Mutations in the human alpha-tectorin gene cuase autosomal dominant non-syndromic hearing impairment. Nat. Genet. 19, 60–62.

Wackym, P.A., Linthicum, F.H., Jr 1985. Diabetes mellitus and hearing loss: Clinical and histopathological relationships. Am. J. Otolaryngol. 7, 176–182.

Weidauer, H., Arnold, W. 1976. Strukturelle verandergunen am hororgan beim Alport syndrome. Z. Laryngol. Rhinol. Otol. 55, 6–16.

Wester, D.C., Atkin, C.L., Gregory, M.C. 1995. Alport syndrome: Clinical update. J. Am. Acad. Audiol. 6, 73–79.

Woolf, N.K., Koern, F.J., Ryan, A.F. 1992. Immunohistochemical localization of fibronectin-like protein in the inner ear of the developing gerbil and rat. Dev. Brain Res. 65, 21–33.

Woolf, N.K., Jaquish, D.V., Koehrn, F.J., Woods, V.J., Jr., Peterson, D.A. 1996. Improved resolution of fibronectin mRNA expression in the inner ear using laser scanning confocal microscopy. J. Histochem. Cytochem. 44, 27–34.

Woolf, N.K., Koehrn, F.J., Ryan, A.F. 1992. Immunohistochemical localization of fibronectin-like protein in the inner ear of the developing gerbil and rat. Dev. Brain Res. 65, 21–33.

Xu, H., Christmas, P., Wu, X-R., Wever, U.M., Engvall, E. 1994. Defective muscle basement membrane and lack of M-laminin in the dystrophic dy/dy mouse. Proc. Natl. Acad. Sci. USA 91, 5572–5576.

Yoo, T.J., Tomoda, K. 1988. Type II collagen distribution in rodents. Laryngoscope 98, 1255–1260.

Yoshiba, K., Yoshiba, N., Aberdam, D., Meneguzzi, G., Perrin-Schmitt, F., Stoetzel, C., Ruch, J.V., Lesot, H. 1998. Expression and localization of laminin-5 subunits during mouse tooth development. Dev. Dyn. 211, 164–176.

Further Reading

Alloisio, N., Morle, L., Bozon, M., Godet, J., Verhoeven, K., Van Camp, G., Plauchu, H., Muller, P., Collet, L., Lina-Granade, G. 1999. Mutation in the zonadhesin-like domain of alpha-tectorin associated with autosomal dominant non-syndromic hearing loss. Eur. J. Hum. Genet. 7, 255–258.

Griffith, A.J., Gebarski, S.S., Shepard, N.T., Kileny, P.R. 2000. Audiovestibular phenotype associated with a COL11A1 mutation in Marshall syndrome. Arch. Otolaryngol. Head Neck Surg. 126, 891–894.

Heaney, D.L., Schulte, B.A. 2003. Dystroglycan expression in the mouse cochlea. Hear. Res. 177, 12–20.

Simon-Assmann, P., Lefebvre, O., Bellissent-Waydelich, A., Olsen, J., Oria-Roussea, V., DeArcangelia, A. 1998. The laminins: Role in intestinal morphogenesis and differentiation. Ann. NY Acad. Sci. 859, 46–64.

Spicer, S.S., Gratton, M.A., Schulte, B.A. 1997. Expression patterns of ion-transport enzymes in spiral ligament fibrocytes change in relation to strial atrophy in the aged gerbil cochlea. Hear Res. 111, 93–102.

Szymko-Bennett, Y.M., Kurima, K., Olsen, B., Seegmiller, R., Griffith, A.J. 2003. Auditory function associated with Col11a1 haploinsufficiency in chondrodysplasia (cho) mice. Hear Res. 175, 178–182.

Thomopolous, G.N., Spicer, S.S., Gratton, M.A., Schulte, B.A. 1997. Age-related thickening of basement membrane in stria vascularis membrane. Hear Res. 111, 31–41.

Tomatis, D., Echtermayer, F., Schober, S., Balzac, F., Retta, S.F., Silengo, L., Tarone, G. 1999. The muscle-specific laminin receptor 7 1 integrin negatively regulates 5 1 fibronectin receptor function. Exp. Cell Res. 246, 421–432.

Whitlon, D.S., Zhang, X., Kusakabe, M. 1999. Tenascin-C in the cochlea of the developing mouse. J. Comp. Neurol. 406, 361–374.

Extracellular matrix and the development of disease: The role of its components in cancer progression

Roy Zent and Ambra Pozzi

Division of Nephrology, Departments of Medicine and Cancer Biology, Vanderbilt University Medical Center and Veterans' Affairs Hospital, Nashville, Tennessee

Contents

1. Introduction

Tumor metastasis is a complex multi-step event that requires tumor cell growth, adhesion, migration, and invasion. The steps involved in tumor invasion and metastasis include the destruction of cell stroma and basement membranes, specialized extracellular matrix (ECM) structures that separate epithelia from the surrounding stroma. The large array of functions that a tumor cell has to fulfill to settle as a metastasis in a distant organ requires

Advances in Developmental Biology
Volume 15 ISSN 1574-3349
DOI: 10.1016/S1574-3349(05)15007-8

cooperative activities between the tumor and the surrounding tissue. Our understanding of the events involved in tumor invasion has increased considerably in the last decades as a result of the identification of a large number of newly discovered components of the ECM, the realization that cleavage products of the ECM can strongly influence tumor cell behavior, and the recognition that malignant cells can and do modify the production, as well as the degradation of ECM components. Here we focus on the role of ECM molecules in the control of tumor progression and angiogenesis.

Cancer is the second leading cause of death in the Western world. Death is usually not due to the primary tumors but rather to metastasis to distant organs. Tumor metastasis is a complex multi-step event that requires tumor cell growth, adhesion, migration, and invasion. During cancer progression, cells become more motile and acquire invasive qualities. The first steps involved in tumor invasion and metastasis include the destruction of cell stroma and basement membranes (Engbring and Kleinman, 2003); the integrity of basement membranes is often used as a marker to determine the degree of malignancy of a tumor (Jones and De Clerck, 1982). Benign tumors are usually uniformly circumscribed by basement membranes, while degradation of basement membrane components around tumors is crucial in the promotion of invasion, migration into blood and/or lymphatic vessels, and distant metastasis. Once tumor cells reach a distal site, they must adhere to the vessel wall, extravasate from the vessel, invade the new tissue, and proliferate to form a secondary tumor. Another key requirement for tumor formation, growth, progression, and metastasis is angiogenesis, the formation of new blood vessels from pre-existing vasculature (McDonnell et al., 2000).

It is well known that cooperative activities between the tumor and the surrounding tissue is required for a tumor cell to settle as a metastasis in a distant site (Bissell et al., 2002; Hsu et al., 2002; Krutovskikh, 2002; Quaranta, 2002; De Wever and Mareel, 2003). Several classes of molecules such as cell-cell adhesion molecules (Okegawa et al., 2004), cell-ECM molecules (Hood and Cheresh, 2002; Engbring and Kleinman, 2003; Goel and Languino, 2004; Jin and Varner, 2004), and matrix-degrading enzymes (Stetler-Stevenson and Yu, 2001; Coussens et al., 2002; Freije et al., 2003; Folgueras et al., 2004) are involved. Our knowledge of the complexity of events implicated in tumor invasion/progression has increased in part due to the identification of new ECM components involved in tumorigenesis, the realization that cleavage products of the ECM can strongly influence tumor cell behavior, and the recognition that malignant cells modify the production and degradation of ECM components.

In this chapter we will focus primarily on ECM molecules that are components of the basement membranes, which control tumor progression and angiogenesis. We will describe: 1) the major components of basement membranes; 2) how these components interact with tumor and stromal cells; 3) the role of intact and/or cleaved basement membrane molecules in the control of specific steps of tumor progression (migration, invasion, and recruitment of blood vessels); and 4) how these ECM components could potentially be used as therapeutic tools for the treatment and prevention of cancer.

2. Basement membrane composition

The ECM consists of a complex mixture of structural and functional macromolecules and is important in tissue and organ morphogenesis as well as in the maintenance of cell and tissue structure and function (Bosman and Stamenkovic, 2003). In pathological conditions such as cancer, both tumor and stromal cells contribute to ECM deposition. The morphology and composition of the ECM in tumors is diverse, resulting in varied influences on tumor cell properties and functions. In addition, ECM can both exert mechanical force and act as a reservoir for signaling molecules, thus facilitating different cellular processes such as adhesion, proliferation, migration, and invasion, which are key events in tumor progression.

Basement membranes are specialized structures that separate epithelial elements from the surrounding stroma (Madri et al., 1984; Leblond and Inoue, 1989). The two major representatives of the components of basement membranes are non-fibrillar collagens and laminins (Madri et al., 1984; Leblond and Inoue, 1989).

Collagen IV, a required component of all basement membranes, belongs to a family that has at least 25 distinct members. Six distinct genes have been identified as belonging to the type IV collagen gene family (Soininen et al., 1987; Hostikka and Tryggvason, 1988; Myers et al., 1990; Leinonen et al., 1994; Mariyama et al., 1994). These genes encode the six chains of collagen IV—$\alpha 1$(IV) through $\alpha 6$(IV)—which are selectively expressed in different basement membranes. Each of these six chains has three domains: a short 7S domain at the N-terminal; a long, collagenous domain that occupies the mid section of the molecule; and a noncollagenous domain (NC1) positioned at the C-terminal (Fig. 1). Despite the many potential permutations, the six chains of collagen IV appear to form only three sets of triple helical molecules called protomers, which are designated as $\alpha 1 \alpha 1 \alpha 2$(IV), $\alpha 3 \alpha 4 \alpha 5$(IV), and $\alpha 5 \alpha 5 \alpha 6$(IV) (Borza et al., 2001, 2002; Boutaud et al., 2000; Timpl et al., 1981) (Fig. 1). These protomers create collagenous networks by uniting two NC1 trimers to form hexamers and uniting four 7S domains to form tetramers with other protomers. The $\alpha 1 \alpha 1 \alpha 2$(IV) network is ubiquitously expressed in all basement membranes, while the $\alpha 3 \alpha 4 \alpha 5$(IV) is restricted to organs such as the kidneys and the lungs, where it is found primarily in pulmonary vessels.

Alteration in collagen IV integrity or assembly has been observed in various tumors. In aggressive basal-cell carcinomas, for example, downregulation of the $\alpha 5$(IV) chain and degradation of the $\alpha 1$(IV) chains have been

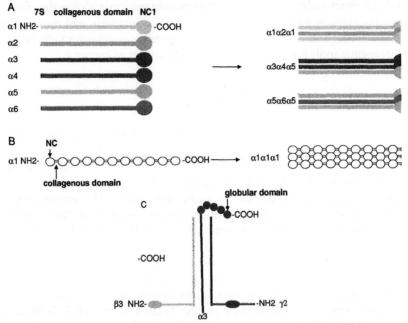

Fig. 1. Structure of laminin-5 and non-fibrillar collagens. (A) Collagen IV is composed by six different chains, α1(IV) to α6(IV), characterized by a N-terminal 7S domain, a central collagenous domain, and a C terminal NC1 domain. The six chains assemble to form the trimeric structures indicated on the right. (B) Collagen XVIII is formed by the assembly of three α1(XVIII) chains. Each chain is composed by 11 non-collagenous and 10 collagenous domains. (C) Laminin-5 is composed of the α3, β3, and γ2 chains. The C-terminal of the α3 chain is characterized by the presence of five globular domains which contains important modules via which laminin molecules interact with cells and other matrix molecules. (See Color Insert.)

observed (Quatresooz et al., 2003). Furthermore, in invasive colorectal cancer, the α1/α2(IV) chains are linearly stained in the basement membranes, while the α5/α6(IV) chains are discontinuously or negatively stained, suggesting that the α5/α6(IV) chain staining may be a diagnostic marker for the invasiveness of colorectal cancer (Hiki et al., 2002).

Other non-fibrillar collagens highly expressed in basement membranes are collagen XV and collagen XVIII. Collagens type XV and XVIII, identified as a chondroitin sulfate and heparan sulfate proteoglycans, respectively (Halfter et al., 1998; Li et al., 2000) are closely related non-fibrillar collagens that define the multiplexin subfamily (multiple triple helix domains with interruptions) (Abe et al., 1993; Oh et al., 1994a,b; Rehn et al., 1994; Rehn and Pihlajaniemi, 1994). Unlike collagen IV, α1(XV) and α1(XVIII) chains organize as homotrimers (Fig. 1). Each chain is divided into three subdomains which, as in collagen IV, includes a C-terminal NC1 domain. Collagen XV is highly expressed in the heart, the skeletal muscles, and the placenta, and moderately expressed in the adrenal gland, the kidney, and the pancreas (Muragaki et al., 1994; Kivirikko et al., 1995). Its expression is associated with vascular, neuronal, mesenchymal, and some epithelial basement membranes, indicating a probable function in adhesion between basement membranes and the underlying connective tissue stroma (Myers et al., 1996). Type XVIII collagen contains 10 collagenous triple-helical domains, separated by 11 non-collagenous regions (Oh et al., 1994a) (Fig. 1). Three distinct variants of this collagen have been described in mice, but only two variants (a short and a long form) have been described in humans (Saarela et al., 1998). Each of the variants shows a characteristic expression pattern, with the highest levels in the liver, the kidneys, and the lungs. In addition, type XVIII collagen is a component of vascular and epithelial basement membranes (Muragaki et al., 1995; Musso et al., 1998). Besides the typical features of a collagen, type XVIII collagen also has properties of a heparan sulfate proteoglycan (HSPG), with several Ser-Gly dipeptides suitable for glycosaminoglycan attachment, and long heparitinase-sensitive carbohydrate chains. Thus, this molecule belongs to the group of basement membrane HSPGs (Halfter et al., 1998), in addition to being a collagen.

In pathological conditions, such as mammary ductal carcinoma, complete degradation of collagen type XV precedes tumor cell migration and invasion, suggesting that the disappearance of collagen XV can be used as an early marker to determine breast cancer aggressiveness (Amenta et al., 2003). Furthermore, the observation that type XV collagen expression is downregulated in the basement membranes of human colonic adenocarcinomas, but increased in the surrounding interstitium, suggests a role for this protein in the invasive process and the possibility that it may provide a sensitive indicator of tumor invasion (Amenta et al., 2000). Consistent with this data, low collagen XVIII expression by hepatocarcinoma cells correlates with large tumor size and invasiveness, while tumors expressing high levels of collagen XVIII are smaller and not invasive (Musso et al., 2001). In contrast, it has been observed that overexpression of collagen XVIII in non-small–cell lung cancer is associated with a poorer outcome compared with tumor tissues that do not express collagen XVIII (Musso et al., 2001).

Laminins, together with type IV collagen, are one of the main components of the basement membranes. They belong to a multigenic family, all of which are the product of closely related genes. The laminins are an adhesive glycoprotein family comprised of high molecular weight disulfide bonded heterotrimers with the chain composition of α, β, and γ subunits, which associate to from a cross-like structure (Fig. 1) (Beck et al., 1990). To date five α, three β, and two γ chains have been described, forming at least 12 distinct laminin heterotrimers. All of the known laminins have large globular (G) domains at the C-terminal of the α subunit (Fig. 1). The G domain of laminins contains important modules via which laminin molecules interact with cells and other matrix molecules.

Laminin-5 is the most studied laminin found in basement membranes that is associated with cancer. It consists of $\alpha3\beta3\gamma2$ chains, and is a component of the cell adhesion complex containing hemidesmosomes and anchoring fibrils (Aumailley et al., 2003) (Fig. 1). This protein is a major constituent of the basement components and has recently proved to be an invasion marker for epithelial cells in many immunohistochemical surveys. Overexpression of laminin-5 and its localization at the invading edges of different epithelial cancer, such lung adenocarcinomas (Niki et al., 2002), endometrial carcinoma (Lundgren et al., 2003), and squamous cell carcinomas (Yamamoto et al., 2001) has been documented (Pyke et al., 1995; Katayama and Sekiguchi, 2004; Skyldberg et al., 1999). In addition, intracellular and/or extracellular accumulation of monomeric $\gamma2$ chain has been widely observed in invasive carcinoma cells, suggesting a role of this chain in promoting invasiveness and metastasis (Pyke et al., 1994; Ono et al., 1999; Kagesato et al., 2001; Yamamoto et al., 2001; Niki et al., 2002; Lundgren et al., 2003). In contrast, downregulation of laminin-5 has been observed in prostate and breast cancer (Hao et al., 1996, 2001, Martin et al., 1998; Sathyanarayana et al., 2003). As laminin-5 has been associated to hemidesmosomes, specialized structures that enhance adhesion of polarized epithelial cell to basement membranes, it has been postulated that its downregulation in certain tumors might be necessary to induce the switch from a stationary to a migratory phenotype.

Besides laminin-5, there are other laminins which are major components of basement membranes, including laminin-1 (composed by $\alpha1\beta1\gamma1$ chains) and laminin-10 and -11 (composed by $\alpha5\beta1\gamma1$ and $\alpha5\beta2\gamma1$ chains, respectively). However, despite the recent observation that patients with metastatic melanoma have increased serum levels of laminin-1 (Burchardt et al., 2003), suggesting a role of this ECM in melanoma progression, the exact role of these laminins in the progression of human cancers has not yet been well characterized.

3. ECM receptors

Cell-ECM interactions play a crucial role in cellular functions such as cell adhesion, migration, proliferation, and apoptosis (Pozzi and Zent, 2003). The heterodimeric molecules integrins, which are type I transmembrane glycoproteins comprised of α and β subunits, mediate these interactions. There are 18 α and 8 β subunits described that combine in a restricted manner to form dimers, each of which exhibits different, although often overlapping, ligand-binding properties (Hynes, 2002). The extracellular domains of integrin subunits constitute the ligand binding domain(s) of these receptors, while the cytoplasmic domains play a critical role in the promotion of cell anchorage as they interact with the cytoskeleton and provide a physical connection between the internal and external environment. In addition to their anchoring function, integrins also mediate cell signaling by transducing multiple pathways through their cytoplasmic tails following activation by ligands ('outside-in' signaling) (Schwartz et al., 1995). The cytoplasmic domains also modulate integrin affinity for their ligands by changing the conformation of the extracellular domains via a process called integrin activation, or 'inside-out' signal transduction. Therefore, integrins are signaling receptors that transmit information in both directions across the plasma membrane and provide an inter section where mechanical forces, cytoskeletal organization, and biochemical signals meet.

The importance of ECM receptors in tumor progression is well documented (Hood and Cheresh, 2002; Felding-Habermann, 2003). Changes in integrin expression have been associated with invasive potential (Gordon et al., 2003). Downregulation of integrin $\alpha 6\beta 4$, the major components of hemidesmosomes and a laminin-5 receptor, has been observed in invasive ovarian carcinoma epithelial cells (Skubitz et al., 1996). In contrast, in invasive breast cancer, increased expression of integrin $\alpha 6\beta 4$ has been associated with enhanced tumor cell invasiveness and poor outcome (Tagliabue et al., 1998). A dual role of integrin $\alpha 3\beta 1$, a second major laminin-5 receptor, in cancer progression has also been documented. While downregulation of this receptor seems to be a key event in the invasive process of small-cell lung cancer (Barr et al., 1998), overexpression of the same integrin in gastric carcinoma specifically enhances peritoneal metastases, as well as the depth of tumor invasiveness (Ura et al., 1998).

Among the collagen-binding receptors, overexpression of integrin $\alpha 1\beta 1$ and $\alpha 2\beta 1$ has been found in hepatocaricoma cell lines and is associated with increased migration and invasion (Yang et al., 2003). In addition, $\alpha 1$ and $\alpha 2$ integrins enhance invasion of mouse mammary carcinoma by promoting protease synthesis (Lochter et al., 1999), and overexpression of integrin $\alpha 2\beta 1$ in gastric carcinoma enhances lymph node and liver metastases of this carcinoma (Ura et al., 1998). In contrast, abrogation of integrin $\alpha 1\beta 1$ function via genetic manipulation (Pozzi et al., 2000, 2002), blocking antibodies (Senger et al., 1997), or specific disintegrins such as obtustatin (Marcinkiewicz et al., 2003) has been shown to decrease tumor growth by preventing growth factor- and tumor-induced angiogenesis.

Another integrin involved in various steps of tumor growth, namely invasion and angiogenesis, is integrin $\alpha v\beta 3$. This receptor, originally described as a vitronectin-binding integrin, is probably the most promiscuous among the ECM-binding receptors as it can bind vitronectin, fibronectin, and cleavage products of collagens IV, XV, and XVIII (see the following text for details). Increased expression of this integrin in both tumor and host cells has been associated with tumor progression and is closely associated with increased cell invasion and metastasis (Felding-Habermann et al., 2002).

Notably, integrin $\alpha v \beta 3$ is expressed on invasive melanoma but not benign nevi or normal melanocytes (Gehlsen et al., 1992). Additionally, increased $\alpha v \beta 3$ expression levels correlate with increased rates of melanoma metastases (Nip et al., 1992). In the host, integrin $\alpha v \beta 3$ has primarily been found to be overexpressed in angiogenic tumor blood vessels (Hamano et al., 2003) and it seems to promote tumor angiogenesis by cooperating with growth factor receptors (Friedlander et al., 1995) and by binding secreted proteases (Brooks et al., 1998) thus enhancing matrix remodeling and endothelial cell migration. Integrin $\alpha v \beta 3$ is considered an excellent anti-angiogenic target as it is expressed only on tumor angiogenic blood vessels (Hamano et al., 2003); anti $\alpha v \beta 3$ antibodies block tumor angiogenesis and tumor growth (Brooks et al., 1994) and binding of collagen-derived cleavage products to this receptor inhibits tumor angiogenesis (see the following text for details).

4. ECM remodeling

The ECM undergoes constant remodeling and degradation under physiological conditions such as wound healing, embryonic development, menstruation, and tissue remodeling; however, this process also represents a crucial step in cancer invasion. In cancer progression, ECM degradation processes involve the synthesis and secretion of proteolytic enzymes from both tumor and stromal cells. One of the best examples of this are the cysteine proteinases, a subclass of thiol-dependent endopeptidases, which have been associated with tumor cell invasion and ECM degradation (Sloane and Honn, 1984; Sloane et al., 1991; Kos and Lah, 1998; Krepela, 2001; Siewinski et al., 2003). Cathepsin B, a lysosomal cysteine proteinases, is the most studied enzyme in this class and it has been shown to facilitate invasion either directly by dissolving ECM barriers like the basement membrane, or indirectly by activating other proteases capable of digesting the ECM (Berquin and Sloane, 1996; Yan et al., 1998; Dohchin et al., 2000; Dennhofer et al., 2003). In addition, cathepin L has been shown to cleave collagen XVIII generating NC1 fragments with anti-angiogenic activity (see the following text for details) (Felbor et al., 2000).

Another group of proteases involved in tumor progression, angiogenesis, and metastasis is the urokinase plasminogen activator (uPA) system, (Mazar et al., 1999; Dohchin et al., 2000; Choong and Nadesapillai, 2003; Sidenius and Blasi, 2003; Behrendt, 2004). Binding of uPA with its receptor (uPAR) initiates a proteolytic cascade that results in the conversion of plasminogen to plasmin. Plasmin has proteolytic activity and degrades a range of extracellular basement membrane components and activates other proteases, such as the metalloproteinases. Overexpression of uPA and its receptor has been associated with breast and gastric tumor progression (Costantini et al., 1996; Lee et al., 2003).

The most studied and well-characterized group of proteolytic enzymes involved in ECM remodeling and tumor progression is the matrix metalloproteinases (MMPs) (Stetler-Stevenson and Yu, 2001; Freije et al., 2003; Stamenkovic, 2003; Folgueras et al., 2004). MMPs, a family of more than 25 distinct zinc-dependent proteolytic enzymes, are the most potent ECM-degrading enzymes. The majority of the MMP family members are secreted as proMMPs and activated upon cleavage. Some members, the membrane type MMPs (MT-MMPs), are anchored to the plasma membranes and are involved in both ECM degradation (Jiang and Pei, 2003; Koshikawa et al., 2004) and activation of soluble MMPs (Jiang and Pei, 2003; Nie and Pei, 2003; Dreier et al., 2004). Both MMPs and MT-MMPs are frequently overexpressed in human tumors and play a role in tumor progression (Sato and Seiki, 1996; Ohnishi et al., 2001; Stetler-Stevenson and Yu, 2001; Freije et al., 2003; Stamenkovic, 2003; Folgueras et al., 2004). Functional studies have shown that MMPs play an important role in the proteolytic destruction of virtually all ECM components and basement membranes, thereby facilitating tumor invasion and metastasis. Moreover, MMPs are upregulated in virtually all human and animal tumors, as well as in most tumor cell lines. Changes in MMP levels can markedly affect the invasive behavior of tumor cells and their ability to metastasize in experimental animal models. MMPs are now known to contribute to multiple steps of tumor progression in addition to invasion, including tumor promotion, angiogenesis, and the establishment and growth of metastatic lesions in distant organ sites. In addition, it is recognized that MMPs are not only synthesized by tumor cells, but are frequently produced by surrounding stromal cells, including fibroblasts, endothelial cells, and infiltrating inflammatory cells. Increased MMP production and consequent ECM degradation was thought to be a key factor in the promotion of tumor cell migration and invasion, and MMP expression was always associated with a bad prognosis for cancer. More recently, however, it has been shown that MMPs can also be beneficial to the host as they can generate cleavage products from both non-ECM (Dong et al., 1997; Patterson and Sang, 1997; Cornelius et al., 1998; Sang, 1998; O'Reilly et al., 1999; Pozzi et al., 2000, 2002) and ECM components (Hamano et al., 2003; Lin et al., 2001) that are able to prevent important events in tumor progression such as tumor cell migration (Martinella-Catusse et al., 2001) and angiogenesis (O'Reilly et al., 1999; Pozzi et al., 2000, 2002; Hamano et al., 2003). Thus, MMPs appear to play a complicated role in tumor progression as on the one hand they degrade ECM allowing for increased tumor growth and spread, while on the other hand they produce cleavage products that have anti-angiogenic properties. Identifying the predominant effects of MMPs in cancer progression will be critical in determining whether blocking the effects of these enzymes will play a therapeutic role.

5. ECM cleaved products and their function in cancer progression

ECM cleavage products raised a great deal of interest in the scientific community after the finding that some ECM fragments exert anti-angiogenic and/or anti-tumorigenic activity. In particular, non-fibrillar collagen-derived fragments were shown to alter specific cell functions, such as tumor cell migration/invasion and endothelial cell proliferation/survival (reviewed in Marneros and Olsen, 2001; Pasco et al., 2004). Table 1 summarizes the role of these cleavage products with respect to their anti-angiogenic and anti-tumorigenic properties.

Cleavage of the C-terminal region of the NC1 domain of collagen XVIII leads to the generation of endostatin, a potent anti-angiogenic peptide (Fig. 2). This fragment inhibits *in vivo* tumor growth, especially melanoma, via its potent anti-angiogenic activity (O'Reilly et al., 1997). Cathepsin L (Felbor et al., 2000), MMP-7 (Lin et al., 2001), and pancreatic elastase (Ferreras et al., 2000) are able to release endostatin from collagen XVIII (Fig. 2). Although it is unknown how endostatin exerts its anti-angiogenic activity, it has been shown that this ECM fragment has a number of biologically important anti-angiogenic activities. Endostatin can compete with b-FGF and VEGF for heparan sulfate chain binding thus preventing the storage of potent angiogenic molecules in the ECM milieu (Kreuger et al., 2002; Eriksson et al., 2003). It can also bind to the catalytic domain of MMP-2 preventing its activity as well as MMP-2–mediated activation of MT1-MMP (Kim et al., 2000; Lee et al., 2002). Moreover, endostatin inhibits endothelial cell proliferation and migration by interacting with $\beta1$-containing integrins (Wickstrom et al., 2004), or via ligation of $\alpha v\beta3$ or $\alpha v\beta5$ integrins, receptors primarily expressed by angiogenic blood vessels (Rehn et al., 2001) (Table 1).

Another fragment, restin, is derived from the NC1 domain of type XV collagen and shows a 60% sequence homology with endostatin (Sasaki et al., 2000; Hohenester and Engel, 2002). Despite this high homology, restin differs from endostatin in its binding properties to ECM macromolecules as well as in its tissue distribution (Sasaki et al., 2000). Moreover, restin exhibits a potent anti-migratory but not anti-proliferative activity on endothelial cells and inhibits b-FGF–induced cell migration and growth in tumor cells (Ramchandran et al., 1999) (Table 1).

The NC1 domains of the different $\alpha(IV)$ chains have also been associated with inhibition of tumor angiogenesis and/or tumor growth. The $\alpha1(IV)$ NC1 domains, also known as arresten, acts as an anti-angiogenic molecule by inhibiting endothelial cell proliferation, migration, and tube formation *in vitro*, as well as primary and metastatic tumor growth *in vivo* (Colorado et al., 2000). The anti-angiogenic properties are mediated by binding to integrin $\alpha1\beta1$, the major collagen IV binding receptor, although the molecular mechanism has not yet been described (Table 1). Canstatin, the NC1 domain

Table 1
The NC1 domains of collagen chains and their effect on tumor progression

NC1	Biological effect/s	Surface receptor/s	Possible mechanisms	References
$\alpha1$(XVIII) endostatin	anti-angiogenic	$\beta1$, $\alpha v \beta3$, and $\alpha v \beta5$ integrins	Competition with angiogenic growth factors for heparan sulfate chain binding. Binding to the catalytic domain of MMP2. Inhibition of cell migration and proliferation	Kim et al., 2000; Rehn et al., 2001; Kreuger et al., 2002; Lee et al., 2002; Eriksson et al., 2003; Wickstrom et al., 2004
$\alpha1$(XV) restin	anti-angiogenic anti-tumorigenic		Anti-migratory on endothelial cells Prevention of bFGF-induced cell migration in tumor cells.	Ramchandran et al., 1999 Petitclerc et al., 2000
$\alpha1$(IV) arresten	anti-angiogenic	$\alpha1\beta1$ integrin		Colorado et al., 2000
$\alpha2$(IV) canstatin	anti-angiogenic	$\alpha v \beta3$ integrin	C-canstatin: inhibition of cell proliferation N-canstatin stimulation of cell apoptosis	
$\alpha3$(IV)	anti-tumorigenic	$\alpha3\beta1$ integrin	Inhibition of cell proliferation	He et al., 2003, 2004
	54–132 fragment (tumstatin)			
	anti-angiogenic	$\alpha v \beta3$ integrin	Inhibition of protein synthesis Inhibition of FAK and Akt phosphorylation.	Maeshima et al., 2000b, 2002
	185–203 fragment			
	anti-tumorigenic	$\alpha v \beta3$ integrin	Prevention of binding and activation of MMP2 by tumor cells. Downregulation of MT1-MMP and $\beta3$ integrin	Han et al., 1997; Shanan et al., 1999; Passco et al., 2000a
$\alpha4$(IV)	not known			
$\alpha5$(IV)	poor anti-angiogenic poor anti-tumorigenic			Petitclerc et al., 2000
$\alpha6$(IV)	anti-angiogenic	$\alpha v \beta3$ integrin		Petitclerc et al., 2000

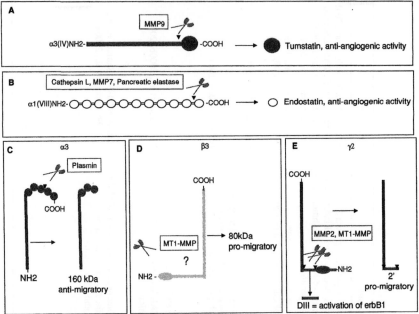

Fig. 2. Cleavage of ECM generates fragment with distinct cell functions. Cleavage of the NC1 domain of the α3(IV) (A) and the α1(XVIII) (B) chains by proteases generates tumstatin and endostatin respectively, both of which have anti-angiogenic activities. The cleavage of the α3 (C), β3 (D), and γ2 (E) chains of laminin-5 can be cleaved by different proteases generating fragments with both anti- and pro-migratory activity. The cleavage site of the different chains is indicated by the scissors.

of the α2(IV) chain has also been shown to have anti-angiogenic and anti-tumorigenic activity via αv and β1 integrin-dependent mechanisms (Kamphaus et al., 2000; Petitclerc et al., 2000). Interestingly, two different domains of canstatin have been identified, the N-canstatin (1–89 aa) and the C-canstatin (157–277aa). Although both of these fragments have been shown to decrease tumor angiogenesis *in vivo*, the C-canstatin seems to exert its functions by specifically inhibiting endothelial cell proliferation (He et al., 2004), whereas N-canstatin enhances endothelial cell apoptosis (He et al., 2003) (Table 1). Increased endothelial cell apoptosis is mainly attributed to the inhibition of Akt, FAK, and eukaryotic initiation factor-4E–binding protein-1 phophorylation, as well as induction of Fas ligand expression, and procaspases 8 and 9 cleavage (Panka and Mier, 2003).

Since observing that this collagen domain is able to inhibit bFGF-driven angiogenesis in chick chorioallantoic membrane assay, particular attention has been given to the biological effects of the NC1 domain of the α3(IV) chain (Petitclerc et al., 2000). Given this, two distinct peptides within the α3(IV) NC1 domain have been identified with specific anti-angiogenic (Maeshima et al., 2000a,b) and anti-tumorigenic properties (Han et al., 1997) (Table 1). The peptide sequence encompassing residues 185–203 of the NC1 α3(IV) domain inhibits melanoma cell proliferation without affecting proliferation of normal dermal fibroblasts or endothelial cells (Han et al., 1997; Shahan et al., 1999). The same fragment also decreases *in vitro* melanoma cell migration by preventing the binding and activation of MMP-2 at the melanoma cell surface, downregulating the expression of MT1-MMP and inhibiting β3 integrin subunit expression (Pasco et al., 2000a). In contrast, the peptide sequence encompassing residues 54–132 of the α3(IV) NC1 domain, (Maeshima et al., 2001a, 2001b) inhibits endothelial cell proliferation, without any effect on tumor cell proliferation (Maeshima et al., 2000a). Like the 185–203 residues, it binds to the αvβ3 integrin, but its binding site on this integrin seems to be distinct from that of the 185–203 fragment (Maeshima et al., 2000a; Pasco et al., 2000b). The α3(IV) chain, generated *in vivo* by MMP-9 cleavage (Hamano et al., 2003) (Fig. 2), inhibits protein synthesis of endothelial cells through a complex transduction pathway involving inhibition of FAK and Akt phosphorylation preventing the dissociation of eukaryotic initiation factor 4E protein from 4E-binding protein 1 (Maeshima et al., 2002).

Besides regulating tumor and endothelial cell functions via its NC1 domains, collagen IV has been shown to alter tumor cell functions via its triple helix domain/s (reviewed in Pasco et al., 2004). Highly metastatic melanoma cells, for example, adhere, spread, and become more motile in response to a triple helix-rich domain, while no effects are observed after stimulation with the NC1 domains (Chelberg et al., 1989). Three triple helix-rich domains within the α1 (IV) chain, namely the CB3 (Chelberg et al., 1989), the 531–543 (Miles et al., 1995), and 1263–1277 (Lauer-Fields et al., 2003) regions, have been shown to affect the functions of different tumor types in an RGD and/or RGD-independent manner (Chelberg et al., 1990). The major receptors transducing the signals initiated by these collagen domains are CD44 (Lauer-Fields et al., 2003), a member of the proteoglycan family, and β1-containing integrins, primarily α1β1, α2β1, and α3β1 (Vandenberg et al., 1991; Eble et al., 1993; Miles et al., 1995). Upon binding to these receptors, specific pathways involved in cell migration and invasion are triggered, including phosphorylation of FAK, paxillin, activation of small G-proteins, PKC and PI3 kinase, and changes in intracellular levels of calcium.

Among the laminin family members, cleavage products derived from laminin-5 are probably the most studied. Laminin-5 is the only laminin to contain a γ2 chain (Fig. 1). As previously mentioned, laminin-5 is abundantly expressed in basement membranes where it promotes adhesion and epithelial cell polarization by binding to integrin α3β1 and α6β4, the major component of hemidesomoses. All three chains of laminin-5 can be cleaved in both physiological (i.e., wound healing, tissue remodeling) and pathological (i.e., tumor progression) events by different proteases, such as plasmin and MMPs (reviewed in Hintermann and Quaranta, 2004) (Fig. 2). Interestingly, proteolytic processing of laminin-5 can both stimulate and downregulate epithelial cell migration, thus playing opposite effects on cell invasion. Post-translational processing of the α3 subunit of laminin-5 by plasmin cleavage has been shown to inhibit migration and enhance adhesion of squamous carcinoma and transformed oral epithelial cell lines (Goldfinger et al., 1998). Tumor cells depositing laminin-5-rich matrix containing unprocessed α3 chain are migratory and do not present hemidesmosome assembly. When the full length (190 kDa) α3 chain is cleaved by the serine protease plasmin, the generated 160 kDa α3 fragment form is able to nucleate hemidesmosome formation, resulting in a decrease in cell motility and enhanced adhesion (Goldfinger et al., 1998; Ghosh and Stack, 2000). In contrast to the α3 chain, cleavage of the β3 chain of laminin-5 by MT1-MMP leads to the generation of an 80 kDa β3 fragment that promotes migration of prostate carcinoma cells (Udayakumar et al., 2003) (Fig. 2). Similarly, cleavage of laminin-5 γ2 chain by MMP-2 and MT1-MMP leads to the exposure of a putative cryptic migratory signal, which induced cell motility in breast epithelial cells (Giannelli et al., 1997; Koshikawa et al., 2000) (Fig. 2). The importance of the γ2 cleavage in tissue remodeling and/or tumor progression also comes from the finding that this fragment is not present in quiescent tissues (Giannelli et al., 1999), while it is abundantly detected at the leading edge of invasive carcinomas (Pyke et al., 1994; Kagesato et al., 2001), where it might support cancer invasion.

It has been recently shown that the fragment of laminin-5 γ2 chain realized upon MMP cleavage (DIII fragment) consists of several EGF-like repeats and can bind to erbB1 triggering chemotaxis on laminin-5 (Schenk et al., 2003) (Fig. 2). Interestingly, stimulation of erbB1 induces downregulation of hemidesmosomes (Mariotti et al., 2001), allowing integrin-mediated cell migration. Moreover, the DIII domain induces selective upregulation of MMP-2 gene expression (Schenk et al., 2003), suggesting a positive feedback loop that tightly controls laminin-5 γ2 degradation and cell migration.

Although laminin-5 is probably the most studied member of the laminin family in the context of its ability to alter tumor cell functions, laminin-1 (α1β1γ1), -10 (α5β1γ1), and -11 (α5β2γ1), also components of basement membranes, can influence cell events, such as migration and invasion. Melanoma cells injected in the presence of the synthetic peptides AG-73 and A-13 from the C-terminal globular region of the laminin α1 chain metastasize to the lungs and/or the liver more aggressively than cells injected in the absence of this peptide (Kim et al., 1998; Kuratomi et al., 1999). The increased metastatic potential is mediated by the interaction of these peptides, at least AG-73, with a specific cell receptor expressed by melanoma cells, namely a heparan sulfate/chondroitin sulfate-containing proteoglycan receptor (Engbring et al., 2002). On the other hand, AG-10 and AG-32, two other peptides derived from the C-terminal globular domain of laminin-1 α1 chain, promote invasion of melanoma cells by interacting with integrin $\alpha6\beta1$ and stimulating lamellopodia formation (Nakahara et al., 1996). Moreover, a peptide derived from the $\gamma1$ chain globular domain of laminin-1 (C-16) significantly enhanced migration and pulmonary metastases of melanoma cells (Kuratomi et al., 2002), as well as vessel sprouting and angiogenesis in the chick chorioallantoic membrane assay (Ponce et al., 1999). Although the molecular mechanism is still unclear, it has been observed that this peptide induced MMP-9 synthesis and secretion, thus promoting ECM degradation (Kuratomi et al., 2002).

While peptides derived from laminin-1 are generally pro-angiogenic and pro-tumorigenic, it has recently been demonstrated that four peptides derived from the globular domain of laminin-10 $\alpha5$ chain inhibited melanoma metastasis and growth, as well as blocked tumor cell migration, invasion, and angiogenesis. Interestingly, one of these four peptides, namely A5G27, binds to the CD44 receptor of melanoma cells via the glycosaminoglycan residues on CD44 and prevents tumor cell migration and invasion in a dominant-negative manner (Hibino et al., 2004).

Unlike fragments of laminin-5 that can be generated *in vivo* and *in vitro* by proteolytic enzymes, the identification of laminin-1 and -10 peptides has been done by systematic screening of synthetic peptides and, at present, it is still unknown whether all or some of these fragments can be generated *in vivo* during both physiological and/or pathological events.

6. The challenges of deriving specific ECM-derived products

The finding that ECM cleavage products inhibit certain steps of tumor progression, such as tumor angiogenesis and tumor invasion, has raised interest in clinical trials. Despite this exciting observation, the potential use of these cleavage products in the clinical trials of cancers has not yet been realized. Treatment with many of these ECM cleavage products does not lead to beneficial effects; it requires a high concentration of the molecule(s) or causes undesirable side effects. A possible explanation for the failure and/or unsatisfactory action of these ECM-derived compounds is that these fragments have more than one specific effect. Endostatin, for example, has been shown to be anti-angiogenic (O'Reilly et al., 1997) and anti-tumorigenic with its ability to block migration and invasion of certain types of tumor cells, namely head and neck squamous cell carcinoma cells (Wilson et al., 2003). Therefore, it is conceivable that the effects of endostatin are tumor-specific and its beneficial effects are observed when both endothelium and tumor cells are targeted at the same time. However, these fragments may also target "healthy" cells resulting in undesirable side effects as they primarily work in extremely high concentrations. For example, endostatin not only binds integrin $\alpha v\beta3$ (Rehn et al., 2001), but also $\beta1$-containing integrins (Wickstrom et al., 2004), thus interfering with the cytoskeleton and impairing cell motility. Unlike $\alpha v\beta3$, primarily expressed on tumor blood vessels, β1-containing integrins are widely expressed in the body. Thus, it is conceivable that endostatin might also affect motility of surrounding "normal" cells. Similarly, it has been observed that the α3(IV) NC1 domain, besides blocking tumor angiogenesis (Maeshima et al., 2001a; Hamano et al., 2003), inhibits the function of stimulated neutrophils by raising intracellular levels of cAMP (Ziaie et al., 1999; Fawzi et al., 2000).

Some ECM-derived fragments act by binding the catalytic domains of certain MMP family members inhibiting their actions. As MMPs play a role during both physiological and pathological events, once again it is possible these ECM fragments also impair normal physiological functions. Moreover, in anti-cancer and anti-angiogenic therapy the activity of a certain drug is dependent upon the time of treatment, the duration of the treatment, and most importantly, at what stage of tumor progression the drug is delivered. In this regard it has been shown that the anti-angiogenic activity of endostatin is sensitive to the timing of treatment (Macpherson et al., 2003), thus explaining why this molecule might be ineffective in patients with advanced cancer.

Another issue related to the use of the ECM-derived products in cancer is that their anti tumor properties are strictly conformation-dependent (Floquet et al., 2004). Thus, it is conceivable that while naturally derived ECM products might have desired and selective effects, exogenously administered fragments might fail due to changes in conformation. Finally, as previously mentioned, some degrading enzymes released during cancer progression might generate anti-angiogenic and anti-tumorigenic ECM-derived products, such as endostatin, canstatin, tumstatin, the α3 chain fragment of laminin-5, but at the same time pro-tumorigenic fragments including fragments of the β3 and the γ2 chains of laminin-5. Thus, the administration of anti-angiogenic and/or anti-tumorigenic compounds might not be sufficient in conditions in which the endogenous production of pro-tumorigenic molecules is in vast excess.

7. Conclusions

The various intact basement membrane components, namely collagens and laminins, play a role in the homeostasis of polarized epithelia by balancing proliferation, migration, survival, and polarization of different cell types. In contrast, cleavage products of these basement membrane components may act in an opposite way, preventing cell proliferation, exacerbating cell migration, or inducing cell apoptosis. In pathological conditions, such as cancer progression, degradation of basement membrane components around tumors is considered critical for the promotion of invasion, migration into blood and/or lymphatic vessels, and traveling to distant sites. Generation of ECM fragments can play opposite effects on cancer progression by enhancing tumor cell migration and by blocking tumor blood supply at the same time. The observation that some of these ECM fragments have anti-angiogenic activity has raised a great deal of interest in the cancer field, making these fragments attractive candidate targets for potential cancer therapies. Treatment of tumor-bearing mice with synthesis peptides or recombinant basement membrane fragments has led to such promising results that several fragments (i.e., endostatin, tumstain) are now being tested in patients (Marx, 2003). Although these treatments induce tumor regression or stabilization in some cases, the preliminary results of the clinical studies suggest that these fragments will not be able to completely abolish and/or block cancer growth. As previously mentioned side effects, lack of selectivity, time of treatment, and three-dimensional conformation might explain the poor efficacy of these ECM fragments in cancer progression. Nevertheless, the discovery of these fragments has led to the hope that they could be administered with other anti-cancer drugs to induce tumor regression.

Acknowledgments

This work was supported by NCI/NIH R01 CA94849–01 (A.P.); NIH/ NIDDK O'Brien Center Grant P50-DK39261–16 (A.P., R.Z.); an Advanced Career Development (R.Z.) and a merit award from the Department of Veterans Affairs (R.Z.); and a Grant-in-Aid from the AHA (R.Z.). RZ is a Clinician Scientist of the National Kidney Foundation of the USA.

References

Abe, N., Muragaki, Y., Yoshioka, H., Inoue, H., Ninomiya, Y. 1993. Identification of a novel collagen chain represented by extensive interruptions in the triple-helical region. Biochem. Biophys. Res. Commun. 196, 576–582.

Amenta, P.S., Briggs, K., Xu, K., Gamboa, E., Jukkola, A.F., Li, D., Myers, J.C. 2000. Type XV collagen in human colonic adenocarcinomas has a different distribution than other basement membrane zone proteins. Hum. Pathol. 31, 359–366.

Amenta, P.S., Hadad, S., Lee, M.T., Barnard, N., Li, D., Myers, J.C. 2003. Loss of types XV and XIX collagen precedes basement membrane invasion in ductal carcinoma of the female breast. J. Pathol. 199, 298–308.

Aumailley, M., El Khal, A., Knoss, N., Tunggal, L. 2003. Laminin 5 processing and its integration into the ECM. Matrix Biol. 22, 49–54.

Barr, L.F., Campbell, S.E., Bochner, B.S., Dang, C.V. 1998. Association of the decreased expression of alpha3beta1 integrin with the altered cell: Environmental interactions and enhanced soft agar cloning ability of c-myc-overexpressing small cell lung cancer cells. Cancer Res. 58, 5537–5545.

Beck, K., Hunter, I., Engel, J. 1990. Structure and function of laminin: Anatomy of a multi-domain glycoprotein. FASEB J. 4, 148–160.

Behrendt, N. 2004. The urokinase receptor (uPAR) and the uPAR-associated protein (uPARAP/ Endo180): Membrane proteins engaged in matrix turnover during tissue remodeling. Biol. Chem. 385, 103–136.

Berquin, I.M., Sloane, B.F. 1996. Cathepsin B expression in human tumors. Adv. Exp. Med. Biol. 389, 281–294.

Bissell, M.J., Radisky, D.C., Rizki, A., Weaver, V.M., Petersen, O.W. 2002. The organizing principle: Microenvironmental influences in the normal and malignant breast. Differentiation 70, 537–546.

Borza, D.B., Bondar, O., Ninomiya, Y., Sado, Y., Naito, I., Todd, P., Hudson, B.G. 2001. The NC1 domain of collagen IV encodes a novel network composed of the alpha 1, alpha 2, alpha 5, and alpha 6 chains in smooth muscle basement membranes. J. Biol. Chem. 276, 28532–28540.

Borza, D.B., Bondar, O., Todd, P., Sundaramoorthy, M., Sado, Y., Ninomiya, Y., Hudson, B. G. 2002. Quaternary organization of the goodpasture autoantigen, the alpha 3(IV) collagen chain. Sequestration of two cryptic autoepitopes by intrapromoter interactions with the alpha4 and alpha5 NC1 domains. J. Biol. Chem. 277, 40075–40083.

Bosman, F.T., Stamenkovic, I. 2003. Functional structure and composition of the extracellular matrix. J. Pathol. 200, 423–428.

Boutaud, A., Borza, D.B., Bondar, O., Gunwar, S., Netzer, K.O., Singh, N., Ninomiya, Y., Sado, Y., Noelken, M.E., Hudson, B.G. 2000. Type IV collagen of the glomerular basement membrane. Evidence that the chain specificity of network assembly is encoded by the noncollagenous NC1 domains. J. Biol. Chem. 275, 30716–30724.

Brooks, P.C., Montgomery, A.M., Rosenfeld, M., Reisfeld, R.A., Hu, T., Klier, G., Cheresh, D.A. 1994. Integrin alpha v beta 3 antagonists promote tumor regression by inducing apoptosis of angiogenic blood vessels. Cell 79, 1157–1164.

Brooks, P.C., Silletti, S., von Schalscha, T.L., Friedlander, M., Cheresh, D.A. 1998. Disruption of angiogenesis by PEX, a noncatalytic metalloproteinase fragment with integrin binding activity. Cell 92, 391–400.

Burchardt, E.R., Hein, R., Bosserhoff, A.K. 2003. Laminin, hyaluronan, tenascin-C and type VI collagen levels in sera from patients with malignant melanoma. Clin. Exp. Dermatol. 28, 515–520.

Chelberg, M.K., McCarthy, J.B., Skubitz, A.P., Furcht, L.T., Tsilibary, E.C. 1990. Characterization of a synthetic peptide from type IV collagen that promotes melanoma cell adhesion, spreading, and motility. J. Cell. Biol. 111, 261–270.

Chelberg, M.K., Tsilibary, E.C., Hauser, A.R., McCarthy, J.B. 1989. Type IV collagen-mediated melanoma cell adhesion and migration: Involvement of multiple, distinct domains of the collagen molecule. Cancer Res. 49, 4796–4802.

Choong, P.F., Nadesapillai, A.P. 2003. Urokinase plasminogen activator system: a multifunctional role in tumor progression and metastasis. Clin. Orthop. S46–S58.

Colorado, P.C., Torre, A., Kamphaus, G., Maeshima, Y., Hopfer, H., Takahashi, K., Volk, R., Zamborsky, E.D., Herman, S., Sarkar, P.K., Ericksen, M.B., Dhanabal, M., Simons, M., Post, M., Kufe, D.W., Weichselbaum, R.R., Sukhatme, V.P., Kalluri, R. 2000. Anti-angiogenic cues from vascular basement membrane collagen. Cancer Res. 60, 2520–2526.

Cornelius, L.A., Nehring, L.C., Harding, E., Bolanowski, M., Welgus, H.G., Kobayashi, D.K., Pierce, R.A., Shapiro, S.D. 1998. Matrix metalloproteinases generate angiostatin: Effects on neovascularization. J. Immunol. 161, 6845–6852.

Costantini, V., Sidoni, A., Deveglia, R., Cazzato, O.A., Bellezza, G., Ferri, I., Bucciarelli, E., Nenci, G.G. 1996. Combined overexpression of urokinase, urokinase receptor, and plasminogen activator inhibitor-1 is associated with breast cancer progression: An immunohistochemical comparison of normal, benign, and malignant breast tissues. Cancer 77, 1079–1088.

Coussens, L.M., Fingleton, B., Matrisian, L.M. 2002. Matrix metalloproteinase inhibitors and cancer: Trials and tribulations. Science 295, 2387–2392.

De Wever, O., Mareel, M. 2003. Role of tissue stroma in cancer cell invasion. J. Pathol. 200, 429–447.

Dennhofer, R., Kurschat, P., Zigrino, P., Klose, A., Bosserhoff, A., van Muijen, G., Krieg, T., Mauch, C., Hunzelmann, N. 2003. Invasion of melanoma cells into dermal connective tissue *in vitro*: Evidence for an important role of cysteine proteases. Int. J. Cancer. 106, 316–323.

Dohchin, A., Suzuki, J.I., Seki, H., Masutani, M., Shiroto, H., Kawakami, Y. 2000. Immunostained cathepsins B and L correlate with depth of invasion and different metastatic pathways in early stage gastric carcinoma. Cancer 89, 482–487.

Dong, Z., Kumar, R., Yang, X., Fidler, I.J. 1997. Macrophage-derived metalloelastase is responsible for the generation of angiostatin in Lewis lung carcinoma. Cell 88, 801–810.

Dreier, R., Grassel, S., Fuchs, S., Schaumburger, J., Bruckner, P. 2004. Pro-MMP-9 is a specific macrophage product and is activated by osteoarthritic chondrocytes via MMP-3 or a MT1-MMP/MMP-13 cascade. Exp. Cell. Res. 297, 303–312.

Eble, J.A., Golbik, R., Mann, K., Kuhn, K. 1993. The alpha 1 beta 1 integrin recognition site of the basement membrane collagen molecule [alpha 1(IV)]2 alpha 2(IV). EMBO J. 12, 4795–4802.

Engbring, J.A., Hoffman, M.P., Karmand, A.J., Kleinman, H.K. 2002. The B16F10 cell receptor for a metastasis-promoting site on laminin-1 is a heparan sulfate/chondroitin sulfate-containing proteoglycan. Cancer Res. 62, 3549–3554.

Engbring, J.A., Kleinman, H.K. 2003. The basement membrane matrix in malignancy. J. Pathol. 200, 465–470.

Eriksson, K., Magnusson, P., Dixelius, J., Claesson-Welsh, L., Cross, M.J. 2003. Angiostatin and endostatin inhibit endothelial cell migration in response to FGF and VEGF without interfering with specific intracellular signal transduction pathways. FEBS Lett. 536, 19–24.

Fawzi, A., Robinet, A., Monboisse, J.C., Ziaie, Z., Kefalides, N.A., Bellon, G. 2000. A peptide of the alpha 3(IV) chain of type IV collagen modulates stimulated neutrophil function via activation of cAMP-dependent protein kinase and Ser/Thr protein phosphatase. Cell Signal 12, 327–335.

Felbor, U., Dreier, L., Bryant, R.A., Ploegh, H.L., Olsen, B.R., Mothes, W. 2000. Secreted cathepsin L generates endostatin from collagen XVIII. EMBO J. 19, 1187–1194.

Felding-Habermann, B. 2003. Integrin adhesion receptors in tumor metastasis. Clin. Exp. Metastasis 20, 203–213.

Felding-Habermann, B., Fransvea, E., O'Toole, T.E., Manzuk, L., Faha, B., Hensler, M. 2002. Involvement of tumor cell integrin alpha v beta 3 in hematogenous metastasis of human melanoma cells. Clin. Exp. Metastasis 19, 427–436.

Ferreras, M., Felbor, U., Lenhard, T., Olsen, B.R., Delaisse, J. 2000. Generation and degradation of human endostatin proteins by various proteinases. FEBS Lett. 486, 247–251.

Floquet, N., Pasco, S., Ramont, L., Derreumaux, P., Laronze, J.Y., Nuzillard, J.M., Maquart, F.X., Alix, A.J., Monboisse, J.C. 2004. The antitumor properties of the alpha3(IV)-(185–203) peptide from the NC1 domain of type IV collagen (tumstatin) are conformation-dependent. J. Biol. Chem. 279, 2091–2100.

Folgueras, A.R., Pendas, A.M., Sanchez, L.M., Lopez-Otin, C. 2004. Matrix metalloproteinases in cancer: From new functions to improved inhibition strategies. Int. J. Dev. Biol. 48, 411–424.

Freije, J.M., Balbin, M., Pendas, A.M., Sanchez, L.M., Puente, X.S., Lopez-Otin, C. 2003. Matrix metalloproteinases and tumor progression. Adv. Exp. Med. Biol. 532, 91–107.

Friedlander, M., Brooks, P.C., Shaffer, R.W., Kincaid, C.M., Varner, J.A., Cheresh, D.A. 1995. Definition of two angiogenic pathways by distinct alpha v integrins. Science 270, 1500–1502.

Gehlsen, K.R., Davis, G.E., Sriramarao, P. 1992. Integrin expression in human melanoma cells with differing invasive and metastatic properties. Clin. Exp. Metastasis. 10, 111–20.

Ghosh, S., Stack, M.S. 2000. Proteolytic modification of laminins: Functional consequences. Microsc. Res. Tech. 51, 238–246.

Giannelli, G., Falk-Marzillier, J., Schiraldi, O., Stetler-Stevenson, W.G., Quaranta, V. 1997. Induction of cell migration by matrix metalloprotease-2 cleavage of laminin-5. Science 277, 225–228.

Giannelli, G., Pozzi, A., Stetler-Stevenson, W.G., Gardner, H.A., Quaranta, V. 1999. Expression of matrix metalloprotease-2-cleaved laminin-5 in breast remodeling stimulated by sex steroids. Am. J. Pathol. 154, 1193–1201.

Goel, H.L., Languino, L.R. 2004. Integrin signaling in cancer. Cancer Treat. Res. 119, 15–31.

Goldfinger, L.E., Stack, M.S., Jones, J.C. 1998. Processing of laminin-5 and its functional consequences: Role of plasmin and tissue-type plasminogen activator. J. Cell. Biol. 141, 255–265.

Gordon, L.A., Mulligan, K.T., Maxwell-Jones, H., Adams, M., Walker, R.A., Jones, J.L. 2003. Breast cell invasive potential relates to the myoepithelial phenotype. Int. J. Cancer 106, 8–16.

Halfter, W., Dong, S., Schurer, B., Cole, G.J. 1998. Collagen XVIII is a basement membrane heparan sulfate proteoglycan. J. Biol. Chem. 273, 25404–25412.

Hamano, Y., Zeisberg, M., Sugimoto, H., Lively, J.C., Maeshima, Y., Yang, C., Hynes, R.O., Werb, Z., Sudhakar, A., Kalluri, R. 2003. Physiological levels of tumstatin, a fragment of collagen IV alpha3 chain, are generated by MMP-9 proteolysis and suppress angiogenesis via alphaV beta3 integrin. Cancer Cell. 3, 589–601.

Han, J., Ohno, N., Pasco, S., Monboisse, J.C., Borel, J.P., Kefalides, N.A. 1997. A cell binding domain from the alpha3 chain of type IV collagen inhibits proliferation of melanoma cells. J. Biol. Chem. 272, 20395–20401.

Hao, J., Jackson, L., Calaluce, R., McDaniel, K., Dalkin, B.L., Nagle, R.B. 2001. Investigation into the mechanism of the loss of laminin 5 (alpha3beta3gamma2) expression in prostate cancer. Am. J. Pathol. 158, 1129–1135.

Hao, J., Yang, Y., McDaniel, K.M., Dalkin, B.L., Cress, A.E., Nagle, R.B. 1996. Differential expression of laminin 5 (alpha 3 beta 3 gamma 2) by human malignant and normal prostate. Am. J. Pathol. 149, 1341–1349.

He, G.A., Luo, J.X., Zhang, T.Y., Hu, Z.S., Wang, F.Y. 2004. The C-terminal domain of canstatin suppresses in vivo tumor growth associated with proliferation of endothelial cells. Biochem. Biophys. Res. Commun. 318, 354–360.

He, G.A., Luo, J.X., Zhang, T.Y., Wang, F.Y., Li, R.F. 2003. Canstatin-N fragment inhibits in vitro endothelial cell proliferation and suppresses in vivo tumor growth. Biochem. Biophys. Res. Commun. 312, 801–805.

Hibino, S., Shibuya, M., Engbring, J.A., Mochizuki, M., Nomizu, M., Kleinman, H.K. 2004. Identification of an active site on the laminin alpha5 chain globular domain that binds to CD44 and inhibits malignancy. Cancer Res. 64, 4810–4816.

Hiki, Y., Iyama, K., Tsuruta, J., Egami, H., Kamio, T., Suko, S., Naito, I., Sado, Y., Ninomiya, Y., Ogawa, M. 2002. Differential distribution of basement membrane type IV collagen alpha1(IV), alpha2(IV), alpha5(IV) and alpha6(IV) chains in colorectal epithelial tumors. Pathol. Int. 52, 224–233.

Hintermann, E., Quaranta, V. 2004. Epithelial cell motility on laminin-5: regulation by matrix assembly, proteolysis, integrins and erbB receptors. Matrix Biol. 23, 75–85.

Hohenester, E., Engel, J. 2002. Domain structure and organisation in extracellular matrix proteins. Matrix Biol. 21, 115–128.

Hood, J.D., Cheresh, D.A. 2002. Role of integrins in cell invasion and migration. Nat. Rev. Cancer 2, 91–100.

Hostikka, S.L., Tryggvason, K. 1988. The complete primary structure of the alpha 2 chain of human type IV collagen and comparison with the alpha 1(IV) chain. J. Biol. Chem. 263, 19488–19493.

Hsu, M.Y., Meier, F., Herlyn, M. 2002. Melanoma development and progression: a conspiracy between tumor and host. Differentiation 70, 522–536.

Hynes, R. 2002. Integrins: bidirectional, allosteric signaling machines. Cell 110, 673–687.

Jiang, A., Pei, D. 2003. Distinct roles of catalytic and pexin-like domains in membrane-type matrix metalloproteinase (MMP)-mediated pro-MMP-2 activation and collagenolysis. J. Biol. Chem. 278, 38765–38771.

Jin, H., Varner, J. 2004. Integrins: Roles in cancer development and as treatment targets. Br. J. Cancer 90, 561–565.

Jones, P.A., De Clerck, Y.A. 1982. Extracellular matrix destruction by invasive tumor cells. Cancer Metastasis Rev. 1, 289–317.

Kagesato, Y., Mizushima, H., Koshikawa, N., Kitamura, H., Hayashi, H., Ogawa, N., Tsukuda, M., Miyazaki, K. 2001. Sole expression of laminin gamma 2 chain in invading tumor cells and its association with stromal fibrosis in lung adenocarcinomas. Jpn. J. Cancer Res. 92, 184–192.

Kamphaus, G.D., Colorado, P.C., Panka, D.J., Hopfer, H., Ramchandran, R., Torre, A., Maeshima, Y., Mier, J.W., Sukhatme, V.P., Kalluri, R. 2000. Canstatin, a novel matrix-derived inhibitor of angiogenesis and tumor growth. J. Biol. Chem. 275, 1209–1215.

Katayama, M., Sekiguchi, K. 2004. Laminin-5 in epithelial tumour invasion. J. Mol. Histol. 35, 277–286.

Kim, W.H., Nomizu, M., Song, S.Y., Tanaka, K., Kuratomi, Y., Kleinman, H.K., Yamada, Y. 1998. Laminin-alpha1-chain sequence Leu-Gln-Val-Gln-Leu-Ser-Ile-Arg (LQVQLSIR) enhances murine melanoma cell metastases. Int. J. Cancer. 77, 632–639.

Kim, Y.M., Jang, J.W., Lee, O.H., Yeon, J., Choi, E.Y., Kim, K.W., Lee, S.T., Kwon, Y.G. 2000. Endostatin inhibits endothelial and tumor cellular invasion by blocking the activation and catalytic activity of matrix metalloproteinase. Cancer Res. 60, 5410–5413.

Kivirikko, S., Saarela, J., Myers, J.C., Autio-Harmainen, H., Pihlajaniemi, T. 1995. Distribution of type XV collagen transcripts in human tissue and their production by muscle cells and fibroblasts. Am. J. Pathol. 147, 1500–1509.

Kos, J., Lah, T.T. 1998. Cysteine proteinases and their endogenous inhibitors: Target proteins for prognosis, diagnosis and therapy in cancer (review). Oncol. Rep. 5, 1349–13461.

Koshikawa, N., Giannelli, G., Cirulli, V., Miyazaki, K., Quaranta, V. 2000. Role of cell surface metalloprotease MT1-MMP in epithelial cell migration over laminin-5. J. Cell. Biol. 148, 615–624.

Koshikawa, N., Schenk, S., Moeckel, G., Sharabi, A., Miyazaki, K., Gardner, H., Zent, R., Quaranta, V. 2004. Proteolytic processing of laminin-5 by MT1-MMP in tissues and its effects on epithelial cell morphology. FASEB J. 18, 364–6.

Krepela, E. 2001. Cysteine proteinases in tumor cell growth and apoptosis. Neoplasma 48, 332–349.

Kreuger, J., Matsumoto, T., Vanwildemeersch, M., Sasaki, T., Timpl, R., Claesson-Welsh, L., Spillmann, D., Lindahl, U. 2002. Role of heparan sulfate domain organization in endostatin inhibition of endothelial cell function. EMBO J. 21, 6303–6311.

Krutovskikh, V. 2002. Implication of direct host-tumor intercellular interactions in non-immune host resistance to neoplastic growth. Semin. Cancer Biol. 12, 267–276.

Kuratomi, Y., Nomizu, M., Nielsen, P.K., Tanaka, K., Song, S.Y., Kleinman, H.K., Yamada, Y. 1999. Identification of metastasis-promoting sequences in the mouse laminin alpha 1 chain. Exp. Cell. Res. 249, 386–395.

Kuratomi, Y., Nomizu, M., Tanaka, K., Ponce, M.L., Komiyama, S., Kleinman, H.K., Yamada, Y. 2002. Laminin gamma 1 chain peptide, C–16 (KAFDITYVRLKF), promotes migration, MMP-9 secretion, and pulmonary metastasis of B16-F10 mouse melanoma cells. Br. J. Cancer 86, 1169–1173.

Lauer-Fields, J.L., Malkar, N.B., Richet, G., Drauz, K., Fields, G.B. 2003. Melanoma cell CD44 interaction with the alpha 1(IV)1263–1277 region from basement membrane collagen is modulated by ligand glycosylation. J. Biol. Chem. 278, 14321–14330.

Leblond, C.P., Inoue, S. 1989. Structure, composition, and assembly of basement membrane. Am. J. Anat. 185, 367–390.

Lee, D.H., Yang, Y., Lee, S.J., Kim, K.Y., Koo, T.H., Shin, S.M., Song, K.S., Lee, Y.H., Kim, Y. J., Lee, J.J., Choi, I., Lee, J.H. 2003. Macrophage inhibitory cytokine-1 induces the invasiveness of gastric cancer cells by up-regulating the urokinase-type plasminogen activator system. Cancer Res. 63, 4648–4655.

Lee, S.J., Jang, J.W., Kim, Y.M., Lee, H.I., Jeon, J.Y., Kwon, Y.G., Lee, S.T. 2002. Endostatin binds to the catalytic domain of matrix metalloproteinase-2. FEBS Lett. 519, 147–152.

Leinonen, A., Mariyama, M., Mochizuki, T., Tryggvason, K., Reeders, S.T. 1994. Complete primary structure of the human type IV collagen alpha 4(IV) chain. Comparison with structure and expression of the other alpha (IV) chains. J. Biol. Chem. 269, 26172–26177.

Li, D., Clark, C.C., Myers, J.C. 2000. Basement membrane zone type XV collagen is a disulfide-bonded chondroitin sulfate proteoglycan in human tissues and cultured cells. J. Biol. Chem. 275, 22339–22347.

Lin, H.C., Chang, J.H., Jain, S., Gabison, E.E., Kure, T., Kato, T., Fukai, N., Azar, D.T. 2001. Matrilysin cleavage of corneal collagen type XVIII NC1 domain and generation of a 28-kDa fragment. Invest. Ophthalmol. Vis. Sci. 42, 2517–2524.

Lochter, A., Navre, M., Werb, Z., Bissell, M.J. 1999. alpha1 and alpha2 integrins mediate invasive activity of mouse mammary carcinoma cells through regulation of stromelysin-1 expression. Mol. Biol. Cell. 10, 271–282.

Lundgren, C., Frankendal, B., Silfversward, C., Nilsson, B., Tryggvason, K., Auer, G., Nordstrom, B. 2003. Laminin-5 gamma2-chain expression and DNA ploidy as predictors of prognosis in endometrial carcinoma. Med. Oncol. 20, 147–156.

Macpherson, G.R., Ng, S.S., Forbes, S.L., Melillo, G., Karpova, T., McNally, J., Conrads, T. P., Veenstra, T.D., Martinez, A., Cuttitta, F., Price, D.K., Figg, W.D. 2003. Anti-angiogenic activity of human endostatin is HIF-1-independent in vitro and sensitive to timing of treatment in a human saphenous vein assay. Mol. Cancer Ther. 2, 845–854.

Madri, J.A., Pratt, B.M., Yurchenco, P.D., Furthmayr, H. 1984. The ultrastructural organization and architecture of basement membranes. Ciba. Found. Symp. 108, 6–24.

Maeshima, Y., Colorado, P.C., Kalluri, R. 2000. Two RGD-independent alpha vbeta 3 integrin binding sites on tumstatin regulate distinct anti-tumor properties. J. Biol. Chem. 275, 23745–23750.

Maeshima, Y., Colorado, P.C., Torre, A., Holthaus, K.A., Grunkemeyer, J.A., Ericksen, M.B., Hopfer, H., Xiao, Y., Stillman, I.E., Kalluri, R. 2000. Distinct antitumor properties of a type IV collagen domain derived from basement membrane. J. Biol. Chem. 275, 21340–21348.

Maeshima, Y., Manfredi, M., Reimer, C., Holthaus, K.A., Hopfer, H., Chandamuri, B.R., Kharbanda, S., Kalluri, R. 2001. Identification of the anti-angiogenic site within vascular basement membrane-derived tumstatin. J. Biol. Chem. 276, 15240–15248.

Maeshima, Y., Sudhakar, A., Lively, J.C., Ueki, K., Kharbanda, S., Kahn, C.R., Sonenberg, N., Hynes, R.O., Kalluri, R. 2002. Tumstatin, an endothelial cell-specific inhibitor of protein synthesis. Science 295, 140–143.

Maeshima, Y., Yerramalla, U.L., Dhanabal, M., Holthaus, K.A., Barbashov, S., Kharbanda, S., Reimer, C., Manfredi, M., Dickerson, W.M., Kalluri, R. 2001. Extracellular matrix-derived peptide binds to alpha(v)beta(3) integrin and inhibits angiogenesis. J. Biol. Chem. 276, 31959–31968.

Marcinkiewicz, C., Weinreb, P.H., Calvete, J.J., Kisiel, D.G., Mousa, S.A., Tuszynski, G.P., Lobb, R.R. 2003. Obtustatin: A potent selective inhibitor of alpha1beta1 integrin in vitro and angiogenesis in vivo. Cancer Res. 63, 2020–2023.

Mariotti, A., Kedeshian, P.A., Dans, M., Curatola, A.M., Gagnoux-Palacios, L., Giancotti, F. G. 2001. EGF-R signaling through Fyn kinase disrupts the function of integrin alpha6beta4 at hemidesmosomes: Role in epithelial cell migration and carcinoma invasion. J. Cell. Biol. 155, 447–458.

Mariyama, M., Leinonen, A., Mochizuki, T., Tryggvason, K., Reeders, S.T. 1994. Complete primary structure of the human alpha 3(IV) collagen chain. Coexpression of the alpha 3(IV) and alpha 4(IV) collagen chains in human tissues. J. Biol. Chem. 269, 23013–23017.

Marneros, A.G., Olsen, B.R. 2001. The role of collagen-derived proteolytic fragments in angiogenesis. Matrix Biol. 20, 337–345.

Martin, K.J., Kwan, C.P., Nagasaki, K., Zhang, X., O'Hare, M.J., Kaelin, C.M., Burgeson, R. E., Pardee, A.B., Sager, R. 1998. Down-regulation of laminin-5 in breast carcinoma cells. Mol. Med. 4, 602–613.

Martinella-Catusse, C., Polette, M., Noel, A., Gilles, C., Dehan, P., Munaut, C., Colige, A., Volders, L., Monboisse, J.C., Foidart, J.M., Birembaut, P. 2001. Down-Regulation of MT1-MMP expression by the alpha3 chain of type IV collagen inhibits bronchial tumor cell line invasion. Lab. Invest 81, 167–175.

Marx, J. 2003. Angiogenesis. A boost for tumor starvation. Science 301, 452–454.

Mazar, A.P., Henkin, J., Goldfarb, R.H. 1999. The urokinase plasminogen activator system in cancer: Implications for tumor angiogenesis and metastasis. Angiogenesis 3, 15–32.

McDonnell, C.O., Hill, A.D., McNamara, D.A., Walsh, T.N., Bouchier-Hayes, D.J. 2000. Tumour micrometastases: the influence of angiogenesis. Eur. J. Surg. Oncol. 26, 105–115.

Miles, A.J., Knutson, J.R., Skubitz, A.P., Furcht, L.T., McCarthy, J.B., Fields, G.B. 1995. A peptide model of basement membrane collagen alpha 1 (IV) 531–543 binds the alpha 3 beta 1 integrin. J. Biol. Chem. 270, 29047–29050.

Muragaki, Y., Abe, N., Ninomiya, Y., Olsen, B.R., Ooshima, A. 1994. The human alpha 1(XV) collagen chain contains a large amino-terminal non-triple helical domain with a tandem repeat structure and homology to alpha 1(XVIII) collagen. J. Biol. Chem. 269, 4042–4046.

Muragaki, Y., Timmons, S., Griffith, C.M., Oh, S.P., Fadel, B., Quertermous, T., Olsen, B.R. 1995. Mouse Col18a1 is expressed in a tissue-specific manner as three alternative variants and is localized in basement membrane zones. Proc. Natl. Acad. Sci. USA 92, 8763–8767.

Musso, O., Rehn, M., Saarela, J., Theret, N., Lietard, J., Hintikka, Lotrian, D., Campion, J.P., Pihlajaniemi, T., Clement, B. 1998. Collagen XVIII is localized in sinusoids and basement membrane zones and expressed by hepatocytes and activated stellate cells in fibrotic human liver. Hepatology 28, 98–107.

Musso, O., Rehn, M., Theret, N., Turlin, B., Bioulac-Sage, P., Lotrian, D., Campion, J.P., Pihlajaniemi, T., Clement, B. 2001. Tumor progression is associated with a significant decrease in the expression of the endostatin precursor collagen XVIII in human hepatocellular carcinomas. Cancer Res. 61, 45–49.

Myers, J.C., Dion, A.S., Abraham, V., Amenta, P.S. 1996. Type XV collagen exhibits a widespread distribution in human tissues but a distinct localization in basement membrane zones. Cell Tissue Res. 286, 493–505.

Myers, J.C., Jones, T.A., Pohjolainen, E.R., Kadri, A.S., Goddard, A.D., Sheer, D., Solomon, E., Pihlajaniemi, T. 1990. Molecular cloning of alpha 5(IV) collagen and assignment of the gene to the region of the X chromosome containing the Alport syndrome locus. Am. J. Hum. Genet. 46, 1024–1033.

Nakahara, H., Nomizu, M., Akiyama, S.K., Yamada, Y., Yeh, Y., Chen, W.T. 1996. A mechanism for regulation of melanoma invasion. Ligation of alpha6beta1 integrin by laminin G peptides. J. Biol. Chem. 271, 27221–27224.

Nie, J., Pei, D. 2003. Direct activation of pro-matrix metalloproteinase-2 by leukolysin/membrane-type 6 matrix metalloproteinase/matrix metalloproteinase 25 at the asn(109)-Tyr bond. Cancer Res. 63, 6758–6762.

Niki, T., Kohno, T., Iba, S., Moriya, Y., Takahashi, Y., Saito, M., Maeshima, A., Yamada, T., Matsuno, Y., Fukayama, M., Yokota, J., Hirohashi, S. 2002. Frequent co-localization of Cox-2 and laminin-5 gamma2 chain at the invasive front of early-stage lung adenocarcinomas. Am. J. Pathol. 160, 1129–1141.

Nip, J., Shibata, H., Loskutoff, D.J., Cheresh, D.A., Brodt, P. 1992. Human melanoma cells derived from lymphatic metastases use integrin alpha v beta 3 to adhere to lymph node vitronectin. J. Clin. Invest. 90, 1413.

Oh, S.P., Kamagata, Y., Muragaki, Y., Timmons, S., Ooshima, A., Olsen, B.R. 1994. Isolation and sequencing of cDNAs for proteins with multiple domains of Gly-Xaa-Yaa repeats identify a distinct family of collagenous proteins. Proc. Natl. Acad. Sci. USA 91, 4229–4233.

Oh, S.P., Warman, M.L., Seldin, M.F., Cheng, S.D., Knoll, J.H., Timmons, S., Olsen, B.R. 1994. Cloning of cDNA and genomic DNA encoding human type XVIII collagen and localization of the alpha 1(XVIII) collagen gene to mouse chromosome 10 and human chromosome 21. Genomics 19, 494–499.

Ohnishi, Y., Tajima, S., Ishibashi, A. 2001. Coordinate expression of membrane type-matrix metalloproteinases-2 and 3 (MT2-MMP and MT3-MMP) and matrix metalloproteinase-2 (MMP-2) in primary and metastatic melanoma cells. Eur. J. Dermatol. 11, 420–423.

Okegawa, T., Pong, R.C., Li, Y., Hsieh, J.T. 2004. The role of cell adhesion molecule in cancer progression and its application in cancer therapy. Acta Biochim. Pol. 51, 445–457.

Ono, Y., Nakanishi, Y., Ino, Y., Niki, T., Yamada, T., Yoshimura, K., Saikawa, M., Nakajima, T., Hirohashi, S. 1999. Clinocopathologic significance of laminin-5 gamma2 chain expression in squamous cell carcinoma of the tongue: immunohistochemical analysis of 67 lesions. Cancer 85, 2315–2321.

O'Reilly, M.S., Boehm, T., Shing, Y., Fukai, N., Vasios, G., Lane, W.S., Flynn, E., Birkhead, J. R., Olsen, B.R., Folkman, J. 1997. Endostatin: an endogenous inhibitor of angiogenesis and tumor growth. Cell 88, 277–285.

O'Reilly, M.S., Wiederschain, D., Stetler-Stevenson, W.G., Folkman, J., Moses, M.A. 1999. Regulation of angiostatin production by matrix metalloproteinase-2 in a model of concomitant resistance [In Process Citation]. J. Biol. Chem. 274, 29568–29571.

Panka, D.J., Mier, J.W. 2003. Canstatin inhibits Akt activation and induces Fas-dependent apoptosis in endothelial cells. J. Biol. Chem. 278, 37632–37636.

Pasco, S., Han, J., Gillery, P., Bellon, G., Maquart, F.X., Borel, J.P., Kefalides, N.A., Monboisse, J.C. 2000. A specific sequence of the noncollagenous domain of the alpha3(IV) chain of type IV collagen inhibits expression and activation of matrix metalloproteinases by tumor cells. Cancer Res. 60, 467–473.

Pasco, S., Monboisse, J.C., Kieffer, N. 2000. The alpha 3(IV)185–206 peptide from noncollagenous domain 1 of type IV collagen interacts with a novel binding site on the beta 3 subunit of integrin alpha Vbeta 3 and stimulates focal adhesion kinase and phosphatidylinositol 3-kinase phosphorylation. J. Biol. Chem. 275, 32999–3007.

Pasco, S., Ramont, L., Maquart, F.X., Monboisse, J.C. 2004. Control of melanoma progression by various matrikines from basement membrane macromolecules. Crit. Rev. Oncol. Hematol. 49, 221–233.

Patterson, B.C., Sang, Q.A. 1997. Angiostatin-converting enzyme activities of human matrilysin (MMP-7) and gelatinase B/type IV collagenase (MMP-9). J. Biol. Chem. 272, 28823–28825.

Petitclerc, E., Boutaud, A., Prestayko, A., Xu, J., Sado, Y., Ninomiya, Y., Sarras, M.P. ,Jr., Hudson, B.G., Brooks, P.C. 2000. New functions for non-collagenous domains of human collagen type IV. Novel integrin ligands inhibiting angiogenesis and tumor growth *in vivo*. J. Biol. Chem. 275, 8051–8061.

Ponce, M.L., Nomizu, M., Delgado, M.C., Kuratomi, Y., Hoffman, M.P., Powell, S., Yamada, Y., Kleinman, H.K., Malinda, K.M. 1999. Identification of endothelial cell binding sites on the laminin gamma 1 chain. Circ. Res. 84, 688–694.

Pozzi, A., Moberg, P.E., Miles, L.A., Wagner, S., Soloway, P., Gardner, H.A. 2000. Elevated matrix metalloprotease and angiostatin levels in integrin alpha 1 knockout mice cause reduced tumor vascularization. Proc. Natl. Acad. Sci. USA 97, 2202–2207.

Pozzi, A., Le Vine, W.F., Gardner, H.A. 2002. Low plasma levels of matrix metalloproteinase 9 permit increased tumor angiogenesis. Oncogene 21, 272–281.

Pozzi, A., Zent, R. 2003. Integrins: Sensors of extracellular matrix and modulators of cell function. Nephron. Exp. Nephrol. 94, e77–e84.

Pyke, C., Romer, J., Kallunki, P., Lund, L.R., Ralfkiaer, E., Dano, K., Tryggvason, K. 1994. The gamma 2 chain of kalinin/laminin 5 is preferentially expressed in invading malignant cells in human cancers. Am. J. Pathol. 145, 782–791.

Pyke, C., Salo, S., Ralfkiaer, E., Romer, J., Dano, K., Tryggvason, K. 1995. Laminin-5 is a marker of invading cancer cells in some human carcinomas and is coexpressed with the receptor for urokinase plasminogen activator in budding cancer cells in colon adenocarcinomas. Cancer Res. 55, 4132–4139.

Quaranta, V. 2002. Motility cues in the tumor microenvironment. Differentiation 70, 590–8.

Quatresooz, P., Martalo, O., Pierard, G.E. 2003. Differential expression of alpha1 (IV) and alpha5 (IV) collagen chains in basal-cell carcinoma. J. Cutan. Pathol. 30, 548–552.

Ramchandran, R., Dhanabal, M., Volk, R., Waterman, M.J., Segal, M., Lu, H., Knebelmann, B., Sukhatme, V.P. 1999. Antiangiogenic activity of restin, NC10 domain of human collagen XV: Comparison to endostatin. Biochem. Biophys. Res. Commun. 255, 735–739.

Rehn, M., Hintikka, E, Pihlajaniemi, T. 1994. Primary structure of the alpha 1 chain of mouse type XVIII collagen, partial structure of the corresponding gene, and comparison of the alpha 1(XVIII) chain with its homologue, the alpha 1(XV) collagen chain. J. Biol. Chem. 269, 13929–13935.

Rehn, M., Pihlajaniemi, T. 1994. Alpha 1(XVIII), a collagen chain with frequent interruptions in the collagenous sequence, a distinct tissue distribution, and homology with type XV collagen. Proc. Natl. Acad. Sci. USA 91, 4234–4238.

Rehn, M., Veikkola, T., Kukk-Valdre, E., Nakamura, H., Ilmonen, M., Lombardo, C., Pihlajaniemi, T., Alitalo, K., Vuori, K. 2001. Interaction of endostatin with integrins implicated in angiogenesis. Proc. Natl. Acad. Sci. USA 98, 1024–1029.

Saarela, J., Rehn, M., Oikarinen, A., Autio-Harmainen, H., Pihlajaniemi, T. 1998. The short and long forms of type XVIII collagen show clear tissue specificities in their expression and location in basement membrane zones in humans. Am. J. Pathol. 153, 611–626.

Sang, Q.X. 1998. Complex role of matrix metalloproteinases in angiogenesis. Cell Res. 8, 171–177.

Sasaki, T., Larsson, H., Tisi, D., Claesson-Welsh, L., Hohenester, E., Timpl, R. 2000. Endostatins derived from collagens XV and XVIII differ in structural and binding properties, tissue distribution and anti-angiogenic activity. J. Mol. Biol. 301, 1179–1190.

Sathyanarayana, U.G., Padar, A., Suzuki, M., Maruyama, R., Shigematsu, H., Hsieh, J.T., Frenkel, E.P., Gazdar, A.F. 2003. Aberrant promoter methylation of laminin-5-encoding genes in prostate cancers and its relationship to clinicopathological features. Clin. Cancer Res. 9, 6395–6400.

Sato, H., Seiki, M. 1996. Membrane-type matrix metalloproteinases (MT-MMPs) in tumor metastasis. J. Biochem. (Tokyo) 119, 209–215.

Schenk, S., Hintermann, E., Bilban, M., Koshikawa, N., Hojilla, C., Khokha, R., Quaranta, V. 2003. Binding to EGF receptor of a laminin-5 EGF-like fragment liberated during MMP-dependent mammary gland involution. J. Cell. Biol. 161, 197–209.

Schwartz, M.A., Schaller, M.D., Ginsberg, M.H. 1995. Integrins: emerging paradigms of signal transduction. Ann. Rev. Cell. Dev. Biol. 11, 549–599.

Senger, D.R., Claffey, K.P., Benes, J.E., Perruzzi, C.A., Sergiou, A.P., Detmar, M. 1997. Angiogenesis promoted by vascular endothelial growth factor: Regulation through alpha1-beta1 and alpha2beta1 integrins. Proc. Natl. Acad. Sci. USA 94, 13612–13617.

Shahan, T.A., Ohno, N., Pasco, S., Borel, J.P., Monboisse, J.C., Kefalides, N.A. 1999. Inhibition of tumor cell proliferation by type IV collagen requires increased levels of cAMP. Connect. Tissue Res. 40, 221–232.

Sidenius, N., Blasi, F. 2003. The urokinase plasminogen activator system in cancer: Recent advances and implication for prognosis and therapy. Cancer Metastasis. Rev. 22, 205–222.

Siewinski, M., Saleh, Y., Ziolkowski, P. 2003. Cysteine peptidases in health and diseases. Folia. Med. Cracov. 44, 169–178.

Skubitz, A.P., Bast, R.C. ,Jr., Wayner, E.A., Letourneau, P.C., Wilke, M.S. 1996. Expression of alpha 6 and beta 4 integrins in serous ovarian carcinoma correlates with expression of the basement membrane protein laminin. Am. J. Pathol. 148, 1445–1461.

Skyldberg, B., Salo, S., Eriksson, E., Aspenblad, U., Moberger, B., Tryggvason, K., Auer, G. 1999. Laminin-5 as a marker of invasiveness in cervical lesions. J. Natl. Cancer. Inst. 91, 1882–1887.

Sloane, B.F., Honn, K.V. 1984. Cysteine proteinases and metastasis. Cancer Metastasis. Rev. 3, 249–263.

Sloane, B.F., Rozhin, J., Krepela, E., Ziegler, G., Sameni, M. 1991. The malignant phenotype and cysteine proteinases. Biomed. Biochim. Acta. 50, 549–554.

Soininen, R., Haka-Risku, T., Prockop, D.J., Tryggvason, K. 1987. Complete primary structure of the alpha 1-chain of human basement membrane (type IV) collagen. FEBS Lett. 225, 188–194.

Stamenkovic, I. 2003. Extracellular matrix remodelling: The role of matrix metalloproteinases. J. Pathol. 200, 448–464.

Stetler-Stevenson, W.G., Yu, A.E. 2001. Proteases in invasion: Matrix metalloproteinases. Semin. Cancer Biol. 11, 143–152.

Tagliabue, E., Ghirelli, C., Squicciarini, P., Aiello, P., Colnaghi, M.I., Menard, S. 1998. Prognostic value of alpha 6 beta 4 integrin expression in breast carcinomas is affected by laminin production from tumor cells. Clin. Cancer Res. 4, 407–410.

Timpl, R., Wiedemann, H., van Delden, V., Furthmayr, H., Kuhn, K. 1981. A network model for the organization of type IV collagen molecules in basement membranes. Eur. J. Biochem. 120, 203–211.

Udayakumar, T.S., Chen, M.L., Bair, E.L., Von Bredow, D.C., Cress, A.E., Nagle, R.B., owden, G.T. 2003. Membrane type-1-matrix metalloproteinase expressed by prostate carcinoma cells cleaves human laminin-5 beta3 chain and induces cell migration. Cancer Res. 63, 2292–2299.

Ura, H., Denno, R., Hirata, K., Yamaguchi, K., Yasoshima, T. 1998. Separate functions of alpha2beta1 and alpha3beta1 integrins in the metastatic process of human gastric carcinoma. Surg. Today 28, 1001–1006.

Vandenberg, P., Kern, A., Ries, A., Luckenbill-Edds, L., Mann, K., Kuhn, K. 1991. Characterization of a type IV collagen major cell binding site with affinity to the alpha 1 beta 1 and the alpha 2 beta 1 integrins. J. Cell. Biol. 113, 1475–1483.

Wickstrom, S.A., Alitalo, K., Keski-Oja, J. 2004. An endostatin-derived peptide interacts with integrins and regulates actin cytoskeleton and migration of endothelial cells. J. Biol. Chem. 279, 20178–20185.

Wilson, R.F., Morse, M.A., Pei, P., Renner, R.J., Schuller, D.E., Robertson, F.M., Mallery, S. R. 2003. Endostatin inhibits migration and invasion of head and neck squamous cell carcinoma cells. Anticancer. Res. 23, 1289–1295.

Yamamoto, H., Itoh, F., Iku, S., Hosokawa, M., Imai, K. 2001. Expression of the gamma(2) chain of laminin-5 at the invasive front is associated with recurrence and poor prognosis in human esophageal squamous cell carcinoma. Clin. Cancer Res. 7, 896–900.

Yan, S., Sameni, M., Sloane, B.F. 1998. Cathepsin B and human tumor progression. Biol. Chem. 379, 113–123.

Yang, C., Zeisberg, M., Lively, J.C., Nyberg, P., Afdhal, N., Kalluri, R. 2003. Integrin alpha1beta1 and alpha2beta1 are the key regulators of hepatocarcinoma cell invasion across the fibrotic matrix microenvironment. Cancer Res. 63, 8312–8317.

Ziaie, Z., Fawzi, A., Bellon, G., Monboisse, J.C., Kefalides, N.A. 1999. A peptide of the alpha3 chain of type IV collagen protects basement membrane against damage by PMN. Biochem. Biophys. Res. Commun. 261, 247–250.

Index

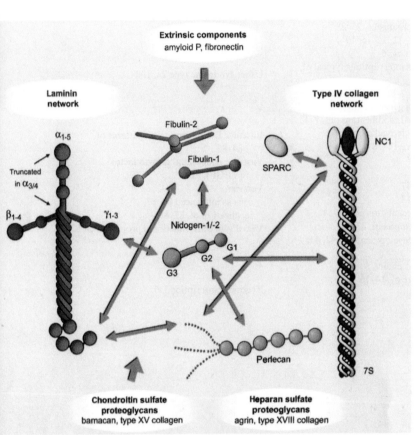

Scott J. Harvey and Paul S. Thorner, Chapter 1, Figure 1. Molecular composition and assembly of basement membranes. All basement membranes have four major intrinsic components: type IV collagen, laminin, nidogen, and heparan sulfate proteoglycans (HSPGs). Type IV collagen, a heterotrimer of 3 α chains, forms a network through self-interaction at both its carboxyl (NC1) and amino (7S) termini. Laminin molecules are "cross-shaped" heterotrimers consisting of an α, β, and γ chain that form a separate network through interactions between globular domains at the termini of its three short arms. These networks are bridged by nidogen, which binds to the γ1 chain of laminin through its G3 domain and also to triple helical region of type IV collagen through its G2 domain. The HSPGs such as perlecan are incorporated into this scaffold by binding to the molecules above via their core protein or heparan sulfate side chains. In some basement membranes, chondroitin sulfate proteoglycans are incorporated through similar interactions. Other intrinsic components include SPARC (which binds to type IV collagen) and members of the fibulin family (which bind to nidogen and laminin). Extrinsic molecules, such as fibronectin and amyloid P component, are not synthesized locally but are deposited in basement membranes through their affinity for the various intrinsic molecules. The components are illustrated roughly twice their scale relative to type IV collagen. Figure adapted from Current Opinion in Cell Biology, Vol. 6, No. 5, Yurchenco & O'Rear,

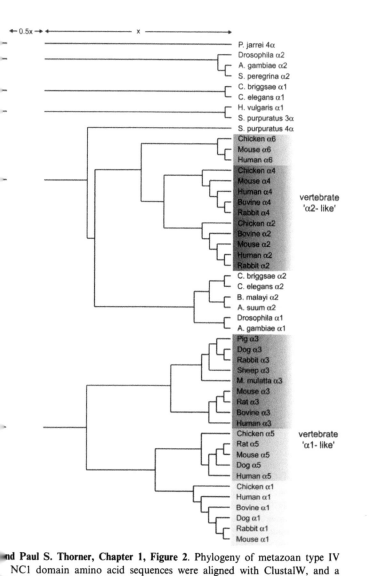

←0.5x→ ←——————— x ———————→

P. jarrei 4α
Drosophila α2
A. gambiae α2
S. peregrina α2
C. briggsae α1
C. elegans α1
H. vulgaris α1
S. purpuratus 3α
S. purpuratus 4α
Chicken α6
Mouse α6
Human α6
Chicken α4
Mouse α4
Human α4
Bovine α4
Rabbit α4
Chicken α2
Bovine α2
Mouse α2
Human α2
Rabbit α2
C. briggsae α2
C. elegans α2
B. malayi α2
A. suum α2
Drosophila α1
A. gambiae α1
Pig α3
Dog α3
Rabbit α3
Sheep α3
M. mulatta α3
Mouse α3
Rat α3
Bovine α3
Human α3
Chicken α5
Rat α5
Mouse α5
Dog α5
Human α5
Chicken α1
Human α1
Bovine α1
Dog α1
Rabbit α1
Mouse α1

vertebrate 'α2- like'

vertebrate 'α1- like'

nd Paul S. Thorner, Chapter 1, Figure 2. Phylogeny of metazoan type IV
NC1 domain amino acid sequences were aligned with ClustalW, and a
derived using the protein parsimony algorithm of the program protpars in the
ware package. The tree was rooted through the sequence of the sponge (P.
onsidered ancestral. Some available sequences that were incomplete were
analysis (10).

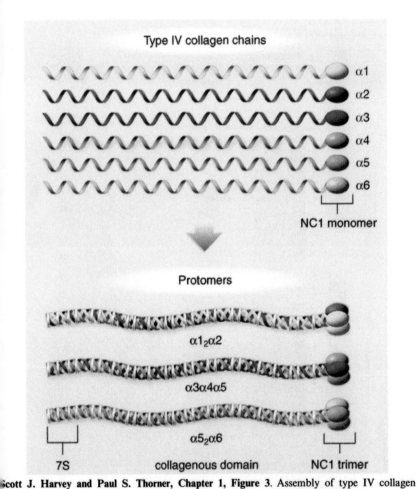

Type IV collagen chains

α1
α2
α3
α4
α5
α6

NC1 monomer

Protomers

α1₂α2

α3α4α5

α5₂α6

7S   collagenous domain   NC1 trimer

Scott J. Harvey and Paul S. Thorner, Chapter 1, Figure 3. Assembly of type IV collagen protomers. The six isoforms of type IV collagen assemble into three triple helical molecules (protomers) that differ in their composition. Each protomer is characterized by an amino-terminal 7S domain ($\sim$140 residues), a main collagenous domain ($\sim$1300 residues) that contains numerous non-collagenous interruptions (white rings), and a carboxyl-terminal non-collage-nous (NC1) domain. Molecular recognition events between the NC1 domains of each α chain govern their triple helical assembly into protomers. The 7S and NC1 domains are involved in cross-linking and aggregation of protomers into networks. Figure reproduced from Hudson, B., Tryggvason, K., Sundaramoorthy, M., Neilson, E. (2003). Alport's syndrome, Goodpasture's syndrome, and type IV collagen. N Engl J Med 348, 2543–2556. Copyright © 2004 Massachusetts Medical Society. All rights reserved (20).

NC1 hexamers

Network assembly

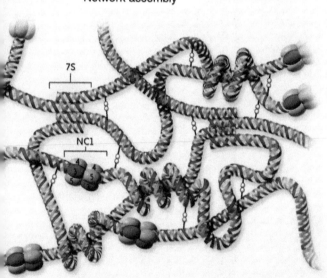

and Paul S. Thorner, Chapter 1, Figure 4. Assembly of type IV collagen
protomers dimerize through their carboxyl termini, which yields the NC1
mer. Three distinct populations of NC1 hexamers have been identified; two
omotypic assembly of $\alpha1_2\alpha2$ or $\alpha3\alpha4\alpha5$ protomers, and a third through
ly of $\alpha1_2\alpha2$ and $\alpha5_2\alpha6$ protomers. The relative arrangement of NC1 domain
hexamer is accurately depicted. The NC1 domain monomers that oppose
the trimer-trimer interface may become covalently cross-linked. (Bottom)
supramolecular assembly into a complex network through aggregation and
eir like ends, and through supercoiling of their triple helical domains. A
of $\alpha3\alpha4\alpha5$ protomers is illustrated. Individual protomers dimerize through
inal NC1 domains (NC1 box) and form tetramers through their overlapping
domains (7S box). The 7S tetramer is depicted in a "stacked" alignment for
it is thought to assume a tetragonal arrangement. The network is stabilized
lisulfide bonds between triple helical domains (white circles) that are also
S tetramer and perhaps the NC1 hexamer (not shown). Further stabilization

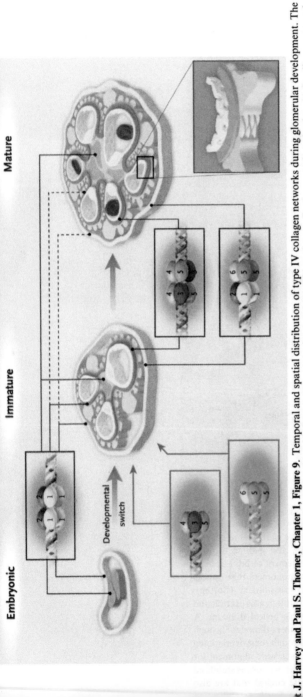

Alport J. Harvey and Paul S. Thorner, Chapter 1, Figure 9. Temporal and spatial distribution of type IV collagen networks during glomerular development. The basement membrane of primitive nephric figures (vesicle and 'comma' shaped nephrons) from which glomeruli arise contain only the α1/α2 collagen network (embryonic). In 'S-shaped' nephrons, the α1/α2/α5/α6 network is deposited in the basement membrane underlying the epithelial component destined to become Bowman's capsule (not shown). Later, in early capillary loop stage glomeruli (immature) the α3/α4/α5 network becomes expressed along with the α1/α2 network in the primitive GBM (purple) that separates visceral epithelium (podocytes) (blue) from endothelium (yellow). At later stages (mature), the α1/α2/α5/α6 and α3/α4/α5 networks appear to replace (dotted line) the α1/α2 network in Bowman's capsule and the GBM, respectively. Figure reproduced from Hudson, B., Tryggvason, K., Sundaramoorthy, M., Neilson, E. (2003). Alport's syndrome, Goodpasture's syndrome, and type IV collagen. N. Engl. J. Med. 348, 2543–2556. Copyright © 2004 Massachusetts Medical Society. All rights reserved (43).

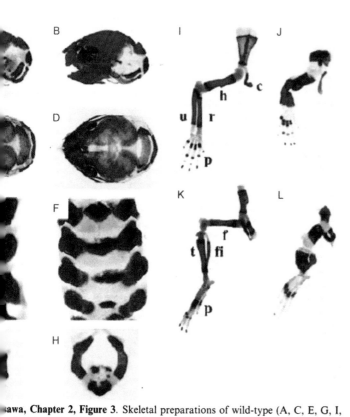

awa, **Chapter 2, Figure** 3. Skeletal preparations of wild-type (A, C, E, G, I, (B, D, F, H, J, L) newborn mice stained with Alcian blue and Alizarin red a et al., 1999).Lateral view of Hspg2−/−skull (B) shows shorter and thicker elayed ossification of bony ossicles of middle ear. Dilated sutures and observed in the skull vault of Hspg2−/− mice (D). Shorter and wider f Hspg2−/− mice are shown. Multiple ossification centers were observed at e vertebral body (H). The scapula, humerus, radius, and ulna of Hspg2−/− um, femur, tibia and fibula of the hindlimb (L) are shorter and broader. The ges in Hspg2−/− mice is relatively normal. c = clavicle; h = humerus; p = is; u = ulna; f = femur; fi = fibula; t = tibia; Hspg 2 = perlecan (70).

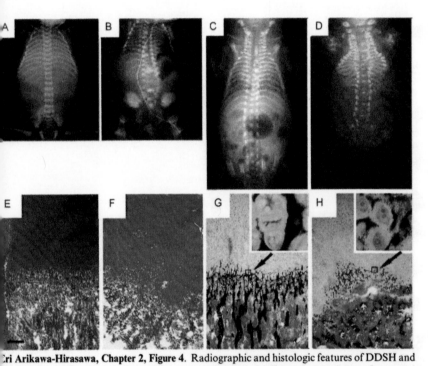

Eri Arikawa-Hirasawa, Chapter 2, Figure 4. Radiographic and histologic features of DDSH and the perlecan-null mouse. (Arikawa-Hirasawa et al., 2001). Radiographs are shown of a normal newborn human (A), a newborn patient with DDSH (B), a normal newborn mouse (C), and an Hspg2-null newborn mouse (D). Note the similarity in the radiographic appearance between the human and the null mouse. In both cases, there is anisospondyly, a small chest, and short, bent long bones of the limbs. Profound dyssegmental ossification is observed in the spine in case 1 and the Hspg2-null mouse. Also shown is the histology of the cartilage of the femur: (E, F) Toluidine blue stain, ×6.25, scale bar, 500 μm; (G, H) von Kossa trichrome stain, ×6.25. Inset, higher magnification, ×50. (E, G), Normal human. (F, H) A patient with DDSH. In DDSH cartilage, reduced matrix staining and endochondral ossification with disorganized columnar structures of hypertrophic chondrocytes are seen. The calcospherites are unfused and show granular appearances in the DDSH hypertrophic zone, whereas they are fused in normal tissue. Hspg 2 = perlecan (72).

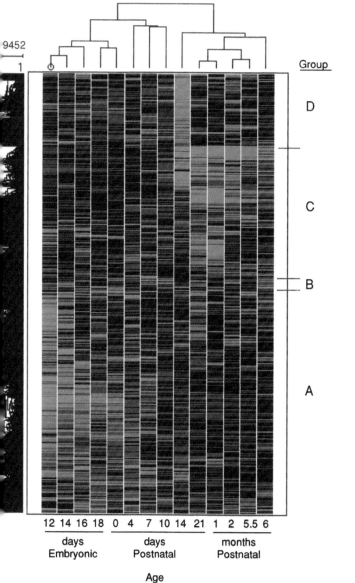

Brigham H. Mecham, Cassandra M. Kelleher, Thomas J. Mariani and Robert

ter 3, **Figure 3**. Hierarchical clustering of 9452 probe sets that changed by at
their average difference value over the course of the time series. Numbers at
te mouse developmental age and letters on the right side indicate functional
m hierarchical clustering of the probe sets. The dendrigram at the top shows
ts cluster into two large groups. The first cluster includes the "early" time
and the second cluster consists of the "late" time points P7 to P6 months.
' cluster, the time points within the respective sub-clusters of E12 and E14

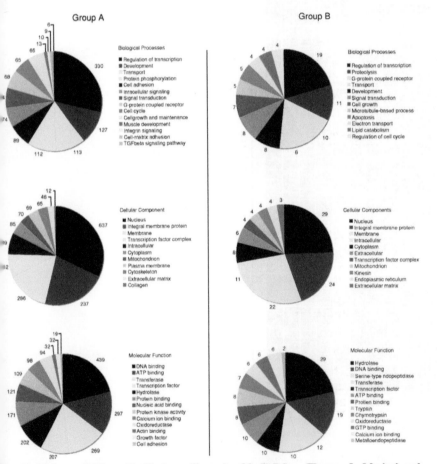

an E. McLean, Brigham H. Mecham, Cassandra M. Kelleher, Thomas J. Mariani and
bert P. Mecham, Chapter 3, Figure 4. (*continued*)

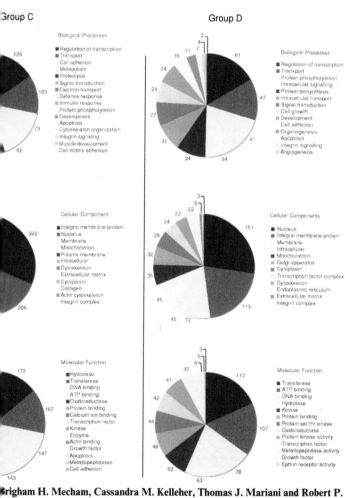

Brigham H. Mecham, Cassandra M. Kelleher, Thomas J. Mariani and Robert P.
B, Figure 4. Functional associations of gene clusters associated with groups A
d C and D (panel B). The Gene Ontology Database (www.geneontology.org)
ize the ontology annotations for each group into three ontologies: Molecular
al Process, and Cellular Components. The numbers indicate how many probe
subcategory (91, 92).

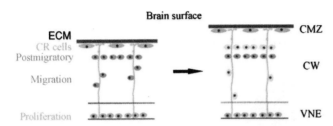

Ventricle

Richard Belvindrah and Ulrich Müller, Chapter 5, Figure 1. Development of the cerebral cortex. Two stages of development are shown. Left picture: neurons (blue) are generated in the ventricular neuroepithelium (VNE) from proliferating precursor cells (dark green). The cells migrate along radial glial fibers (dark green) into the cortical wall (CW). The cells terminate migration close to the cortical marginal zone (CMZ) where the Cajal Retzius (CR) cells (orange) are situated below the meningeal basement membrane (ECM). The migrating neurons detach from the glial fibers and are assembled into cortical layers. Right picture: a second wave of migrating neurons (light green) bypasses the earlier born neurons and forms a layer close to the CR cell layer. Subsequent waves of neurons (not shown) will similarly bypass earlier born neurons leading to the formation of the multilayered cerebral cortex (155).

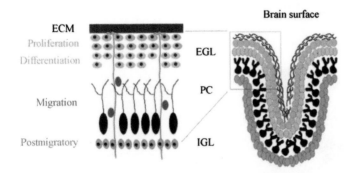

Cerebellar Folium

Richard Belvindrah and Ulrich Müller, Chapter 5, Figure 2. Development of the cerebellar cortex. A cerebellar folium is shown to the right, and an enlargement of a small area of a folium (boxed) to the left. Granule cell precursors (green) proliferate in the external granule cell layer (EGL) that contacts the meningeal ECM. The cells give rise to premigratory neurons (orange) that subsequently differentiate further and migrate (blue cells) along Bergman glial fibers (orange) through the Purkinje cell layer (PC) (black) into the internal granule cell layer (IGL) (157).

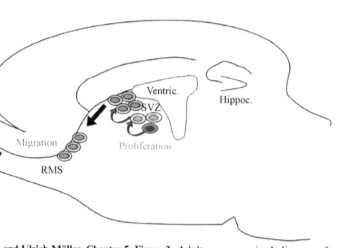

and **Ulrich Müller, Chapter 5, Figure 3**. Adult neurogenesis. A diagram of a
on is shown. Neuroblasts (yellow) that are generated from stem cells (blue) in
zone (SVZ) migrate along the rostral migratory stream (RMS) toward the
3) where they differentiate (163).

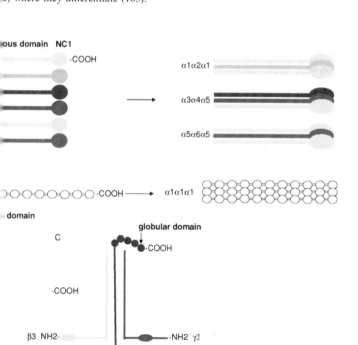

bra **Pozzi, Chapter 7, Figure 1**. Structure of laminin-5 and non-fibriillar
agen IV is composed by six different chains, α1(IV) to α6(IV), characterized
domain, a central collagenous domain, and a C terminal NC1 domain. The
to form the trimeric structures indicated on the right. (B) Collagen XVIII is

Printed and bound by CPI Group (UK) Ltd, Croydon, CR0 4YY

08/05/2025

01864966-0006